Gibberellins

The three editors of this volume: Professors Nobutaka Takahashi (right), Bernard O. Phinney (center), and Jake MacMillan (left) (On July 23, 1989).

Nobutaka Takahashi   Bernard O. Phinney
Jake MacMillan
Editors

Gibberellins

With 176 Illustrations

Springer-Verlag
New York Berlin Heidelberg London
Paris Tokyo Hong Kong Barcelona

Nobutaka Takahashi
Department of Agricultural Chemistry
The University of Tokyo
Bunkyo-ku
Tokyo 113 Japan

Bernard O. Phinney
Department of Biology
University of California, Los Angeles
405 Hilgard Avenue
Los Angeles, CA 90024 USA

Jake MacMillan
School of Chemistry
University of Bristol
Cantock's Close
Bristol BS8 1TS England

Library of Congress Cataloging-in-Publication Data
Gibberellins/Nobutaka Takahashi, Bernard O. Phinney, Jake MacMillan,
   editors.
     p. cm.
   Papers from a symposium held at The University of Tokyo July 20–23,
1989.
   Includes bibhographical references.
   1. Gibberellins—Congresses. I. Takahashi, Nobutaka, 1930–   .
II. Phinney, Bernard O.   III. MacMillan, J. (Jake)   IV. Tōkyō
Daigaku.
QK898.G45G49   1990
581.3'1—dc20   90-9538
                                                                   CIP

Printed on acid-free paper

© 1991 Springer-Verlag New York Inc.
All rights reserved. This work may not be translated or copied in whole or in part without the written permission of the publisher (Springer-Verlag New York, Inc., 175 Fifth Avenue, New York, NY 10010, USA), except for brief excerpts in connection with reviews or scholarly analysis. Use in connection with any form of information and retrieval, electronic adaptation, computer software, or by similar or dissimilar methodology now known or hereafter developed is forbidden.
The use of general descriptive names, trade names, trademarks, etc., in this publication, even if the former are not especially identified, is not to be taken as a sign that such names, as understood by the Trade Marks and Merchandise Marks Act, may accordingly be used freely by anyone.
Permission to photocopy for internal or personal use, or the internal or personal use of specific clients, is granted by Springer-Verlag New York, Inc. for libraries registered with the Copyright Clearance Center (CCC), provided that the base fee of $0.00 per copy, plus $0.20 per page is paid directly to CCC, 21 Congress St., Salem, MA 01970, USA. Special requests should be addressed directly to Springer-Verlag New York, 175 Fifth Avenue, New York, NY 10010, USA.
ISBN 0-387-97259-5/1991 $0.00 + 0.20

Typeset by Asco Trade Typesetting Ltd., Hong Kong
Printed and bound by Edwards Brothers, Ann Arbor, Michigan
Printed in the United States of America.

9 8 7 6 5 4 3 2 1

ISBN 0-387-97259-5 Springer-Verlag New York Berlin Heidelberg
ISBN 3-540-97259-5 Springer-Verlag Berlin Heidelberg New York

Preface

The cultivation of rice in Japan has suffered from damage caused by bakanae disease, in which rice seedlings show abnormal growth (elongation) as the result of infection by a plant pathogen. Investigation of the taxonomy of this pathogen led to the commencement of gibberellin (GA) research among Japanese plant pathologists, who later identified it as *Gibberella fujikuroi*, its other name being *Fusarium moniliforme*. In 1926, Kurosawa demonstrated the occurrence of an active principle in the culture media of fungus that showed the same symptoms as those of the rice disease. In 1938, this finding was followed by the successful isolation of the active principles as crystals from the culture filtrate. This was achieved by the Japanese agricultural chemists Yabuta and Sumiki, of The University of Tokyo, who named these active principles gibberellins A and B. Following World War II, this discovery attracted the interest of scientists around the world, and research on GA was pursued on a worldwide scale.

One of the most outstanding discoveries in GA research after the isolation of GA as the metabolite of the plant pathogen must be the isolation and characterization of GAs from tissues of higher plants by the MacMillan group, West and Phinney, and the Tokyo University group in 1958 and 1959. Thus, GAs have been recognized as one of the most important classes of plant hormones.

Up to now, seventy-nine free GAs and more than 10 GA conjugates have been chemically characterized as naturally occurring. It must be emphasized that the identification of such a large number of GAs could not have been achieved without the combination of the technology developed in plant physiology and natural product chemistry. More especially, the recent advancement in the methodology of isolation and characterization is remarkable. Without the use of modern techniques such as high-performance liquid chromatography (HPLC) and combined gas chromatography—mass spectrometry (GC-MS), such a large number of GAs would not have been characterized. On the other hand, such technology can also be enhanced through intensive GA research. It is in this sense that GA research and advancement in the methodology of isolation and characterization are complementary.

After lengthy chemical research carried out by groups from The University of Tokyo and Imperial Chemical Industry (ICI), the dierpenoidal structure of GA was finally established by the ICI group. In this research, new methodologies such as nuclear magnetic resonance (NMR), mass spectroscopy (MS), and X-ray analysis had been widely introduced and extensively used. They have showed their power in the structural determination of minute amounts of bioactive compounds. The total synthesis of GA_3 has been achieved by the Corey group, and further important total and relay syntheses of other GAs have been conducted by organic synthetic chemists. Thus, rare GAs, such as GA_{19}, which were obtained in very small amounts from natural sources have been synthesized and used in the identification and physiological study of GAs.

Basic plant physiological research has been intensively conducted, and much notable progress has been achieved. The scope of research has extended to the molecular mechanism of the action of GA, and investigations on the purification of the enzyme participating in the biosynthesis of GAs and on the genetic background of the enzyme are now in progress. As the result of intensive applied research, GAs continue to maintain a leading position among plant hormones.

Fifty years have passed since GA was isolated. This memorable discovery was conducted on the campus of The University of Tokyo in 1938. To mark the fiftieth anniversary of the birth of GA, one of the editors of this book (N.T.) thought that it was timely to organize a GA symposium, in which the recent progress in GA research could be summarized. Fortunately, the University decided to help financially and spiritually to organize the symposium. At the same time, the Society for Chemical Regulation of Plants proposed to support the symposium.

Thus, the Gibberellin Symposium Tokyo 1989 was organized under the auspices of The University of Tokyo and The Society for Chemical Regulation of Plants and held from July 20 to 23, 1989, on the campus of The University of Tokyo. Almost 100 scientists from more than ten countries attended the meeting. All participants wish to express their thanks to the University and to the Society for the deep understanding and financial support shown toward GA research.

It is my great pleasure to be able to publish this volume, which contains all thirty-nine lectures presented in this symposium.

I wish to express my sincere thanks as the editor to Professor J. Macmillan, University of Bristol, and Professor B.O. Phinney, University of California at Los Angeles (UCLA) for their kind help in reviewing the manuscript. Thanks are also due to the members of the organizing committee, the staff of The University of Tokyo, Professor N. Murofushi, Drs. T. Yokota, I. Yamaguchi, and H. Yamane, and the staff of The Institute of Physical and Chemical Research (RIKEN), Drs. A. Sakurai, S. Yoshida, and Y. Kamiya, for their enthusiasm and hard work in preparing the symposium. I am also thankful to Dr. C.R. Spray of UCLA for his help in the editorial work of this volume. I also express my thanks to the following organizations and companies for their financial support: Japan Agronomy Association; Japan Gibberellin Research Association; BASF Japan Ltd.; Chugai Pharmaceutical Co. Ltd.; Imperial Chemical Industry Japan Ltd.; Japan Tobacco Inc.; Kumiai Chemical Industry Co. Ltd.; Kyowa Hakko Kogyo Co. Ltd.; Nihon Tokushu Noyaku Seizo K.K.; Sumitomo Chemical Co. Ltd.

I thank Springer-Verlag for their interest and help in publishing the proceedings of the symposium.

Nobutaka Takahashi

Contents

Preface	v
Contributors	xv

1. **Historical Aspects of Gibberellins** — 1
   S. Tamura

I. Organ Specificity and Dwarfism

2. **Organ-Specific Gibberellins in Rice: Roles and Biosynthesis** — 9
   N. Takahashi and M. Kobayashi

3. **Gibberellin Metabolism in Maize: Tissue Specificity** — 22
   B.O. Phinney, C.R. Spray, Y. Suzuki, and P. Gaskin

4. **Gibberellins from the Tassel, Cob, 'Seed,' Silk, and Pollen of Maize** — 32
   N. Murofushi, I. Honda, R. Hirasawa, I. Yamaguchi, N. Takahashi, and B.O. Phinney

5. **Gibberellin Mutants in *Pisum* and *Lathyrus*** — 40
   J.B. Reid and J.J. Ross

II. Biosynthetic Enzymes

6. **The Relationship of Different Gibberellin Biosynthetic Pathways in *Cucurbita maxima* Endosperm and Embryos and the Purification of a C-20 Oxidase from the Endosperm** — 51
   J.E. Graebe, T. Lange, S. Pertsch, and D. Stöckl

7. **Enzymatic 3β-Hydroxylation of Gibberellins A_{20} and A_5** — 62
   V.A. Smith, K.S. Albone, and J. MacMillan

8. **Partial Characterization of the Gibberellin 3β-Hydroxylase** 72
   from Immature Seeds of *Phaseolus vulgaris*
   Y. Kamiya and S-S. Kwak

9. **Gibberellin Metabolism in Cell-Free Preparations** 83
   from *Phaseolus coccineus*
   A. Crozier, C.G.N. Turnbull, J.M. Malcolm, and J.E. Graebe

10. **Gibberellin Biosynthetic Enzymes and the Regulation of** 94
    Gibberellin Concentration
    P. Hedden

III. Molecular Aspects

11. **Gibberellin A_3-Regulated α-Amylase Synthesis and Calcium** 106
    Transport in the Endoplasmic Reticulum of Barley Aleurone Cells
    D.S. Bush, L. Sticher, and R.L. Jones

12. **Rice α-Amylase and Gibberellin Action—A Personal View** 114
    T. Akazawa, J. Yamaguchi, and M. Hayashi

13. **Gibberellin Production and Action During Germination of Wheat** 125
    J.R. Lenton and N.E.J. Appleford

14. **Probing Gibberellin Receptors in the *Avena fatua* Aleurone** 136
    R. Hooley, M.H. Beale, and S.J. Smith

15. **A Minireview on the Immunoassay for Gibberellins** 146
    I. Yamaguchi and E.W. Weiler

IV. Physiology and Metabolism

16. **Physiology of Gibberellins in Relation to Floral Initiation and** 166
    Early Floral Differentiation
    R.P. Pharis

17. **Role of Endogenous Gibberellins During Fruit and Seed** 179
    Development
    G.W.M. Barendse, C.M. Karssen, and M. Koornneef

18. **Correlations Between Apparent Rates of *ent*-Kaurene** 188
    Biosynthesis and Parameters of Growth and Development in
    Pisum sativum
    T.C. Moore and R.C. Coolbaugh

19. Gibberellins and the Regulation of Shoot Elongation in 199
    Woody Plants
    O. Juntilla

20. The Gibberellin Control of Cell Elongation 211
    M. Katsumi and K. Ishida

21. The role of Gibberellin in the Formation of Onion Bulbs 220
    H. Shibaoka

22. Gibberellin Requirement for the Normal Growth of Roots 229
    E. Tanimoto

23. Effects of Gibberellin A_3 on Growth and Tropane Alkaloid 241
    Synthesis in Ri Transformed Plants of *Datura innoxia*
    H. Kamada, T. Ogasawara, and H. Harada

24. Biochemical and Physiological Aspects of Gibberellin Conjugation 249
    G. Sembdner, W. Schliemann, and G. Schneider

25. Metabolism of [^{3}H]Gibberellin A_4 and [^{2}H]Gibberellin A_4 in 264
    Cell Suspension Cultures of Rice, *Orygza sativa* cv. Nihonbare
    M. Koshioka, E. Minami, H. Saka, R.P. Pharis, and
    L.N. Mander

V. Light Effects

26. Stem Growth and Gibberellin Metabolism in Spinach in 273
    Relation to Photoperiod
    J.A.D. Zeevaart, M. Talon, and T.M. Wilson

27. Phytochrome Mediation of Gibberellin Metabolism and 280
    Epicotyl Elongation in Cowpea, *Vigna sinensis* L.
    N. Fang, B. A. Bonner, and L. Rappaport

28. Role of Gibberellins in Phytochrome-Mediated Lettuce 289
    Seed Germination
    Y. Inoue

VI. Growth Retardants

29. Inhibitors of Gibberellin Biosynthesis: Applications in 296
    Agriculture and Horticulture
    W. Rademacher

30. Studies on the Action of the Plant Growth Regulators 311
    BX-112, DOCHC, and DOCHC-Et
    I. Nakayama, T. Miyazawa, M. Kobayashi, Y. Kamiya,
    H. Abe, and A. Sakurai

31. Studies on Sites of Action of the Plant Growth Retardant 4'- 320
    Chloro-2'-(α-Hydroxybenzyl)isonicotinanilide (Inabenfide) in
    Gibberellin Biosynthesis
    T. Miki, T. Ichikawa, Y. Kamiya, M. Kobayashi, and A. Sakurai

32. Effects of the Growth Retardant Uniconazole-P on Endogenous 330
    Levels of Hormones in Rice Plants
    K. Izumi and H. Oshio

33. Inconsistency Between Growth and Endogenous Levels of 339
    Gibberellins, Brassinosteroids, and Sterols in *Pisum sativum*
    Treated with Uniconazole Antipodes
    T. Yokota, Y. Nakamura, N. Takahashi, M. Nonaka,
    H. Sekimoto, H. Oshio, and S. Takatsuto

VII. Applied Aspects

34. Gibberellin Increases Cropping Efficiency in Sour Cherry 350
    (*Prunus cerasus* L.)
    M.J. Bukovac and E. Yuda

35. Prospects for Gibberellin Control of Vegetable Production 361
    in the Tropics
    C.G. Kuo

36. Gibberellin-Induced Flowering and Morphological Changes in 370
    Taro Plants
    N. Katsura, K. Takayanagi, T. Sato, T. Nishijima,
    and H. Yamaji

VIII. Antheridiogens

37. Antheridiogens of Schizaeaceous Ferns: Structures, 378
    Biological Activities, and Biosynthesis
    H. Yamane

38. Antheridiogen, Gibberellin, and the Control of Sex Differentiation 389
    in Gametophytes of the Fern *Lygodium japonicum*
    K. Takeno

39. Synthetic Pathways to Fern Antheridiogens from Gibberellins   398
    M. Furber, L.N. Mander, and G.L. Patrick

Index   417

Contributors

H. Abe Life Science Research Institute, Kumiai Chemical Industry Co., Ltd., Kikukawa, Shizuoka 439, Japan

T. Akazawa Research Institute for Biochemical Regulation, School of Agriculture, Nagoya University, Aichi 464, Japan

K.S. Albone School of Chemistry, University of Bristol, Bristol BS8 1TS, UK

N.E.J. Appleford Department of Agricultural Sciences, University of Bristol, AFRC Institute of Arable Crops Research, Long Ashton Research Station, Long Ashton, Bristol BS18 9AF, UK

G.W.M. Barendse Department of Experimental Botany, University of Nijmegen, Toernooiveld, 6526 ED Nijmegen, The Netherlands

M.H. Beale School of Chemistry, University of Bristol, Bristol BS8 1TS, UK

B.A. Bonner Department of Botany, University of California, Davis, CA 95616, USA

M.J. Bukovac Department of Horticulture, Michigan State University, East Lansing, MI 48824, USA

D.S. Bush Department of Plant Biology, University of California, Berkeley, CA 94720, USA

R.C. Coolbaugh Department of Botany and Plant Pathology, Purdue University, West Lafayette, IN 47907, USA

A. Crozier Department of Botany, University of Glasgow, Glasgow G12 8QQ, UK

S. Dertsch Pflanzen physiologisches Institut, Untere Karspuele 2, D-3400 Gottingen, FRG

N. Fang Department of Vegetable Crops, Plant Growth Laboratory, University of California, Davis, CA 95616, USA

M. Furber Research School of Chemistry, Australian National University, GPO Box 4, Canberra, A.C.T. 2601, Australia

R. Gaskin School of Chemistry, University of Bristol, Bristol BS8 1TS, UK

J.E. Graebe Pflanzenphysiologisches Institut, Untere Karspüle 2, D-3400 Göttingen, FRG

H. Harada Gene Experiment Center, University of Tsukuba, Tsukabashi, Ibaraki 305, Japan

M. Hayashi Research Institute for Biochemical Regulation, School of Agriculture, Nagoya University, Aichi 464, Japan

P. Hedden Department of Agricultural Sciences, University of Bristol, AFRC Institute of Arable Crops Research, Long Ashton Research Station, Long Ashton Bristol BS18 9AF, UK

R. Hirasawa Department of Agricultural Chemistry, The University of Tokyo, Bunkyo-ku, Tokyo 113, Japan

I. Honda Department of Agricultural Chemistry, The University of Tokyo, Bunkyo-ku, Tokyo 113, Japan

R. Hooley Department of Agricultural Sciences, University of Bristol, AFRC Institute of Arable Crops Research, Long Ashton Research Station, Long Ashton, Bristol BS18 9AF, UK

T. Ichikawa Chugai Pharmaceutical Co., Ltd., 41-8, Takada 3-chome, Toshima-ku, Tokyo 171, Japan

Y. Inoue Department of Botany, Faculty of Science, The University of Tokyo, Hongo, Tokyo 113, Japan

K. Ishida Biology Department, International Christian University, Osawa, Mitaka City, Tokyo 181, Japan

K. Izumi Agricultural Science Research Laboratory, Takarazuka Research Center, Sumitomo Chemical Co., Ltd., Takarazuka, Hyogo 665, Japan

R.L. Jones Department of Plant Biology, University of California, Berkeley, CA 94720, USA

O. Junttila Department of Plant Physiology and Microbiology, University of Tromsø, Tromsø, Norway

H. Kamada Gene Experiment Center, University of Tsukuba, Tsukabashi, Ibaraki 305, Japan

Y. Kamiya The Institute of Physical and Chemical Research, Wako-shi, Saitama 351-01, Japan

C.M. Karrsen Department of Plant Physiology, Agricultural University, Arboretumlaan 4, 6703 BD Wageningen, The Netherlands

M. Katsumi Biology Department, International Christian University, Osawa, Mitaka City, Tokyo 181, Japan

N. Katsura National Research Institute of Vegetables, Ornamental Plants and Tea, Ano, Mie 514-23, Japan

M. Kobayashi Institute of Physical and Chemical Research, Wako-shi, Saitama 351-01, Japan

M. Koornneef Department of Genetics, Agricultural University, Generaal Foulkesweg 53, 6703 BM Wageningen, The Netherlands

M. Koshioka National Institute of Agro-Environmental Sciences, Tsukuba, Ibaraki 305, Japan

C.G. Kuo The Asian Vegetable Research and Development Center, P.O. Box 42, Shanhua, Tainan, Taiwan 74199

S.-S. Kwak The Institute of Physical and Chemical Research, Hirosawa, Wako-shi, Saitama 351-01, Japan

T. Lange Pflanzenphysiologisches Institut, Untere Karspüle 2, D-3400 Göttingen, FRG

J.R. Lenton Department of Agricultural Sciences, University of Bristol, AFRC Institute of Arable Crops Research, Long Ashton Research Station, Long Ashton, Bristol BS18 9AF, UK

J. MacMillan School of Chemistry, University of Bristol, Bristol BS8 1TS, UK

J.M. Malcolm Department of Botany, University of Glasgow, Glasgow G12 8QQ, UK

L.N. Mander Research School of Chemistry, Australian National University, GPO Box 4, Canberra, A.C.T. 2601, Australia

T. Miki Chugai Pharmaceutical Co., Ltd., 41-8, Takada 3-chome, Toshima-ku, Tokyo 171, Japan

E. Minami National Institute of Agro-Environmental Sciences, Tsukuba, Ibaraki 305, Japan

T. Miyazawa Life Science Research Institute, Kumiai Chemical Industry Co., Ltd., Kikugawa Shizuoka 439, Japan

N. Murofushi Department of Agricultural Chemistry, The University of Tokyo, Bunkyo-ku, Tokyo 113, Japan

T.C. Moore Department of Botany and Plant Pathology, Oregon State University, Corvallis, OR 97331, USA

Y. Nakamura Department of Agricultural Chemistry, The University of Tokyo, Bunkyo-ku, Tokyo 113, Japan

I. Nakayama Life Science Research Institute, Kumiai Chemical Industry Co., Ltd., Kikugawa, Shizuoka 439, Japan

T. Nishijima National Research Institute of Vegetables, Ornamental Plants and Tea, Ano, Mie 514-23, Japan

M. Nonaka National Research Institute of Vegetables, Ornamental Plants and Tea, Kurume 830, Japan

T. Ogasawara Gene Experiment Center, University of Tsukuba, Tsukaba-shi, Ibaraki 305, Japan

H. Oshio Agricultural Science Research Laboratory, Takarazuka Research Center, Sumitomo Chemical Co., Ltd., Takarazuka, Hyogo 665, Japan

S. Pertsch Pflanzenphysiologisches Institut, Untere Karspule 2, D-3400 Göttingen, FRG

G.L. Patrick Research School of Chemistry, Australian National University, GPO Box 4, Canberra, A.C.T. 2601, Australia

R.P. Pharis Department of Biological Sciences, University of Calgary, Calgary, Alberta T2N 1N4, Canada

B.O. Phinney Department of Biology, University of California, Los Angeles, CA 90024-1606, USA

W. Rademacher BASF Agricultural Research Station, D-6703 Limburgerhof, FRG

L. Rappaport Department of Vegetable Crops, Plant Growth Laboratory, University of California, Davis, CA 95616

J.B. Reid Department of Plant Science, University of Tasmania, Hobart, Tasmania 7001, Australia

J.J. Ross Department of Plant Science, University of Tasmania, Hobart, Tasmania 7001, Australia

H. Saka National Institute of Agrobiological Resources, Tsukuba, Ibaraki 305, Japan

A. Sakurai The Institute of Physical and Chemical Research, Wako-shi, Saitama 351-01, Japan

T. Sato National Research Institute of Vegetables, Ornamental Plants and Tea, Ano, Mie 514-23, Japan

W. Schliemann Institute of Plant Biochemistry, Academy of Sciences of the German Democratic Republic, Weinberg 3, DDR-4050 Halle/Saale, GDR

G. Schneider Institute of Plant Biochemistry, Academy of Sciences of the German Democratic Republic, Weinberg 3, DDR-4050 Halle/Saale, GDR

H. Sekimoto Takarazuka Research Center, Sumitomo Chemical Co., Ltd., Takarazuka, Hyogo 665, Japan

G. Sembdner Institute of Plant Biochemistry, Academy of Sciences of the German Democratic Republic, Weinberg 3, DDR-4050 Halle/Saale, GDR

H. Shibaoka Department of Biology, Faculty of Science, Osaka University, Toyonaka, Osaka 560, Japan

S.J. Smith Department of Agricultural Sciences, University of Bristol, AFRC Institute of Arable Crops Research, Long Ashton Research Station, Long Ashton, Bristol BS18 9AF, UK

V.A. Smith School of Chemistry, University of Bristol, Bristol BS8 1TS, UK

C.R. Spray Department of Biology, University of California, Los Angeles, CA 90024-1606, USA

L. Sticher Department of Plant Biology, University of California, Berkeley, CA 94720, USA

D. Stöckl Pflanzenphysiologisches Institut, Untere Karspüle 2, D-3400 Göttingen, FRG

Y. Suzuki Department of Biology, University of California, Los Angeles, CA 90024-1606, USA

N. Takahashi Department of Agricultural Chemistry, The University of Tokyo, Bunkyo-ku, Tokyo 113, Japan

S. Takatsuto Joetsu University of Education, Joetsu City, Niigata 943, Japan

K. Takayanagi National Research Institute of Vegetables, Ornamental Plants and Tea, Ano, Mie 514-23, Japan

K. Takeno Laboratory of Horticultural Science, Faculty of Agriculture, Tohoku University, Sendai 981, Japan

S. Tamura c/o Prof. N. Murofushi, Department of Agricultural Chemistry, The University of Tokyo, Bunkyo-ku, Tokyo 113, Japan

E. Tanimoto Biology Department, College of General Education, Nagoya City University, Mizuho-ku, Mizuho-cho, Nagoya Aichi 467, Japan

M. Talon MSU-DOE Plant Research Laboratory, Michigan State University, East Lansing, MI 48824, USA

C.G.N. Turnbull Department of Botany, University of Glasgow, Glasgow G12 8QQ, UK

E.W. Weiler Department of Plant Physiology, Ruhr Universität Bochum, Postfach 102148, D-4630 Bochum 1, FRG

T.M. Wilson MSU-DOE Plant Research Laboratory, Michigan State University, East Lansing, MI 48824, USA

I. Yamaguchi Department of Agricultural Chemistry, The University of Tokyo, Bunkyo-ku, Tokyo 113, Japan

J. Yamaguchi Research Institute for Biochemical Regulation, School of Agriculture, Nagoya University, Aichi 464, Japan

H. Yamaji National Research Institute of Vegetables, Ornamental Plants and Tea, Ano, Mie 514-23, Japan

H. Yamane Department of Agricultural Chemistry, The University of Tokyo, Bunkyo-ku, Tokyo 113, Japan

T. Yokota Department of Agricultural Chemistry, The University of Tokyo, Bunkyo-ku, Tokyo 113, Japan

E. Yuda Department of Horticulture, Michigan State University, East Lansing, MI 48824, USA
(On leave from University of Osaka Prefecture, Sakai, Osaka, 591, Japan)

J.A.D. Zeevaart MSU-DOE Plant Research Laboratory, Michigan State University, East Lansing, MI 48824, USA

CHAPTER 1
Historical Aspects of Gibberellins

S. Tamura

The early history of gibberellin studies, especially the contribution of Japanese scientists to this subject, will be described here.

Like auxin, gibberellin has been definitely recognized as a class of plant hormones. However, the process of discovery was quite different. Further, it is quite strange that, although gibberellin was isolated in the crystalline state almost at the same time as auxin, the former did not attract attention outside Japan until the end of the Second World War.

Auxin was identified by the successive work of European naturalists and plant physiologists based on the fundamental ideas of Charles and Francis Darwin on the growth of plants. To elucidate the mechanism of phototropism, Charles Darwin and his son conducted a great many experiments on the responses of plants to light and suggested in 1880 that when the tip of a coleoptile is exposed to light, a stimulus formed in the tip is transported to the lower part to cause curvature toward the light. In 1919, A. Paál confirmed that, irrespective of exposure to light, the tip secretes a substance that promotes the growth of the part below the tip. Further, F.W. Went established the method known as the "*Avena* coleoptile test" in 1928 to measure this growth substance. F. Kögl finally isolated the active principle from human urine, identified it as indole-3-acetic acid (IAA) in 1934, and gave it the trivial name heteroauxin. Indole-3-acetic acid is now known to be the main auxin of higher plants.

The presence of gibberellin was detected through the efforts of Japanese plant pathologists working to prevent "bakanae" disease caused by *Gibberella fujikuroi*, which seriously lowered the yield of rice in Japan as well as throughout the Asian continent.

From ancient times, it had been recognized that, for unknown reasons, rice seedlings in paddy fields occasionally showed extreme elongation resulting in sterility or reduction in the yield of rice. Japanese farmers in different localities called these plants by various common names because of their symptoms: *bakanae* (foolish seedling), *ahonae* (stupid seedling), *yurei* (ghost), *somennae* (thin noodle seedling), *yarikatsugi* (spear bearer), *otokonae* (male seedling), *onnanae* (female seedling), and so on. These names

Fig. 1. The cover of "Agricultural Experiment Station Record", 12(1), 1898, in which the first report of Hori on "Bakanae" fungus appeared.

reflect the sense of humor of the farmers. At present, the term *bakanae* is generally accepted.

Bakanae show several characteristic features: elongated seedlings are pale yellow and bear slender leaves together with stunted roots; severely diseased plants gradually die in the paddy fields; plants with slight symptoms survive and form seeds, but the grains in the seeds are either absent or poorly ripened. Nowadays, the outbreaks of bakanae are easily prevented by the sterilization of rice seeds with fungicides prior to sowing.

The first scientific paper on the cause of bakanae was published in 1898 by Shotaro Hori[1] of the Agricultural Experiment Station of the Ministry of Agriculture, Tokyo, who demonstrated that the phenomenon was induced by infection of a fungus belonging to the genus *Fusarium*, probably *Fusarium heterosporium* Nees (Fig. 1). Up to that time, the reason for the occurrence of bakanae was unknown. Hori himself considered at first that bakanae might originate in some physiological disorder in growth points or radicles of sprouting seeds, but all of his experiments carried out from this standpoint ended in failure. In 1897, he collected bakanae in the neighboring paddy fields and conducted a thorough investigation. He found that the color of the stems adjacent to the seeds had changed to brown and a large number of fungal hyphae had penetrated into the tissue of the stems. Further, Hori was able to create bakanae by inoculating healthy seedlings with the microbes isolated from diseased plants.

Fig. 2. Dr. Eiichi Kurosawa.

After the publication of Hori's report, there was much controversy among plant pathologists on the nomenclature of bakanae fungus. The problem was resolved by the 1930s when the imperfect stage of the fungus was named *Fusarium moniliforme* (Sheldon) and the perfect stage, *Gibberella fujikuroi* (Saw.) Wr. by H. W. Wollenweber.

The damage to rice crops due to bakanae disease was especially serious in Taiwan, which was under the control of Japan prior to the Second World War, and the loss in yield was estimated to reach upwards of 40% occasionally. As a result, studies on the prevention of bakanae disease were actively conducted in Taiwan. "*Fujikuroi*" and "Saw." in the term *Gibberella fujikuroi* (Saw.) Wr. were derived from the names of two Japanese plant pathologists, Yosaburo Fujikuro and Kenkichi Sawada, who worked at the Agricultural Experiment Station of the Japanese Government House in Taiwan.

In 1912, Sawada published a paper in the *Formosan Agricultural Review* entitled "The Diseases of Crops in Taiwan" in which he suggested that the elongation in rice seedlings infected with bakanae fungus might be due to the stimulus of fungal hyphae. However, Eiichi Kurosawa (Fig. 2) should be considered the first plant pathologist to disclose the chemical mechanism of the elongation in bakanae. On the basis of irrefutable evidence, he published a paper in 1926 entitled "Experimental Studies on the nature of the substance secreted by bakanae fungus (Preliminary Report)"[2] in the *Transactions of the Natural History Society of Formosa*.

Kurosawa graduated from Chiba Prefectural High School of Horticulture in 1916 and then worked as an assistant at the same school. In 1919, he moved to Sawada's laboratory in Taiwan to work on methods for the control of bakanae disease. During the study with Sawada, Kurosawa was greatly interested in the mechanism responsible for the elongation of rice seedlings infected with bakanae fungus. Originally, he was a biologist and not a chemist; nevertheless, he developed the idea that some "chemical action" due to the infection must have caused the bakanae symptoms. In 1925, he set up his own research program to obtain evidence for this position.

Kurosawa grew the fungus on a liquid medium prepared by extracting dried rice seedlings with Knop's solution at 100°C for 2 h. The culture filtrate thus obtained, even after being heated for 2–3 h at 100°C, was found to cause marked elongation in healthy rice seedlings. In another experiment, the fungus was grown for 2.5 months on a semisolid medium composed of equal amounts of crushed rice grains and distilled water. The culture was ground in a mortar, left to stand with an appropriate volume of water, and then filtered through a Chamberland filter (a sterile filter). The resulting filtrate caused similar elongation in rice plants. Furthermore, these culture filtrates also stimulated elongation in seedlings of oat, millet, barnyard millet, maize, and sesame. Summarizing his experimental results, Kurosawa concluded that bakanae fungus secretes a kind of toxin which stimulates elongation, inhibits chlorophyll formation, and suppresses root growth and that the toxin does not lose its activity even after being kept at 100°C for 4 h. This clearly showed the participation of a heat-resistant substance introduced by the fungus into rice seedlings, the bakanae disease.

After the publication of Kurosawa's report, a number of Japanese plant pathologists, including T. Henmi and F. Seto of Kyoto University, S. Ito and S. Shimada of Hokkaido University, and T. Takahashi of Mie Agricultural College, tried to isolate the active principle. All of their attempts were unsuccessful.

In 1931, Teijiro Yabuta became interested in the problem of isolation. He was trained in organic chemistry, having received his doctoral degree at The University of Tokyo in 1916. He then obtained additional training for two years (1921–1923) in the United Kingdom, France, and the United States, accumulating experience in the isolation and structure elucidation of natural products, especially fungal metabolites. In 1924, Yabuta was appointed Professor in the Department of Agricultural Chemistry at The University of Tokyo. He held the additional post of Engineer at the Agricultural Experiment Station, Ministry of Agriculture, Tokyo. This latter appointment was highly significant for his future work because Kurosawa moved to the Experiment Station from Taiwan in 1933.

Yabuta initiated work on the isolation of the active principle using the

> 研 究 速 報
>
> 稲馬鹿苗病菌の生化學(抜粋)
> 植物を徒長せしむる作用ある物質 Gibberellin の結晶に就いて
> 農學博士 藪田貞治郎, 農學博士 住木諭介
> (東京帝國大學農學部)
> 昭和 13 年 11 月 26 日受理
>
> 稲馬鹿苗病菌 Gibberella fujikuroi をグリセリン, KH₂PO₄, NH₄Cl よりなる培養液に培養し培養液を活性炭にて處理して有効成分を吸着せしめ之をメタノール, アンモニヤにて溶出し, 更に弱酸性陽イオン樹脂等にて處理して得たる粗有効成分の粉末 (Crude Gibberellin と命名, 分解點約 60°) を醋酸エステルリグロイン又はアルコールリグロインの混合溶劑にて處理し, 2 種の有効成分を結晶狀に分離するを得たり. 之を Gibberellin A, Gibberellin B と命名す. Gibberellin A は長柱狀結晶, 融點 194~6°, 稲苗地上部を徒長せしむるのみならず根の發育を促進せしむる作用あり. C% = 75.16, H% = 7.60. M.W = 250 (ラスト法). Gibberellin B は短柱狀結晶, 分解點 245~6°. 稲苗地上部を徒長せしむる作用並に強く根の發育は阻害す. C% = 65.05, H% = 7.12, M.W = 386 (ラスト法), 357 (滴定: モノカルボキシリック酸として), [α]_D = +36.13° (メタノール, C = 4.290). (昭和 13 年 9 月東京支部講演會にて發表)

Fig. 3. The preliminary report of Yabuta and Sumiki on the crystallization of gibberellins A and B [J. Agric. Chem. Soc. Japan, 14, 1526 (1983)].

fungal strains provided by Kurosawa. In 1934, Yabuta isolated a crystalline compound with biological activity from the culture filtrate. However, it showed inhibitory effects on rice seedlings at all concentrations tested. This inhibitory compound was shown to be the same as the inhibitory "substance" found by Kurosawa in 1930. Yabuta named it fusaric acid and determined its structure to be 5-n-butylpicolinic acid.

On the advice of Kurosawa, Yabuta changed the composition of the medium, using glycerol as the carbon source instead of glucose; he also made the medium strongly acidic (pH 2.2–2.4) to suppress the formation of fusaric acid. As a result, he obtained a noncrystalline solid from the culture filtrate that had high promotive activity. He named it gibberellin in 1935. This was the first use of the term "gibberellin" in the scientific literature.

In 1938, Yabuta and his associate Yusuke Sumiki finally succeeded in crystallizing the pale yellow solid to yield gibberellin A and gibberellin B (Fig. 3).[3] (The names were interchanged in 1941: the original gibberellin A was found afterward to be inactive.) Determination of the structure of the active gibberellin was begun immediately by Yabuta and Sumiki, but they constantly had the problem of a shortage of pure crystalline sample because the productivity of their fungal strain was extremely poor on the

Fig. 4. Dr. Teijiro Yabuta, at the time he was awarded the Order of Cultural Merits, Japan (in 1964).

medium described above. Also, they did not know that the "crude" gibberellin A used for the chemical work was not pure, but a mixture of structurally related gibberellins. This made it difficult for them to analyze their chemical data. Unfortunately, the outbreak of war made the continuation of their gibberellin research increasingly difficult.

The war came to an end in 1945, and Sumiki succeeded Yabuta as Professor at The University of Tokyo in 1946. Under adverse circumstances, he renewed studies on gibberellin, but the progress was extremely slow. In 1951, Sumiki visited the United States and met Frank H. Stodola at the USDA Northern Regional Research Laboratories in Peoria, Illinois. In 1953, he visited the Akers Research Laboratories of Imperial Chemical Industries (ICI) in the United Kingdom. It was like a thunderbolt from a clear sky for Sumiki to learn that foreign scientists were already working on the chemistry and biology of fungal gibberellin on a large scale.

With the recovery of the economy, studies in Japan on gibberellins made rapid progress. In 1952, Sumiki's group began to reexamine the purity of gibberellin A produced by the fungal strains from their laboratory. In 1955, Takahashi et al.,[4] members of the group, succeeded in separating the

Fig. 5. Dr. Yusuke Sumiki (in 1969).

methyl ester of gibberellin A into three components, from which corresponding free acids were obtained and named gibberellins A_1, A_2, and A_3, respectively. In 1957, Takahashi et al.[5] isolated a new gibberellin named gibberellin A_4 as a minor component from the culture filtrate. Comparing these gibberelins with those supplied by Stodola and the ICI group, Takahashi confirmed the identity of gibberellin A_1 with Stodola's gibberellin A and of gibberellin A_3 with Stodola's gibberellin X and ICI's gibberellic acid. There were no counterparts for gibberellins A_2 and A_4.

In the 1960s, the number of papers reporting the isolation and identification of new gibberellins from fungal and plant origins rapidly increased, and there was occasional confusion about the nomenclature of these new gibberellins. In 1968, J. MacMillan and N. Takahashi[6] reached an agreement that all gibberellins should be assigned numbers as gibberellin A_1–x, irrespective of their origin.

Acknowledgment. I visited Dr. Frank H. Stodola at the USDA Northern Regional Research Laboratories, Peoria, Illinois, in 1960, when he kindly presented me with a copy of *Source Book on Gibberellin* (1928–1957) published in 1958. The book was very helpful to me in the preparation of this paper on the early history of gibberellin research in Japan. Here, I wish to express my sincerest respect to Dr. Stodola, who expended great effort in editing the valuable *Source Book*.

References

1. Hori S. Some observations on "Bakanae" disease of the rice plant. Mem Agric Res Sta (Tokyo). 1898; 12:110–119.
2. Kurosawa E. Experimental studies on the nature of the substance secreted by the "bakanae" fungus. Nat Hist Soc Formosa. 1926; 16:213–227.
3. Yabuta T, Sumiki Y. On the crystal of gibberellin, a substance to promote plant growth. J Agric Chem Soc Japan. 1938; 14:1526.
4. Takahashi N, Kitamura H, Kawarada A, Seta Y, Takai M, Tamura S, Sumiki Y. Isolation of gibberellins and their properties. Bull Agric Chem Soc Japan. 1955; 19:267–277.
5. Takahashi N, Seta Y, Kitamura H, Sumiki Y. A new gibberellin, gibberellin A_4. Bull Agric Chem Soc Japan. 1957; 21:396–398.
6. MacMillan J, Takahashi N. Proposed procedure for the allocation of trivial names to the gibberellins. Nature. 1968; 217:170–171.

CHAPTER 2

Organ-Specific Gibberellins in Rice: Roles and Biosynthesis

N. Takahashi and M. Kobayashi

1 Introduction

The analysis of plant hormones in the regulation of the life cycle of higher plants should include the following aspects:

1. Isolation and chemical characterization of endogenous plant growth regulators responsible for the physiological phenomenon.
2. Examination of higher plants at various stages of growth for any fluctuations in the levels of endogenous plant growth regulators.
3. Investigation of not only the chemical but also the biochemical aspects of the biosynthetic and metabolic pathways of the endogenous plant growth regulators.
4. Examination of the responses of various higher plants to endogenous plant growth regulators so that their physiological roles within the tissues or intact plants may be understood.
5. Molecular and cell biological investigation of the mechanism of action of endogenous plant growth regulators.

This type of approach can be achieved by the combination of biological and chemical methodologies.

Rice is known to be one of the most important and productive crops in the world, and the agronomic aspects of its life cycle have been extensively investigated by Japanese scientists. Seeds germinate in nursery beds or boxes. The young seedlings at about the 6th leaf stage, 30 to 40 days after germination, are transplanted into a paddy field. New buds (tillers) emerge at the bottom of the leaf sheath. This is the start of the tillering stage. The most vigorous tillering is observed at the 9th leaf stage. At the 11th leaf stage, tillering stops and panicles initiate at the part of the plant that is located between the bottom of the leaf sheath and the root. This is the panicle initiation stage. At this point, the vegetative growth stage is switched into the reproductive growth stage. Within a one-month period, panicles develop to form ears. Then internodes start elongating, and the ears are pushed out of the leaf sheath to cause heading. Anthesis starts at the same time as the initiation of heading. It takes several days to complete flowering. Seeds yielded by pollination and fertilization develop and ripen one and a half months after anthesis.

In the early 1970s, despite much being known about the life cycle of rice, very little was known about the endogenous plant growth regulators in rice, especially their roles in the regulation of its life cycle. In 1975, we started an investigation to identify endogenous plant growth regulators in rice. Herein we describe the identification and quantification of endogenous gibberellin (GA) in rice. Special emphasis is put on the organ-specific GAs in terms of their biosynthesis and physiological roles.

2 Plant Materials and Analytical Methods for GAs

We used *Oryza sativa* cv. Nihonbare (normal, japonica type), Tan-ginbozu (dwarf, japonica type), Waito-C (dwarf, japonica type), and Tong-il (dwarf, a hybrid between japonica and indica types). They were grown in a paddy field or in pots. Various tissues, namely, leaves, shoots, roots, and ears, were harvested at various appropriate stages of growth. If necessary, individual organs, such as spikelet, rachis, lemma, and anther were separated from the ears.

Plant materials were extracted by acetone or methanol in the usual way, and the extracts were purified according to the procedures established by our group.[1] The final identification was conducted by combined gas chromatography–mass spectrometry (GC–MS) with a full mass scan and the quantification by combined gas chromatography–selected ion monitoring (GC–SIM) using internal standards.

3 Identification and Quantification of Endogenous GAs in the Normal cv. Nihonbare

The GAs identified and the amounts found in various tissues are shown in Tables 1 and 2, respectively, and their structures are summarized in Fig. 1.[2-5] The highest levels of GA_{19} were found in the shoot and leaf, 2–7 ng/g fr wt. The GA_{19} content was observed to peak at the vigorous tillering stage, but its level was rather low at the internode elongation stage.

It has been shown that GA_{19} has a very weak effect on the growth of the dwarf cv. Waito-C, in which the conversion of GA_{19} to GA_1 appears to be genetically blocked,[6] but it promotes the growth of the normal cv. Nihonbare and the dwarf cv. Tan-ginbozu. In contrast, GA_1, which is found at the end of the GA biosynthetic pathway, promotes growth in all three cultivars. These observations indicate that GA_{19} itself is an inactive GA at

Table 1. Gibberellins identified in rice

Tissue[a]	13-Hydroxy-GAs	Non-13-hydroxy-GAs
Vegetative		
Shoot	GA_1, DA_8[b], GA_{19}, GA_{20}, GA_{29}, GA_{53}	GA_4, GA_{34}
Reproductive		
Flowering ear	GA_1, GA_8[c], GA_{19}, GA_{20}, GA_{29}, GA_{44}	GA_4, GA_9, GA_{24}, GA_{34}, GA_{51}
Rachis	GA_1, GA_{19}, GA_{20}	GA_4
Spikelet	GA_1, GA_{17}, GA_{19}, GA_{20}	GA_4, GA_{34}, GA_{51}
Lemma	GA_1, GA_{19}, GA_{20}	GA_4, GA_9, GA_{34}, GA_{51}
Anther	GA_1, GA_{19}, GA_{20}, GA_{29}	GA_4, GA_9, GA_{12}, GA_{24}, GA_{34}, GA_{51}
Immature seed	GA_1[c], GA_{19}, GA_{20}, GA_{29}, GA_{44}, GA_{53}	GA_{34}, GA_{51}

[a] GAs were identified in plant materials harvested in 1981–1984
[b] Tentatively identified in seedlings
[c] Tentatively identified

a vegetative growth stage, even though in quantity it is a major GA, while GA_1 is an active GA, although in quantity it is only a minor GA. It is suggested that GA_{19} is converted to GA_1 in the tissues of rice to exhibit biological activity. The fact that the GA_1 levels were highest at the internode elongation stage (0.63 ng/g fr wt) also suggests that GA_1 actually regulates the growth of the internode.[5]

It should be noted that most GAs identified in vegetative tissues carry a C-13 hydroxyl group; these include GA_1, GA_{19}, GA_{20}, and GA_{29}. However, in the shoot and leaf at the flowering stage, low levels of GAs lacking the C-13 hydroxyl group, such as GA_4 and GA_{34}, were detected. On the other hand, in the whole ear and organs within the ear, 13-hydroxy-GAs such as GA_1, GA_{17}, GA_{19}, GA_{20}, GA_{29}, GA_{44}, and GA_{53} and non-13-hydroxy-GAs such as GA_4, GA_9, GA_{12}, GA_{24}, GA_{34}, and GA_{51} were identified. It is noteworthy that high levels of GA_{19} were detected but GA_1 was either not detected or only present at very low levels. Furthermore, GA_4 was one of the most abundant GAs in the ear, and it was found in extremely high concentrations within the anther. The occurrence of many non-13-hydroxy-GAs at fairly high levels in the whole ear and organs within the ear is in contrast to the fact that they are not present in vegetative tissues.

This finding strongly indicates that GA_4 is both an organ-specific GA in the anther and another type of active GA in rice which may play a significant role in the regulation of the physiology of the anther, especially of pollen. This may also suggest that the biosynthetic site of non-13-hydroxy-GAs is localized in the anther.

Table 2. Content (ng/g fr wt) of GAs in various tissues of normal cv. Nihonbare

Tissue[a] and stage	13-Hydroxy-GAs									Non-13-hydroxy-GAs					
	GA_1	GA_8	GA_{17}	GA_{19}	GA_{20}	GA_{29}	GA_{44}	GA_{53}		GA_4	GA_9	GA_{12}	GA_{24}	GA_{34}	GA_{51}
Shoot															
6th leaf	0.19	—[b]		4.5	0.8	—		+[c]		—	—	—	—	—	—
Maximum tillering	0.14	—		6.9	1.1	0.3		+		—	—	—	—	—	—
Panicle initiation	+	—		+	0.2	0.1		+		—	—	—	—	—	—
Anthesis	0.21	—		1.7	0.6	0.3		+		0.3	—	—	—	1.0	—
Ear	0.27	+[d]		13	0.6	0.2	+	—		32	0.6	—	4.3	0.4	2.0
Rachis	0.05	—	—	8	0.1	—	—	—		13	—	—	—	—	—
Spikelet	0.23	—	+[e]	5	0.3	—	—	—		33	1	—	—	0.7	1.0
Lemma	0.13	—	—	15	1.1	—	—	—		24	—	—	—	0.6	1.0
Anther	10	—	—	42	0.3	0.4	—	—		3700	32	+	150	9	29
Immature seed[e]	—[b]			0.43	0.02	0.23	+	—		—	—	—	—	2.6	5.6

[a] Content of GAs was measured using plant materials harvested in 1983
[b] —, Not detected
[c] +, Identified but not quantified
[d] Tentatively identified
[e] Identified in the sample harvested in 1984

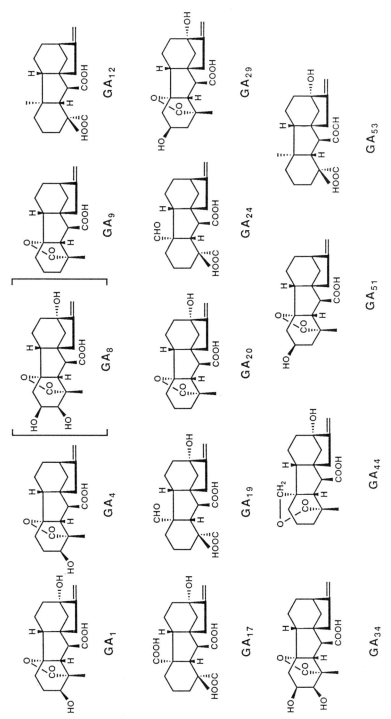

Fig. 1. Structures of endogenous GAs in rice. [ ] Tentatively identified

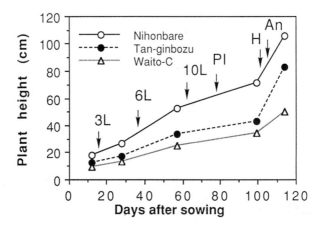

Fig. 2. Growth curves of rice cultivars Nihonbare, Tan-ginbozu, and Waito-C. 3L, 3rd leaf stage; 6L, 6th leaf stage; 10L, 10th leaf stage; PI, panicle initiation stage; H, heading stage; An, anthesis stage

4 Comparison of the Fluctuation Patterns of GA Levels Between the Normal and Dwarf Cultivars

In our previous work,[3] the levels of GAs in the shoot and leaf of the normal cv. Nihonbare and the dwarf cvs. Tan-ginbozu and Tong-il were compared. It has been shown that the levels of endogenous GAs in Nihonbare, especially GA_{19}, were much higher than those found in the dwarf cultivars. On the other hand, the fluctuation patterns of GA levels in the ear of Nihonbare and Tong-il were rather similar, but they were different from those of Tan-ginbozu, in which GA_{19} was not detected.

Recently, more comprehensive analyses of GA levels at various stages of growth have been conducted using cvs. Nihonbare, Tan-ginbozu, and Waito-C.[7]

In Fig. 2 the growth curves of the three cultivars are shown. The plant height of Tan-ginbozu and Waito-C 28 days after sowing, as compared with that of Nihonbare, was found to be 64% and 50%, respectively. The growth curves of the dwarf cultivars were linear during the vegetative stage, while a rapid increase in the plant height was observed during the reproductive stage, especially in Tan-ginbozu.

Figure 3 shows the fluctuation of the endogenous levels of GAs in the shoot at the 3rd, 6th, and 10th leaf stages and the panicle initiation and anthesis stages of 1986. Gibberellins A_1 and A_{20} were identified in the shoot of Tan-ginbozu, and GA_1, GA_{19}, and GA_{20} were identified in Waito-C and Nihonbare. As shown in Fig. 4, the levels of GA_{19} in Waito-C were almost the same as those in Nihonbare while those in Tan-ginbozu were so low that the occurrence of this GA could not be confirmed. The levels of

2. Organ-Specific GAs in Rice 15

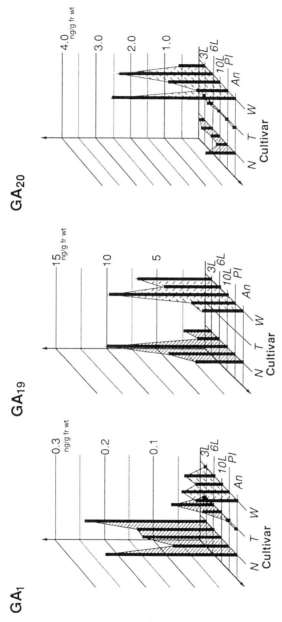

Fig. 3. Fluctuation patterns of endogenous levels of GAs in the shoot and leaf of cvs. Nihonbare, Tan-ginbozu, and Waito-C. N, Nihonbare; T, Tan-ginbozu; W, Waito-C. For definitions of abbreviations, see legend to Fig. 2

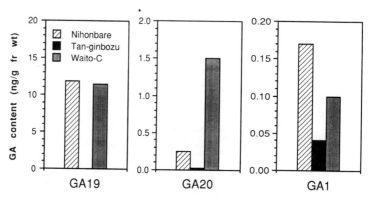

Fig. 4. Content of GA_1, GA_{19}, and GA_{20} in the shoot and leaf at 10th leaf stage of cvs. Nihonbare, Tan-ginbozu, and Waito-C

Table 3. Content of GA_4, GA_{19}, and GA_{20} in the ear of cvs. Nihonbare. Tan-ginbozu, and Waito-C

Cultivar	Content (ng/g fr wt)		
	GA_4	GA_{19}	GA_{20}
Nihonbare	6.3	19	2.1
Tan-ginbozu	14	—[a]	0.4
Waito-C	22	11	3.8

[a]—, Not detected

GA_{20} in Waito-C were much higher than those in Nihonbare while those in Tan-ginbozu were much lower. Although the levels of GA_{19} and GA_{20} were different in the two dwarf cultivars, the levels of GA_1 in both the dwarf cultivars were lower than those in the normal cultivar.

Gibberellin A_4, the major GA in the reproductive tissue of Nihonbare, was also detected in dwarf cultivars at the anthesis stage. Table 3 shows the levels of GA_4, GA_{19}, and GA_{20} in the ear of the three cultivars. Although the levels of 13-hydroxy-GAs were generally low, GA_4 was detected at higher levels in Tan-ginbozu and Waito-C than in Nihonbare. Our previous analysis of GA_4 in the reproductive organs of Nihonbare showed that GA_4 was localized in anther.[5] In our more recent study, the levels of GA_4 in the anther of the dwarf cultivars were found to be 860 ng/g fr wt in Tan-ginbozu and 1200 ng/g fr wt in Waito-C.[7]

Based on the types of GAs identified in the vegetative and reproductive tissues of the normal cv. Nihonbare and the fluctuation of their levels, it is suggested that two independent biosynthetic pathways, early-13-hydroxylation and early-non-hydroxylation, are involved in the life cycle of

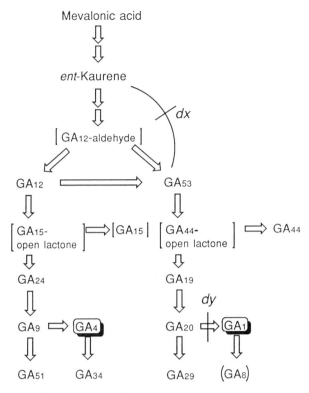

Fig. 5. Biosynthetic pathways of GA in rice. ( ) Tentatively identified; [ ] not identified; ⇨ blocking site in the dwarf cultivars; ⬤ active GAs

rice. These pathways are shown in Fig. 5. The former operates in both vegetative and reproductive tissues while the latter only operates in the reproductive tissues. The comparison among the three cultivars of the fluctuation patterns of GA levels in the shoot and the types of GAs present in the ear indicates that some of the steps of the early-13-hydroxylation pathway are blocked in the vegetative tissues of dwarf cultivars. The comparison has also shown, by the occurrence of very high levels of GA_4 in the anther, that the early-non-hydroxylation pathway should operate in the reproductive tissues of all three cultivars.

In Tan-ginbozu carrying the dx gene, the GA biosynthesis appears to be blocked at a rather early stage. Moore et al.[8] reported that endogenous levels of *ent*-kaurene and *ent*-isokaurene in Tan-ginbozu were almost the same as those in Nihonbare. This evidence, together with the fluctuation of GA levels in Tan-ginbozu, indicates that the dx gene controls the step between *ent*-kaurene and GA_{53}.

Murakami[6] suggested that in Waito-C carrying the dy gene the conversion of GA_{20} to GA_1 was blocked. This suggestion was based on the finding that the biological activity of GA_{20} on the shoot elongation of Waito-C was about one hundred times lower than that of GA_1, indicating that the *dy* gene controls the above step. Our finding that levels of GA_{20} in Waito-C were higher than those in Nihonbare supports Murakami's suggestion.

It must be emphasized that the comparison of endogenous GAs among the three cultivars clearly reveals that GA_4, together with GA_1, is an important and active GA in the anther. Gibberellin A_4, which is one of the organ-specific GAs in rice, has physiological roles which are still unclear but may concern the regulation of heading, anthesis, and/or maturation of pollen.

It is interesting that the GA biosynthesis is not blocked in the anther of the dwarf cultivars, and our results indicate that GA biosynthesis in the shoot is under different genetical controls from those in the anther.

5 Conversion of GAs by Cell-free Extracts from Rice Cultivars

To obtain more evidence for the occurrence of 13-hydroxy-GAs and non-13-hydroxy-GAs in the reproductive tissues, cell-free extracts were prepared from the anther of the normal cv. Nihonbare and the dwarf cv. Waito-C.[9]

Ears were collected at the heading stage, and anthers were picked out from spikelets. Anthers with an equal amount of potassium phosphate buffer (0.05 M, pH 8) were homogenized and extracted. The extract was filtered through cheesecloth and centrifuged (2000 × g for 10 min at 4°C). The supernatant (S-2) was further centrifuged at 200,000 × g for 60 min at 4°C, giving the supernatant (S-200), which was dialyzed against a phosphate buffer (0.05 M, pH 8) containing 0.05 M dithiothreitol (DTT), 0.2 M sucrose, and 30% glycerol. After the dialysis, the enzyme solution was stored at −80°C.

Radioisotope-labeled GAs were incubated with 88 μl of purified cell-free extract in the presence of NADPH, 2-ketoglutarate, ascorbate, and $FeSO_4$ (final concentrations were 2 mM, 10 mM, 10 mM, and 1 mM, respectively) at 30°C for 2 h. The incubation mixture was acidified and extracted with ethyl acetate to give an ethyl acetate-soluble fraction. After prepurification with a Bond Elut DEA cartridge column, the ethyl acetate-soluble fraction was subjected to high-performance liquid chromatography (HPLC) analysis using columns of ODS and Nucleosil 5 $N(CH_3)_2$. Carbon-14-labeled GAs were detected with an on-line radioanalyzer. The radioactivity in collected fractions was measured by a liquid scintillation counter in order to detect the presence of any ³H-labeled GAs.

Gas chromatographic–mass spectrometric analysis was carried out to identify the products. Samples were trimethylsilylated with *N*-methyl-(*N*-trimethylsilyl) trifluoroacetamide at 80°C for 30 min. The GC–MS analysis

Table 4. Conversion of labeled GAs by the cell-free extracts of the anther of normal and dwarf cultivars

Substrate	Products	
	Waito-C enzyme	Nihonbare enzyme
[^{14}C]GA$_{12}$	GA$_{15}$,[a] GA$_{34}$[a]	GA$_{15}$, GA$_{34}$
[^{14}C]GA$_9$	GA$_{34}$[a]	
[^{14}C]GA$_{53}$	GA$_{19}$, GA$_{20}$, GA$_{44}$	GA$_{19}$, GA$_{44}$
[^{13}C,^{3}H]GA$_{20}$	GA$_1$,[a] GA$_8$,[a] GA$_{29}$[a]	

[a] Confirmed by full-scan GC–MS

was conducted with an INCOS 50 mass spectrometer equipped with a capillary column.

[^{14}C]Gibberellin A$_{12}$, [^{14}C]GA$_9$, [^{14}C]GA$_{53}$, and [^{13}C^{3}H]GA$_{20}$ were used as substrates in incubations with the extract from Waito-C. The HPLC–radiochromatogram of the product of GA$_{12}$ incubation showed two major peaks with t_R values of 23 min and 31 min. The first peak was identified as [^{14}C]GA$_{34}$ by GC–MS analysis, and the second as [^{14}C]GA$_{15}$. The first peak at t_R 23 min was also detected in the GA$_9$ incubation experiment.

In the incubation of [^{14}C]GA$_{53}$, GA$_{44}$ was shown to be a major product by HPLC analysis, and GA$_{19}$ and GA$_{20}$ were identified as minor products. Since no 3β-hydroxylated products were detected, [^{13}C^{3}H]GA$_{20}$ was incubated to analyze 3β-hydroxylation activity. The occurrence of two radioactive fractions was indicated by HPLC analysis of the incubation product, and the first fraction was identified by GC–MS as a mixture of [^{13}C]GA$_8$ and GA$_{29}$, while the second was identified as [^{13}C]GA$_1$.

[^{14}C]Gibberellin A$_{12}$ and [^{14}C]GA$_{53}$ were also incubated with the extracts from Nihonbare. Although the efficiency of conversion in each substrate was low, HPLC analysis gave results similar to those mentioned above.

Table 4 shows a summary of the results of incubation. These finally confirmed that non-13-hydroxy-GAs that had accumulated in the ears at the heading stage were produced in the anther. These also demonstrated that both 13-hydroxy-GA and non-13-hydroxy-GA pathways (Fig. 5) operate in the anther and that non-13-hydroxy-GAs are metabolized more efficiently than 13-hydroxy-GAs. This is consistent with the fact that, in the anther, the amount of GA$_4$ present was much higher than that of GA$_1$.

The cell-free extract of the anther from Waito-C showed as much 3β-hydroxylation activity as that of Nihonbare. This result is in accordance with the fact that the anther of dwarf cultivars contained a large amount of GA$_4$. Since it has been shown that 3β-hydroxylation is blocked in the shoot of Waito-C, it is clearly shown that the regulation of GA biosynthesis is specific to the organs. The mechanism of this regulation is still not clear.

6 Conclusion

In the course of our study to clarify the regulatory mechanism of the life cycle of rice, the identification and quantitation of endogenous GAs in various tissues at various stages of growth was conducted using the normal cv. Nihonbare and the dwarf cvs. Tong-il, Tan-ginbozu, and Waito-C.

13-Hydroxy-GAs such as GA_1, GA_{17}, GA_{19}, GA_{20}, GA_{29}, GA_{44}, and GA_{53} were identified in most types of tissue, namely the shoot, leaf, ear, and anther. However, non-13-hydroxy-GAs such as GA_4, GA_9, GA_{12}, GA_{24}, GA_{34}, and GA_{51} were identified mainly in the reproductive tissues, namely, the anther and pollen. It has been suggested that GA_1 is the active GA in vegetative tissue such as the shoot and leaf, based on its biological activity in various dwarf mutants of rice and on biosynthetic considerations. On the other hand, GA_4 may be the active GA in the pollen and anther, based on its high concentration in these tissues.

The comparison among the GA levels of the three cultivars in the course of the life cycle indicates that the production of GA_1, one type of active GA, is inhibited in the vegetative tissues of dwarf cultivars, but that GA_4, another type of active GA, is produced in the reproductive tissues of all three cultivars.

Cell-free preparations from anthers of both the normal cv. Nihonbare and the dwarf cv. Waito-C showed the ability to convert GA_{12} and GA_9 to GA_{34}. The conversion of GA_{20} to GA_1, GA_8, and GA_{29} was also observed to a lesser extent.

This evidence clearly shows the presence of organ-specific GAs. This can be explained by the fact that (1) there exist two independent GA biosynthetic pathways, namely, the early-13-hydroxylation and the early-nonhydroxylation pathways; and (2) the former pathway operates throughout the life cycle of rice, while the latter operates only in the reproductive stage and not in the vegetative stage.

Acknowledgments. This research has been conducted in collaboration with Dr. S. Kurogochi, Dr. N. Murofushi, Mr. Y. Suzuki, Dr. I. Yamaguchi, and Dr. T. Yokota at The University of Tokyo, Dr. A. Sakurai and Dr. Y. Kamiya at The Institute of Physical and Chemical Research, and Dr. Y. Ota and Dr. H. Saka at the National Institute of Agrobiological Resources, to whom we wish to express our thanks.

References

1. Takahashi N, Yamaguchi I. Analyses of endogenous plant hormone level throughout the life cycle of higher plants. Acta Hort. 1986; 179:47–57.
2. Kurogochi S, Murofushi N, Ota Y, et al. Identification of gibberellins in the rice plant and quantitative changes of gibberellin A_{19} throughout its life cycle. Planta. 1979; 146:185–191.
3. Suzuki Y, Kurogochi S, Murofushi N, et al. Seasonal changes of GA_1, GA_{19} and abscisic acid in three rice cultivars. Plant Cell Physiol. 1981; 22:1085–1093.
4. Kobayashi M, Yamaguchi I, Musofushi N, et al. Endogenous gibberellins in immature seeds and flowering ear of rice. Agric Biol Chem. 1984; 48: 2725–2729.
5. Kobayashi M, Yamaguchi I, Murofushi N, et al. Fluctuation and localization of endogenous gibberellin in rice. Agric Biol Chem. 1988; 52:1189–1194.
6. Murakami Y. Dwarfing genes in rice and their relation to gibberellin biosynthesis. In: Carr DJ, ed. Plant growth substances 1970. Berlin: Springer-verlag, 1972: p. 166–174.
7. Kobayashi M, Sakurai A, Saka H, et al. Quantitative analysis of endogenous gibberellins in normal and dwarf cultivars of rice. Plant Cell Physiol. 1989; 30:963–969.
8. Moore TC, Yamane H, Murofushi N, et al. Concentrations of *ent*-kaurene and squalene in vegetative rice shoots. J Plant Growth Regul. 1988; 7:145–151.
9. Kobayashi M, Kamiya Y, Sakurai A, et al. Metabolism of gibberellins in cell-free extracts of anthers from normal and dwarf rice. Plant Cell Physiol. 1990; 31:289–293.

CHAPTER 3

Gibberellin Metabolism in Maize: Tissue Specificity

B.O. Phinney, C.R. Spray, Y. Suzuki, and P. Gaskin

1 Introduction

The single-gene dwarf mutants *d1*, *d2*, *d3*, and *d5* have been used to analyze the early-13-hydroxylation pathway of gibberellin (GA) biosynthesis in the shoots of maize (*Zea mays* L.). These dwarf mutants are GA mutants in that they exhibit normal growth in response to exogenous GA; they do not respond to any other plant hormones or growth regulators. The location of the specific steps blocked by the mutants has been based on the following information: (1) the identification of the GAs native to maize shoots[1]; (2) the relative bioactivities of the maize GAs on each of the four mutants[2]; (3) the quantification of the GAs present in normal, *d1*, *d2*, *d3*, and *d5* maize shoots[1]; and (4) the metabolic fate of the label after feeds of double-labeled GAs and their precursors to normal and mutant maize shoots[3] and to cell-free systems.[4] The results clearly establish the early-13-hydroxylation pathway as the major pathway of GA biosynthesis in maize shoots; the results clearly define the specific steps controlled by two of the mutants (*d1* and *d5*). The above studies have led to the conclusion that GA_1 is the only GA bioactive per se in the pathway controlling stem elongation in maize shoots[5,6]; the other GAs in the pathway are active through their metabolism to GA_1. The recent identification of GA_3 from maize seedling shoots[7] together with the demonstration of its biosynthetic origin from GA_{20} via GA_5[8] raises the question of the relative roles of GA_1 and GA_3 as terminal active GAs controlling shoot growth in maize. The answer will depend on a knowledge of the enzymes which control the conversion of GA_{20} to GA_1, GA_{20} to GA_5, and GA_5 to GA_3. The answer is also dependent on the binding properties of GA_1 (and GA_3?) to the purported and elusive GA receptor.

It is not known *how* the *d1* gene controls the conversion of GA_{20} to GA_1. There are several possibilities, depending upon the nature of the mutation. For example, control could be due to the production of an altered enzyme (3β-hydroxylase?); it could be due to the production of an inhibitor of the enzyme; it could be due to intracellular compartmentation. Again, the question of the control by the *dl* lesion will remain unanswered until we have a knowledge and understanding of the enzyme (3β-hydroxylase?) controlled by this gene.

It is the purpose of this report to present preliminary results on (1) the presence and distribution of GAs in specific regions of the maize vegetative shoot and (2) the ability of these specific regions (tissues) to metabolize GA_{20}. The goal is to find a source material for future enzymological studies. The studies are ultimately directed toward information on the molecular biology of the genes that control shoot elongation in maize.

Preliminary studies with Dr. Valerie Smith of the School of Chemistry, University of Bristol, suggested that homogenates obtained from young stems of maize could convert $[^3H]GA_{20}$ to a $[^3H]GA_1$-like compound (unpublished data, 1988). As a result, the series of experiments reported herein was designed to look for GA_{20} metabolism using maize internodes and leaves. The purpose was to find a convenient tissue source from which to purify the maize 3β-hydroxylase.

2 Materials and Methods

2.1 Plant Material

The material used in this study came from normal (tall) plants of the genetic background CC5/L317. For GA identification and quantification, the plants were grown in the UCLA Botanic Garden; for enzyme analysis, they were grown in the UCLA greenhouse. Plants were approximately 6–8 weeks old at the time of harvest.

Data on the identification and quantification of GAs were obtained from three kinds of plant material: the cortical regions from immature internodes (1–3 cm in length); the cortical regions from mature internodes (6–8 cm in length); and immature leaf sheaths (1–3 cm in length). The epidermal regions from the internodes were removed from the internodes and discarded. The leaf sheath material included both epidermal and cortical regions. Blocks of tissue (200 g fr wt) were removed from the stem and petiole with a razor blade and processed immediately.

Data on GA_{20} metabolism were obtained from ten kinds of plant material: cortical and epidermal regions from both the upper and lower portions of immature internodes (3–5 cm in length); cortical and epidermal regions from both the upper and lower portions of mature internodes (6–8 cm in length); and immature and mature leaf sheaths (1–3 cm in length). The epidermal region (green outermost layer) was removed from the internodes with a razor blade and diced. One-millimeter cubes were obtained from the cortical tissues. All material was used immediately for metabolic studies.

2.2 Identification and Quantification of GAs

a Extraction and Purification

Each of the three tissue samples (ca. 200 g fr wt) was homogenized in a blender with MeOH:H_2O (80:20, v/v; 500 ml). After 20 min the mixture was filtered and the residue was reextracted twice in MeOH:H_2O (80:20, v/v; 500 ml), for 14 h and 6 h, at 5°C. The filtrates from the three extractions were combined, and the following ten internal standards were added: [$^{13}C,^{3}H$]GA_1 (2.11 GBq/mmol, 30 ng); [$^{13}C,^{3}H$]GA_3 (0.74 GBq/mmol, 20 ng); [$^{13}C,^{3}H$]GA_5 (1.51 GBq/mmol, 10 ng); [$^{13}C,^{3}H$]GA_8 (1.48 GBq/mmol, 40 ng); [^{2}H]GA_{19} (102 ng); [$^{13}C,^{3}H$]GA_{20} (1.79 GBq/mmol, 100 ng); [$^{13}C,^{3}H$]GA_{29} (1.58 GBq/mmol, 40 ng); [^{14}C]GA_{44} (4.29 GBq/mmol, 89 ng); [^{14}C]GA_{53} (4.29 GBq/mmol, 158 ng); [$^{13}C,^{3}H$]GA_{12}-aldehyde (1.40 GBq/mmol, 100 ng). All the internal standards were at least 99.5% pure, as determined by gas chromatography–mass spectrometry (GC–MS). [^{14}C]Gibberellin A_{44} and [^{14}C]GA_{53} were gifts from Drs. J.E. Graebe and P. Hedden; [^{2}H]GA_{19} was a gift from Dr. L.N. Mander; [$^{13}C,^{3}H$]GA_3 was prepared by Dr. C.L. Willis (manuscript in preparation); [$^{13}C,^{3}H$]GA_{12}-aldehyde was prepared by Dr. H. Yamane (the preparation will be described in a future paper); the other internal standards were prepared as described by Fujioka et al.[1] The MeOH was evaporated in vacuo at 30°C to give an aqueous residue (ca. 100 ml). Polyvinylpolypyrrolidone (PVPP, 4 g) was added to the aqueous residue, and the mixture was stirred overnight at 5°C. The PVPP was removed by filtration through a bed of Celite. The Celite/PVPP bed was washed with H_2O (ca. 100 ml), and the filtrate volume was brought to 200 ml with H_2O). The filtrate was solvent partitioned as described by Fujioka et al.[1,7] to give aqueous (AQ), neutral ethyl acetate (NE), and acidic ethyl acetate (AE) fractions. The AQ and NE fractions were not analyzed. The AE fraction was dissolved in MeOH (1 ml) and loaded onto a column of DEAE Sephadex A-25. The column (20 ml) was eluted with MeOH, 0.25 N acetic acid (HOAc) in MeOH, 0.50 N HOAc in MeOH, 0.75 N HOAc in MeOH, 1.0 N HOAc in MeOH, 3.0 N HOAc in MeOH, and 5.0 N HOAc in MeOH (20 ml each). Aliquots of each fraction were radiocounted. Radioactive fractions were combined and passed through a Sep-Pak C_{18} cartridge eluted with HOAc:H_2O (0.1:99.9, v/v; 7 ml) followed by MeOH:H_2O (80:20, v/v; 7 ml). The MeOH:H_2O fraction was concentrated (no radioactivity was detected in the aqueous HOAc eluate) and purified by high-performance liquid chromatography (HPLC) on Nucleosil NMe_2. The column (100 × 10 mm) was eluted with MeOH containing 0.1%

HOAc, at a flow rate of 3.0 ml/min. Fractions were collected every minute, and an aliquot of each fraction was radiocounted. Based on the radiocounting results, selected radioactive fractions were combined to give fraction 1 [retention time (Rt): 4–5 min], fraction 2 (Rt: 9–13 min), fraction 3 (Rt: 13–16 min), and fraction 4 (Rt: 16–19 min).

b Derivatization and GC–MS

The four fractions from each tissue sample were analyzed by GC–selected ion monitoring (SIM) for the endogenous presence of GA_{53}, GA_{44}, GA_{19}, GA_{29}, GA_{20}, GA_5, GA_3, GA_8, and GA_1. The M^+ ion cluster was monitored for the MeTMSi derivatives of all GAs except for GA_{19}, for which the M^+-28 ion cluster was used. The procedures for derivatization, GC–SIM, and quantitation were identical to those described by Fujioka et al.,[1] except that the GC temperature program was from 50°C to 200°C at 2°C per minute, then to 300°C at 15°C per minute.

2.3 [^{13}C, ^{3}H]Gibberellin A_{20} Metabolism

Each set of diced tissues (1 g fr wt) was transferred to a glass vial containing 2 ml of incubation medium (Murashige and Skoog medium supplemented with 2% sucrose). [^{13}C,^{3}H]Gibberellin A_{20} (0.448 µg, 1.28 × 10^5 dpm) was added, and the mixture was incubated with gentle shaking, at 25°C for 24 h. The incubation medium was removed from the plant material with a pipette, and the plant material was transferred to a glass homogenizer. The plant material was pulverized in MeOH (5 ml) and then filtered through glass wool. The filtrate was combined with the incubation medium, and the MeOH was removed by evaporation in vacuo. The aqueous residue was acidified with 6 N HCl and extracted with ethyl acetate (EtOAc) (3 × 1.5 ml). The combined EtOAc extracts were evaporated to dryness, redissolved in MeOH:H$_2$O:HOAc (1 ml; 50:49.95:0.05; v/v/v) and injected onto a Lichrosorb C_{18} reverse-phase HPLC column (250 × 10 mm). The column was eluted isocratically with MeOH:H$_2$O:HOAc (50:49.95:0.05; v/v/v), at a flow rate of 2.5 ml/min. Fractions were collected every minute, and an aliquot of each fraction (total of 40 fractions) was analyzed for radioactivity by liquid scintillation counting. The assignment of GA-like metabolites to radioactive fractions was based on the occurrence of radioactivity at the retention time of authentic standards for the expected metabolites of GA_{20}.

Since the above HPLC conditions did not separate GA_{20} from GA_5 (a possible metabolite of GA_{20}), the fractions containing presumptive GA_{20} and GA_5 were combined, concentrated in vacuo, and rechromatographed by HPLC on a Nucleosil NMe$_2$ column (100 × 10 mm) eluted with MeOH containing 0.1% HOAc, at a flow rate of 3.0 ml/min. Fractions were collected every minute and radiocounted.

Table 1. Percent incorporation of isotopic label for each internal standard, determined by GC–SIM, used in the quantitation of GAs from maize stem tissues. The amounts of each internal standard added to the aqueous methanol extracts are also given

Gibberellin	% Isotope incorporation	Amount added (ng/100 g fr wt)		
		MI[a]	II[b]	ILS[c]
GA_{12}-aldehyde	$90.5\,^{13}C_1 : 9.5\,^{13}C_0$	48	48	49
GA_{53}	$38.4\,^{14}C_4 : 61.6\,^{14}C_0$	76	76	78
GA_{44}	$39.4\,^{14}C_4 : 60.6\,^{14}C_0$	43	43	44
GA_{19}	$89.7\,^{2}H_2 : 4.8\,^{2}H_1 : 5.4\,^{2}H_0$	49	49	50
GA_{20}	$91.5\,^{13}C_1 : 8.5\,^{13}C_0$	48	48	49
GA_{29}	$91.8\,^{13}C_1 : 8.2\,^{13}C_0$	19	19	20
GA_1	$91.5\,^{13}C_1 : 8.5\,^{13}C_0$	14	15	15
GA_8	$92.0\,^{13}C_1 : 8.0\,^{13}C_0$	19	19	20
GA_5	$91.5\,^{13}C_1 : 8.5\,^{13}C_0$	4.8	4.8	4.9
GA_3	$93.6\,^{13}C_1 : 6.4\,^{13}C_0$	9.6	9.7	9.9

[a] MI, Mature internode
[b] II, Immature internode
[c] ILS, Immature leaf sheath

Table 2. Isotope ratios, obtained by GC–SIM analysis, for the GAs in the HPLC-purified AE fractions from maize stem tissues. Data for the M$^+$ ion cluster of the MeTMSi derivatives

Gibberellin	HPLC fraction number	Isotope ratios		
		Mature cortex	Immature cortex	Immature leaf sheath
		$^{13}C_1 : ^{13}C_0$	$^{13}C_1 : ^{13}C_0$	$^{13}C_1 : ^{13}C_0$
GA_{12}-aldehyde	1	—[a]	—[a]	—[a]
GA_{20}	3	33.7 : 66.3	53.4 : 46.6	31.8 : 68.2
GA_{29}	3	78.6 : 21.4	61.6 : 38.4	66.2 : 33.8
GA_1	2	47.5 : 52.5	36.4 : 63.6	21.0 : 79.0
GA_8	2,3	56.8 : 43.2	14.6 : 85.4	22.8 : 77.2
GA_5	4	64.3 : 35.7	65.0 : 35.0	54.4 : 45.5
GA_3	2	79.3 : 20.7	48.3 : 51.7	33.3 : 66.7
		$^{14}C_4 : ^{14}C_0$	$^{14}C_4 : ^{14}C_0$	$^{14}C_4 : ^{14}C_0$
GA_{53}	1	19.5 : 80.5	1.7 : 98.3	4.0 : 96.0
GA_{44}	2	8.8 : 91.2	3.2 : 96.8	7.2 : 92.8
		$^{2}H_2 : ^{2}H_1 : ^{2}H_0$	$^{2}H_2 : ^{2}H_1 : ^{2}H_0$	$^{2}H_2 : ^{2}H_1 : ^{2}H_0$
GA_{19}[b]	2	14.7 : 1.3 : 84.0	3.9 : 1.3 : 94.8	3.1 : 4.0 : 92.9

[a] GA_{12}-aldehyde was not detected in the AE–HPLC fractions; however, GA_{12} was present in the samples [ions detected at m/z 328 (M$^+$ − 28), 300, 285, 241, and 240; KRI 2353]. The NE fraction, expected to contain GA_{12}-aldehyde, was not examined by GC–SIM
[b] M$^+$ − 28 ion cluster

Table 3. Amounts of endogenous GAs in maize stem tissues

Tissue	Amount (ng/100 g fr wt)								
	GA_{12}-ald	GA_{53}	GA_{44}	GA_{19}	GA_{20}	GA_{29}	GA_1	GA_8	GA_5	GA_3
Mature internode	—[a]	75	155	250	86	4	13	12	2	2
Immature internode	—[a]	1700[b]	500[b]	875[b]	36	10	22	106	2	9
Immature leaf sheath	—[a]	680[b]	200	625[b]	94	8	50	61	3	16

[a] The NE fraction, expected to contain GA_{12}-aldehyde, was not examined by GC–SIM
[b] These figures are very approximate due to the high dilution of label in the shoot samples (Table 2)

3 Results

3.1 Endogenous GAs

Data are given in Table 1 on the amounts of internal standards added to each extract. These amounts were based on the levels of endogenous GAs previously reported for normal maize seedling shoots by Fujioka et al.[1] In addition, Table 1 shows the percent incorporations of isotopic label obtained by GC–SIM analysis of the M$^+$ ion clusters of the derivatized internal standards for [^{13}C,^{3}H]GA_{12}-aldehyde, [^{14}C]GA_{53}, [^{14}C]GA_{44}, [^{13}C,^{3}H]GA_{20}, [^{13}C,^{3}H]GA_{29}, [^{13}C,^{3}H]GA_1, [^{13}C,^{3}H]GA_8, [^{13}C,^{3}H]GA_5, and [^{13}C,^{3}H]GA_3. The M$^+$ -28 ion cluster was analyzed for [^{2}H]GA_{19} MeTMSi.

Table 2 shows the isotope ratios for GA_{53}, GA_{44}, GA_{19}, GA_{20}, GA_{29}, GA_1, GA_8, GA_5, and GA_3. These values were obtained by GC–SIM analysis. The calculated levels for each endogenous GA are shown in Table 3. The calculations are based on the methods developed by the MacMillan group as described in Fujioka et al.[1]

3.2 [^{13}C, ^{3}H]Gibberellin A_{20} Metabolism

The results of the incubations of radiolabeled GA_{20} with tissue dices are shown in Table 4. All preparations metabolized radiolabeled GA_{20} to a GA_{29}-like compound (major metabolite) and to a GA_1-like compound (minor metabolite). In some incubations using epidermal regions and leaf sheaths, a radioactive peak was detected that corresponded to the retention time of GA_8. All preparations metabolized radiolabeled GA_{20} to a metabolite with an HPLC retention time that did not correspond to that of any of the expected metabolites (see Fig. 1). In addition, all preparations metabolized radiolabeled GA_{20} to a highly polar radiolabeled compound(s). This compound(s) was not extractable by EtOAc from the aqueous phase at pH 2.5.

Table 4. [^{13}C, ^{3}H]Gibberellin A_{20} metabolism from 1-mm cubes obtained from selected maize tissue regions. The data are expressed as percent total recovered radioactivity. The two values for each material represent the results of replicate incubations

Material	GA_{20}-like	GA_{8}-like	GA_{29}-like	GA_{1}-like	Unknown	AQ fraction
Internode—upper						
Cortical region						
Immature	34.6	—[a]	4.3	0.2	9.3	51.8
	64.2	—	9.7	0.2	2.8	22.7
Mature	4.2	—	12.1	<0.1[b]	11.9	31.9
	29.2	—	10.5	<0.1	15.6	44.8
Epidermal region						
Immature	16.5	—	8.8	0.5	2.1	72.1
	35.1	0.1	22.1	0.7	0.9	41.2
Mature	28.7	—	30.5	2.3	0.7	37.8
	16.3	—	21.1	2.4	0.9	59.3
Internode—lower						
Cortical region						
Immature[c]	12.6	—	7.3	0.3	2.5	77.0
Mature	32.7	—	2.1	0.3	4.3	60.7
	42.3	—	2.6	0.5	13.7	40.7
Epidermal region						
Immature	8.4	0.1	4.7	0.6	0.6	85.7
	20.7	0.2	18.2	0.2	0.5	60.4
Mature	37.0	<0.1	23.7	0.4	0.9	38.0
	19.7	—	15.7	2.9	1.0	60.8
Leaf sheath						
Immature	30.2	0.1	3.3	0.3	1.2	65.0
	49.0	<0.1	0.5	<0.1	1.9	48.6
Mature	63.3	—	4.3	<0.1	1.6	30.8
	71.8	0.1	4.9	<0.1	1.5	22.1

[a] —, No metabolite observed
[b] <, Amount of metabolite too low to measure
[c] Only one set of values obtained for this material due to loss of sample.

4 Discussion

4.1 Endogenous GAs

Our initial goal was to identify and quantify the GAs of the early-13-hydroxylation pathway from specific maize tissues. This would provide a basis for expecting the presence in these tissues of the enzyme (3β-hydroxylase) that converts GA_{20} to GA_{1}. Previous studies on the endogenous GAs from normal maize shoots (entire seedlings) have quantified the presence of the eight GAs GA_{53}, GA_{44}, GA_{19}, GA_{20}, GA_{29}, GA_1, GA_8, and GA_5, using GC–SIM and labeled internal standards.[1,7] In the studies reported herein, we used as standards [^{14}C]GA_{53}, [^{14}C]GA_{44}, [^{2}H]GA_{19}, [^{13}C,^{3}H]GA_{20}, [^{13}C,^{3}H]GA_{29}, [^{13}C,^{3}H]GA_1, [^{13}C,^{3}H]GA_8, [^{13}C,^{3}H]GA_5, [^{13}C,^{3}H]GA_{12}-aldehyde, and [^{13}C,^{3}H]GA_3. (GA_3 was previously identified from normal maize seedling shoots by full-scan GC–MS and Kovats retention index[7]; however, since labeled GA_3 was not available, the levels of GA_3 could be estimated only, based on total ion current response data.)

The identification of specific endogenous GAs in a tissue sample, reported here, is based on dilution of the isotope label of the internal standards by nonlabeled endogenous GA (see Table 2). It is concluded that all three types of tissues contained GA_{53}, GA_{44}, GA_{19}, GA_{20}, GA_{29}, GA_1, GA_8, GA_5, and GA_3 (Table 3). These GAs are members of the early-13-hydroxylation pathway.

The levels of the nine endogenous GAs from mature internodes were similar to the levels reported for seedling shoots of normal maize.[1,7] The endogenous GA_3 level from seedling shoots was estimated to be 2 ng/100 g fr wt based on total ion current data[7] (see above). In the present study, we used [17-^{13}C,^{3}H]GA_3 as the internal standard and obtained values of 2 ng/100 g fr wt from mature internodes, 9 ng/100 g fr wt from immature internodes, and 16 ng/100 g fr wt from immature leaf sheaths.

The values for GA_1 are 13 ng/100 g fr wt from mature inernodes, 22 ng/g fr wt from immature internodes, and 50 ng/100 g fr wt from immature leaf sheaths. These values compare to 0, 12, and 22 ng/100 g fr wt from seedling shoots of normal maize.[1,7] Thus, GA_1 is apparently present at higher levels than GA_3 in all tissues studied to date.

4.2 [^{13}C, ^{3}H]Gibberellin A_{20} Metabolism

All preparations examined converted GA_{20} to GA_1-like and GA_{29}-like compounds. In addition, a GA_8-like compound was obtained in preparations from the epidermal regions of upper and lower internodes and from the leaf sheaths. The identification of metabolites was based upon HPLC–radiocounting only. The definitive identification of these metabolites from specific tissue regions will appear in a subsequent paper. It is interesting that the highest metabolism of GA_{20} to GA_1 and GA_8 was by cubes obtained from the epidermal regions.

We also examined the possible metabolism of GA_{20} to GA_5. Since the reverse-phase HPLC conditions did not separate GA_{20} from GA_5, the fractions containing the presumptive GA_{20} and GA_5 were rechromatographed by HPLC on Nucleosil NMe$_2$. No evidence was found (by radiocounting) for the presence of a GA_5-like compound from five separate incubations (data not shown). This is especially interesting in view of the occurrence of GA_5 in maize stems and our findings that intact young maize seedlings metabolize GA_{20} to both GA_1 and GA_5.[8] There are reports that partially purified 3β-hydroxylase preparations from *Phaseolus*

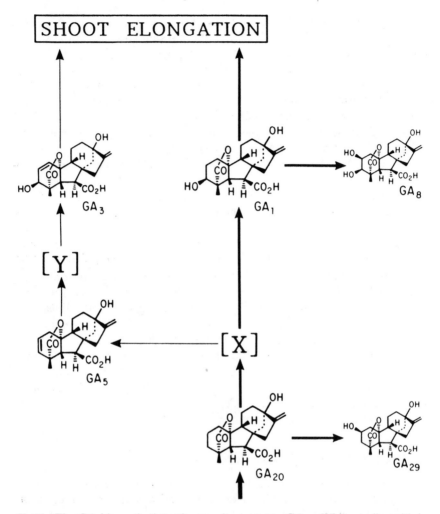

Fig. 1. The GA biosynthetic pathway subsequent to GA_{20}. While studies with intact seedlings show conversion of GA_{20} to GA_5,[8] no evidence was obtained for this conversion in the incubations reported here using internodes and leaf sheaths. [X] and [Y] are presumptive unstable intermediates

vulgaris seeds* catalyze the conversion of GA_{20} to GA_5 (possibly via an intermediate common to the formation of both GA_1 and GA_5; see Fig. 1).

In conclusion, maize internodes represent a convenient and plentiful tissue source from which to attempt the purification of the GA_{20} 3β-hydroxylase. These results are very encouraging since Gilmour et al.[12]

---

*References 9–11 and Chapter 7 of this volume.

have previously referred to unpublished and unsuccessful attempts to obtain GA oxidase activity from the leaves of several plant species and concluded that the isolation of these low-abundance and labile GA oxidases from vegetative tissue is a difficult problem.

Acknowledgments. We wish to thank the following for their generous gifts of labeled GAs used in this work: Drs. P. Hedden and J.E. Graebe for [^{14}C]GA$_{53}$ and [^{14}C]GA$_{44}$, Dr. L.N. Mander for [^{2}H]GA$_{19}$, and Dr. C.L. Willis for [^{13}C,^{3}H]GA$_3$. Finally, we are most grateful to Dr. V.M. Smith for her valuable advice.

References

1. Fujioka S, Yamane H, Spray CR, et al. Qualitative and quantitative analyses of gibberellins in vegetative shoots of normal, *dwarf-1*, *dwarf-2*, *dwarf-3*, and *dwarf-5* seedlings of *Zea mays* L. Plant Physiol. 1988; 88:1367–1372.
2. Phinney BO, Spray CR. Chemical genetics and the gibberellin pathway in *Zea mays* L. In: Wareing PF, ed. Plant growth substances 1982. London: Academic Press, 1982: p. 101–110.
3. Spray CR, Phinney BO, Gaskin P, et al. Internode length in *Zea mays* L. The *dwarf-1* mutation controls the 3β-hydroxylation of gibberellin A$_{20}$ to gibberellin A$_1$. Planta. 1984; 160:464–468.
4. Hedden P, Phinney BO. Comparison of *ent*-kaurene and *ent*-isokaurene synthesis in cell-free systems from etiolated shoots of normal and *dwarf-5* maize seedlings. Phytochemistry. 1979; 18:1475–1479.
5. Phinney BO. Gibberellin A$_1$, dwarfism and shoot elongation in higher plants. Biologia Plantarum (Prague). 1985; 27:47–53.
6. Phinney BO, Spray CR. Gibberellins (GAs), gibberellin mutants and their future in molecular biology. In: Randall DD, Blevins DG, Larson RL, eds. Proc. 4th Ann. Plant Biochem. and Physiol. Symp. Columbia: University of Missouri, 1985: p. 67–74.
7. Fujioka S, Yamane H, Spray CR, et al. The dominant non-gibberellin-responding dwarf mutant (*D8*) of maize accumulates native gibberellins. Proc Natl Acad Sci USA. 1988; 85:9031–9035.
8. Fujioka S, Yamane H, Spray CR, et al. Gibberellin A$_3$ is biosynthesized from gibberellin A$_{20}$ via gibberellin A$_5$ in shoots of *Zea mays* L. Plant Physiol. 1990 (in press).
9. Albone KS, Gaskin P, MacMillan J, et al. Enzymes from seeds of *Phaseolus vulgaris* L.: Hydroxylation of gibberellins A$_{20}$ and A$_1$ and 2,3-dehydrogenation of gibberellin A$_{20}$. Planta. 1989; 177:108–115.
10. Kwak S-S, Kamiya Y, Takahashi M, et al. Metabolism of [^{14}C]GA$_{20}$ in a cell-free system from developing seeds of *Phaseolus vulgaris* L. Plant Cell Physiol. 1988; 29:707–711.
11. Smith VA, Gaskin P, MacMillan J. Partial purification and characterization of the GA$_{20}$ 3β-hydroxylase from seeds of *Phaseolus vulgaris*. Plant Physiol. 1990 (submitted).
12. Gilmour SJ, Bleeker AB, Zeevaart JAD. Partial purification of gibberellin oxidase from spinach leaves. Plant Physiol. 1987; 85:87–90.

CHAPTER 4

Gibberellins from the Tassel, Cob, 'Seed,' Silk, and Pollen of Maize

N. Murofushi, I. Honda, R. Hirasawa, I. Yamaguchi,
N. Takahashi, and B.O. Phinney

1 Introduction

The endogenous gibberellins (GAs) of maize have been extensively studied by Phinney and his co-workers.[1-4] These investigations were based on the use of single-gene dwarf mutants that control specific steps in the GA biosynthetic pathway for maize. The studies provided definitive data for the identification of the native GAs present in the vegetative tissues as well as information on the biosynthetic pathway for these GAs. The control points in the pathway were also defined for each of the mutants *d1*, *d2*, *d3*, and *d5*, thus providing a biochemical basis for the expression of each mutant. The studies in maize focused on the control of shoot elongation, with vegetative tissues as the experimental material. The purpose of this report is to present evidence for the presence of GAs in maize cell and tissue regions associated with reproduction—the young "seed" (fruit), the embryo, the cob (the female inflorescence), the tassel (the male inflorescence), the silk, and pollen.

Previous studies on the endogenous GAs from the ear (terminal inflorescence) of rice have shown the presence of several non-13-hydroxylated GAs, in particular, a high concentration of GA_4.[5,6] In contrast, vegetative tissues of rice contain 13-hydroxylated GAs with relatively high concentrations of GA_{19}.[7,8] The above results implicate the presence of at least two major pathways in rice, the early-non-hydroxylation pathway for the ear (see Chapter 2 by Takahashi and Kobayashi in this volume) and the early-13-hydroxylation pathway for the vegetative shoot.

Plant tissues associated with reproduction must have different physiological properties to explain their obvious differences in growth and form as well as differences associated with sporogenesis and embryogenesis. Information is accumulating which suggests that there are qualitative differences between the major GAs present in vegetative tissues as compared to tissues associated with reproduction (see the review by Pharis and King[9]).

2 Analytical Method

Since the levels of GAs in plant tissues are very low, it is necessary to establish and use analytical methods for the detection of minute amounts of the GAs present in these tissues. Combined gas chromatography–mass spectrometry (GC–MS) has become a powerful tool for this purpose.[10] A systematic analytical method has also been devised by Yamaguchi et al.,[11] in which GC–MS or GC–selected ion monitoring (SIM) is combined with fractionation by high-performance liquid chromatography (HPLC). It is also very useful for the analysis of GAs in plants. The reliability of the quantitative data obtained from the above methods is markedly improved by the use of internal standards, which are now available and are being used for the quantification of GAs. For example, The University of Tokyo group has synthesized deuterium-labeled GA_1, GA_4, and GA_{20} for use in the quantification of the same GAs found to be endogenous to specific plants.[12,13] When deuterium-labeled standards are not available for the quantification of other endogenous GAs, these GAs are first quantified by GC–SIM data; the values are then adjusted upwards based on the percentage recovery of the deuterated GA_1, GA_4, and/or GA_{20} added during purification procedures.

In the experiments reported here, the original identification and quantification of specific GAs was by GC–MS or GC–SIM. In subsequent analyses with similar plant material, identifications and quantifications were based on immunoassay data only. In the latter cases, the presence of specific GAs was ascertained on the basis of retention times (t_R) on two different HPLC columns and reactivity toward antisera. Immunoassays are very useful for the analysis of plant hormones because of their convenience and sensitivity. It should be emphasized that, although the presence of specific GAs can be suggested by immunoassay in combination with HPLC, the only absolute identification is by GC–MS or GC–SIM together with Kovats retention indices (KRI). The identification of GAs made by the combination of HPLC and immunnoassay is reliable but not conclusive. All immunoassays used in the studies reported here employed anti-GA_1-antiserum[14] and anti-GA_{20}-antiserum.[15]

3 Survey of GAs in Maize Seed[16]

Although the levels of GAs are fairly high in the immature seed, especially among members of the Leguminosae, Convolvulaceae, and Cucurbitaceae, these levels are relatively low in the seed of rice.[5] To analyze the GAs present in maize seed, plants of a hybrid dent-type maize, Ko-No. 7, were grown in the field. The tassels and ears were covered with waterproof paper bags after full expansion of the terminal inflorescence and before

Table 1. Content of GAs identified in the "seed" of maize (Ko-No. 7)[16]

GA	Content (ng/g fr wt)	Derivatives[a] and ions used for quantification by GC–SIM	
GA_1	0.7	TMSi–TMSi	564 (M^+)
GA_4[b]	0.4	TMSi–TMSi	476 (M^+)
GA_8[c]	0.2	TMSi–TMSi	652 (M^+)
GA_9	0.06	TMSi	388 (M^+)
GA_{17}	—[e]		
GA_{19}[c]	0.1	Me–TMSi	434 ($M^+ - 28$)
GA_{20}	6.8	TMSi–TMSi	476 (M^+)
GA_{29}	0.5	TMSi–TMSi	564 (M^+)
GA_{34}[d]	—[e]		
GA_{44}	—[e]		
GA_{53}[c]	0.2	TMSi–TMSi	564 (M^+)

Seeds were harvested 7 days after pollination. Identification was achieved by full scan GC–MS or by GC–SIM
[a] TMSi: TMSi ester; TMSi–TMSi: TMSi ester–TMSi ether; Me–TMSi: Me ester–TMSi ether
[b] Identified by GC–SIM and HPLC–immunoassay
[c] Identified by GC–SIM only
[d] Only M^+ was observed in a full-scan GC–MS with the expected KRI
[e] Identified but not quantified

emergence of the silk from the ear. Pollinations were done by hand, and the immature seeds were collected from the ears (female inflorescences) seven days later. The seeds were extracted and purified using methods described by Yamaguchi et al.,[11] with modifications. Identification and quantification were carried out by GC–MS and GC–SIM (Table 1).[16] Gibberellins A_1, A_9, A_{17}, A_{20}, A_{29}, and A_{44} were identified by full-scan GC–MS and KRI. Gibberellins A_4, A_8, A_{19}, and A_{53} were identified by GC–SIM. The definite identification of GA_4 by full-scan GC–MS was unsuccessful, but its identification was supported by HPLC–immunoassay. The occurrence of GA_{34} was also suggested by the ion of m/z 506 (M^+) with the same KRI as authentic GA_{34} methyl ester TMSi ether (GA_{34}-Me TMSi) in a full-scan GC–MS. The kinds of GAs are similar to those reported for rice.[5] These results suggest that the early-13-hydroxylation pathway operates in maize seed. The presence of GA_9 and GA_4 suggests the presence of at least one additional biosynthetic pathway in this material.

In rice, GA_{19} is the major GA in the shoot and ear, the level being much higher than that of other GAs.[8] In normal maize, the level of GA_{19} is also high in the vegetative tissue.[4] However, in maize seed, GA_{20} is the major GA.

If GA_{19} is a "pool GA" in rice, its conversion into GA_{20} could be a regulatory step controlling shoot growth.[7] Likewise, the high level of GA_{20} in maize seed suggests that GA_{20} could also be a "pool GA" and that its conversion to GA_1 regulates seed growth in maize.

Table 2. Content of GA_1, GA_4, GA_9, and GA_{20} in the embryo and the endosperm and fruit wall of the "seed" (Ko-No. 7), harvested 2 weeks after pollination[16]

Organ	Content (ng/g fr wt)			
	GA_1	GA_4[a]	GA_9[b]	GA_{20}
Embryo	0.2 ± 0.03	—[c]	2 ± 0.3	4 ± 0.3
Other part	0.05 ± 0.01	0.06 ± 0.006	1 ± 0.1	10 ± 2

Identification and quantification was achieved based on the t_R for HPLC on Nucleosil NMe_2 and ODS and immunoreactivity to anti-GA_1-antiserum and anti-GA_{20}-antiserum. Each value represents the mean of results from three replicates with the standard error.
[a] Quantified based on the cross-reactivity (48.1%) to anti-GA_1-antiserum
[b] Quantified based on the cross-reactivity (25%) to anti-GA_{20}-antiserum
[c] Not detectable

4 Localization of GAs in the Maize Seed[16]

The identifications of the GAs in maize seed reported here were based on extractions from the entire seed. In order to study the kinds and levels of GAs in the embryo, seeds were separated into embryo and "rest of seed"; each part was then extracted, purified, and analyzed for GAs (Table 2). Only the presence of GA_1, GA_4, GA_{20}, and GA_9 was investigated, because analyses were based on immunoassays using anti-GA_1-antiserum and anti-GA_{20}-antiserum; the presence of GA_4 and GA_9 was determined by cross-reactivity.[14,15] The seeds used for these analyses were collected fourteen days after pollination rather than seven days, because the younger material contained GAs at levels too low for identification. As shown in Table 2, the levels of GA_1 and GA_9 were higher in the embryo than in the rest of the seed, while the level of GA_{20} was higher in the nonembryo part of the seed.

5 Comparison of GA Levels in the Seed of Tall and Dwarf Mutants[16]

Dwarfism in maize is expressed primarily in the shoot. While mature dwarf plants are one-quarter to one-fifth the height of normals, the seeds from F1 plants (with genotypes +/+, +/d1, d1/d1, segregating in the ratio 1:2:1) are indistinguishable from each other in size and shape.[17] Thus, in order to compare the endogenous GAs of the tall (+/+) seed with those of the dwarf (d1/d1) seed, it is necessary to use seeds from plants of known homozygous genotypes. Seeds known to be homozygous normal were obtained by the self-pollination of the tall plants from a line (also a dent-type maize) segregating for the dominant dwarf mutant, D8.[18] Since D8 is a dominant mutation, all segregating tall plants from this line would be

Table 3. Content (ng/g fr wt) of GA_1, GA_4, GA_9, and GA_{20} from the seed of dwarf and tall maize harvested 1, 2, 3, and 4 weeks after pollination[16]

GA		Weeks after pollination			
		1	2	3	4
GA_1	Dwarf	0.04 ± 0.006	0.02 ± 0.002	—[a]	—
	Tall	0.1 ± 0.01	0.07 ± 0.001	—	—
GA_4	Dwarf	0.02 ± 0.004	0.1 ± 0.02	0.04 ± 0.006	0.04 ± 0.008
	Tall	0.2 ± 0.004	0.1 ± 0.02	0.04 ± 0.008	0.04 ± 0.006
GA_9	Dwarf	0.4 ± 0.02	0.4 ± 0.03	1 ± 0.09	2 ± 0.2
	Tall	0.4 ± 0.04	0.3 ± 0.04	2 ± 0.09	3 ± 0.4
GA_{20}	Dwarf	4 ± 0.1	3 ± 0.3	10 ± 1	10 ± 1
	Tall	2 ± 0.1	1 ± 0.2	8 ± 0.9	10 ± 2

For identification and quantification, see the note in Table 2.
[a]—, Detected but not quantified because of low peak intensity and overlapping

normal ($+/+$). Seeds of genotype ($d1/d1$) were obtained by the self-pollination of homozygous $d1$ plants. Seeds of the two genotypes ($d1/d1$ and $+/+$) were harvested 1, 2, 3, and 4 weeks after pollination. The seeds were extracted, purified, and analyzed by immunoassay for the presence of GA_1, GA_4, GA_9, and GA_{20}. As shown in Table 3, there were differences in GA levels in the seed between the tall and dwarf genotypes during development. At maturity the levels appear to be similar. Also, the results from 4-week-old seed are quite different from those found for the vegetative tissues of normal and $d1$ seedlings.[4] The data from the embryos suggest that the $d1$ lesion is expressed *only* during the early developmental stage of embryo growth. The subject obviously warrants further study in terms of gene expression during development.

6 Endogenous GAs in Pollen[14]

Extracts from the anthers of rice (which contained mature pollen) contain non-13-hydroxylated GAs.[6] These GAs were not found in vegetative tissues of the plant, suggesting the presence of a pathway(s) in the pollen not present in the vegetative shoot. Difficulties in obtaining sufficient quantities of pollen that were free from anthers precluded studies with rice pollen per se. However, copious amounts of pollen were obtained from normal maize tassels by bagging the ears and collecting the pollen from these bags.

In the studies reported here, pollen was collected from normal maize in the morning (at the time of initial pollen shedding), then extracted with methanol, and purified by HPLC. The fractions containing presumptive GAs were analyzed by immunoassay and by GC–SIM using an internal standard of deuterated GA_1 for quantification. The quantification data for GAs other than GA_1 by GC–SIM were adjusted based on the recovery of

Table 4. Gibberellins identified from pollen of dent-type maize[14]

	Content (ng/g fr wt)	
	GC–SIM	Radioimmunoassay[a]
GA_1[b]	0.9	0.65 ± 0.03
GA_4[c]	nd[d]	0.03 ± 0.005
GA_9[b]	nd	0.08 ± 0.01
GA_{17}[e]	0.9	ne[f]
GA_{19}[e]	2.1	ne
GA_{20}[b]	47.6	41.5 ± 3.5
GA_{29}[e]	12	ne
GA_{44}[e]	—[g]	ne
GA_{53}[e]	—[h]	ne

[a] Anti-GA_1 antiserum and anti-GA_{20} antiserum were used (see the note in Table 2)
[b] Identified by full-scan GC–MS and HPLC–immunoassay
[c] Mixed spectrum in full-scan GC–MS
[d] nd, Not determined because of interfering peaks
[e] Identified only by GC–SIM
[f] ne, Not examined
[g] Identified but not quantified
[h] Molecular ion was detected but other ions were overlapped

deuterated GA_1. Gibberellins A_1, A_9, and A_{20} were identified by full-scan GC–MS and KRI. The presence of GA_4 was strongly suggested by the combined HPLC and immunoassay analyses and by the full-scan GC–MS of the immunoreactive fraction, which gave a mixed spectrum containing the ions m/z 418 and 386 with the same KRI as the authentic GA_4-Me-TMSi. Table 4 shows the kinds and levels of the GAs identified from maize pollen. The GAs identified from maize pollen implicate the presence of at least two biosynthetic pathways.

7 Gibberellins in Ears and Tassels[16]

It is well known that GAs control sex differentiation in members of the Cucurbitaceae, *Spinacia oleracea*, and *Cannabis sativa*. In *Zea mays*, exogenous application of GA_3 increases femaleness in the tassel (male inflorescence)[19-21]; in dwarf mutants of maize, a lowered level of endogenous GA_1 is correlated with the presence of well-developed anthers in the ears (i.e., male organs in the female cob).[17] Thus, the available evidence clearly suggests a correlation between exogenously applied GA and femaleness/maleness. (For a more detailed examination of GAs in relation to reproduction, see the review by Pharis and King.[9]) There is no definitive information in the literature on the kinds and levels of endogenous GAs found in the reproductive structures of maize. The one report concerning maize tassels used very young material in which the floret inter-

Table 5. Content of GA_1, GA_3 and GA_4 in tassels (male flowers), ears, and silk of maize (Ko-No 7)[16]

	Content (ng/g fr wt)		
	GA_1	GA_3	GA_4
Tassel	+[a]	—[b]	—[c]
Ear	0.06 ± 0.01	1 ± 0.07	0.4 ± 0.04
Silk	0.2 ± 0.05	4 ± 0.2[d]	0.6 ± 0.1

Quantification was conducted by immunoassay (see the note in Table 2).
[a] +, Identified by HPLC–immunoassay but not quantified
[b] —, Not detectable
[c] —, Only weak peak by HPLC–immunoassay
[d] Identified by full-scan GC–MS

nodes were rapidly elongating. The study was thus related more to shoot elongation than to sexuality per se.[22]

In the work reported here, tassels from the hybrid Ko-No. 7 (length ca. 20 cm) were collected just after heading. At this stage, elongation of the tassel had ceased; the anthers contained well-developed pollen. Nonpollinated ears (ca. 25 cm in length, 80 g fresh weight per ear) of the same hybrid were collected seven days after the appearance of silk: the silk (9 g fr wt total) were separated from the seed (fruit). The purified extracts from the tassels, ears, and silk were analyzed for the presence of GA_1, GA_4, and GA_3 by immunoassay using anti-GA_1-antiserum. The results are shown in Table 5. Evidence was obtained for the presence of GA_1 and GA_4 from the tassels, ears, and silk; the data suggest the presence of at least two biosynthetic pathways in tassels, silk, and ears. Evidence was also obtained for the presence of GA_3 from the ears and silk only. The possible presence of GA_3 as a contaminant was minimized by the use of careful laboratory procedures. In addition, the experiment with tassels was run at the same time, in which no evidence was found for the presence of GA_3.

References

1. Phinney BO, Spray C. Chemical genetics and the gibberellin pathway in Zea mays L. In: Wareing PF, ed. Plant growth substances 1982. London: Academic Press, 1982: p. 101–110.
2. Phinney BO, Spray C. Gibberellin biosynthesis in Zea mays. The 3-hydroxylation step to GA_{20} to GA_1. In: Miyamoto J, Kearney PC, eds. Pesticide chemistry, human welfare and environment, Vol. 2. Oxford: Pergamon Press, 1983: p. 81–86.
3. Spray CR, Phinney BO, Gaskin P, et al. Internode length in Zea mays L. The dwarf-1 mutation controls the 3-hydroxylation of gibberellin A_{20} to gibberellin A_1. Planta. 1984; 160:464–468.
4. Fujioka S, Yamane H, Spray CR, et al. Qualitative and quantitative analysis of

gibberellins in vegetative shoots of normal, dwarf-1, dwarf-2, dwarf-3 and dwarf-5 seedlings of Zea mays L. Plant Physiol. 1988; 88:1367–1372.
5. Kobayashi M, Yamaguchi I, Murofushi N, et al. Endogenous gibberellins in immature seeds and flowering ears of rice. Agric Biol Chem. 1984; 48:2725–2729.
6. Kobayashi M, Yamaguchi I, Murofushi N, et al. Fluctuation and localization of endogenous gibberellins in rice. Agric Biol Chem. 1988; 52:1189–1194.
7. Kurogochi S, Murofushi N, Ota Y, et al. Identification of gibberellins in the rice plant and quantitative changes of gibberellin A_{19} throughout its life cycle. Planta. 1979; 146:185–191.
8. Suzuki Y, Kurogochi S, Murofushi N, et al. Seasonal changes of GA_1, GA_{19} and abscisic acid in three rice cultivars. Plant Cell Physiol. 1981; 22:1085–1093.
9. Pharis RP, King RW. Gibberellins and reproductive development in seed plants. Ann Rev Plant Physiol. 1985; 36:517–568.
10. Binks R, MacMillan J, Pryce RJ. Plant hormones—VIII. Combined gas chromatography–mass spectrometry of the methyl esters of gibberellins A_1 to A_{24} and their trimethylsilyl ethers. Phytochemistry. 1969; 8:271–284.
11. Yamaguchi I, Fujisawa S, Takahashi N. Qualitative and semiquantitative analysis of gibberellins. Phytochemistry. 1969; 8:271–284.
12. Endo K, Yamane H, Nakayama M, et al. Endogenous gibberellins in the vegetative shoots of tall and dwarf cultivars of Phaseolus vulgaris. L. Plant Cell Physiol. 1989; 30:137–142.
13. Nakayama M, Yamane H, Yamaguchi I, et al. Endogenous gibberellins in the shoots of normal- and bush-types of Cucumis sativus L. J Plant Growth Regul. 1989; 8:237–247.
14. Yamaguchi I, Nakazawa H, Nakagawa R, et al. Identification and quantification of gibberellins in the pollen of Zea mays by immunoassay and GC/MS. Plant Cell Physiol. in press.
15. Yamaguchi I, Nakagawa R, Kurogochi S, et al. Radioimmunoassay of gibberellins A_5 and A_{20}. Plant Cell Physiol. 1988; 28:815–824.
16. Murofushi N, Honda I, Hirasawa R, et al. Gibberellins from the tassel, cob, 'seed' and silk of maize. Submitted to Agric Biol Chem.
17. Coe EH, Neuffer MG. The genetics of corn. In: Sprague GF, ed. Corn and corn improvement. Madison, Wisconsin: American Society of Agronomy, p. 111–223.
18. Fujioka S, Yamane H, Spray CR, et al. The dominant non-gibberellin-responding dwarf mutant (D8) of maize accumulates native gibberellins. Proc Natl Acad Sci USA. 1988; 85:9031–9035.
19. Nickerson NH. Sustained treatment with gibberellic acid of maize plants carrying one of the dominant genes teopod and corn-grass. Amer J Bot. 1960; 47:809–815.
20. Hansen DJ, Bellman SK, Sacher RN. Gibberellic acid-controlled sex expression of corn tassels. Crop Sci. 1976; 16:371–374.
21. Kano S. On the feminization of the tassel induced by gibberellin in Zea mays. 1. Effects of gibberellin applied at different stages of growth and the morphology of the female spikelets induced by gibberellin. Proc Crop Sci Soc Japan. 1975; 44:199–204.
22. Hedden P, Phinney BO, Heupel RH, et al. Hormones of young tassels of Zea mays. Phytochemistry. 1982; 21:391–393.

CHAPTER 5

Gibberellin Mutants in *Pisum* and *Lathyrus*

J.B. Reid and J.J. Ross

1 Introduction

Over the last few years, our group has characterized internode length mutants in the garden pea (*Pisum sativum* L.) and the related species *Lathyrus odoratus* L. (sweet pea) in order to examine the partial processes controlling stem elongation. In the garden pea, we have identified fifteen loci which exert a marked effect on the synthesis of gibberellin A_1 (GA_1) or the response to applied GA_1.[1] We possess three alleles at six of these loci. In sweet peas, two loci have been identified. We will give a brief overview of our findings before concentrating on new work relating to GA biosynthesis in *Lathyrus odoratus* and the nature of the end-of-day far-red (FR) response in *Pisum*.

2 Gibberellin Synthesis Mutants in Peas

Genes at four loci, *le*, *na*, *ls*, and *lh*, block the synthesis of GA_1.[2,3] All four genes are leaky since either the double mutants (e.g., *lh ls*, *ls na*, and *lh na*) are considerably shorter than the single mutants[4] or low levels of GA_1 may be found in the mutant (e.g., *le*). Multiple mutational events have been recorded at all four loci, sometimes resulting in distinctly different degrees of leakiness (e.g., *le* and *le^d*).[5] Gene *le* partially blocks the 3β-hydroxylation of GA_{20} to GA_1[6] while gene *na* possibly blocks the conversion of *ent*-7α-hydroxykaurenoic acid to GA_{12}-aldehyde.[7] Application data suggest that genes *lh* and *ls* block GA biosynthesis prior to kaurene[7] (Fig. 1). Across all genotypes, elongation correlates only with the level of GA_1, not with the level of other biologically active GAs.[2,5,6] This suggests that the other GAs may possess activity only because they are precursors of GA_1.

The genes *na* and *Le* show pronounced tissue specificity. For example, gene *Le* is not expressed in developing seeds[8,9] and possibly not in mature stem tissue[10] while gene *na* does not block GA biosynthesis in developing

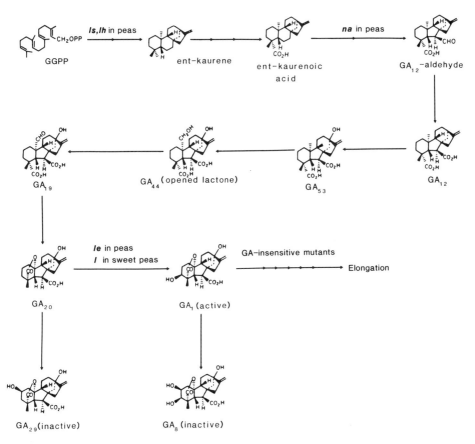

Fig. 1. The possible GA biosynthetic pathway in *Pisum salivum* (peas) and possibly *Lathyrus odoratus* (sweet peas). The site of action of the GA-biosynthesis genes in peas, *le*, *na*, *lh*, and *ls*, and sweet peas, *l*, is also indicated

seeds.[9] However, whether this action of gene *na* implies that it is a tissue-specific regulatory gene or that an alternative gene family controls GA biosynthesis in the seed has not been examined.

3 Gibberellins in Sweet Peas and the Action of Gene *l*

A quantitative relationship between GA_1 levels and elongation has only been established in a relatively small number of species. Most of these are monocots (e.g., maize,[11,12] wheat,[13] and rice[14]). Among dicots, only peas[5,6] and recently *Brassica* species[15] have been examined in detail. We have therefore looked at the gene *l* in sweet peas[16] to see if this mechanism also operates in other species of the Leguminosae. Using combined gas chromatography–mass spectrometry (GC–MS) or combined GC–selected ion monitoring (GC–SIM), evidence has been obtained for the presence of the C_{19} GAs GA_{20}, GA_{29}, 2-epi-GA_{29}, GA_1, and GA_8 in the young shoots of wild-type tall sweet peas.[17] This suggests that the early-13-hydroxylation pathway (Fig. 1) may be the dominant pathway for GA biosynthesis in this species as it is in the garden pea.[18]

The endogenous levels of GA_1 and GA_{20} were measured by GC–SIM using deuterated internal standards. The GA_1 level in the apical portion of wild-type (L) sweet peas was 3.7 ng/g fr wt compared with 0.4 ng/g fr wt in the dwarf (l) plants. By comparison, the GA_{20} levels were somewhat lower in the L plants (19.8 ng/g fr wt) compared with the l plants (38.1 ng/g fr wt). This suggests that gene l may partially block the 3β-hydroxylation of GA_{20} to GA_1. Application data support this interpretation since if GA_1 is fed to tall (L) and dwarf (l) plants, shortened with the GA-synthesis inhibitor paclobutrazol, elongation is strongly promoted in both genotypes. GA_{20} has almost the same effect on L plants as GA_1 but is substantially less effective on l plants (Fig. 2). These data also support the hypothesis that GA_1 levels may control stem elongation in sweet peas, as has been previously shown in other species.[6,12]

Feeds with [^{3}H, ^{13}C]GA_{20} confirm this action of gene l. Extracts from L plants contained four major metabolites. The 3β-hydroxylated GAs, GA_1 and GA_8, were identified by GC-MS in addition to the 2-hydroxylated GAs, GA_{29} and 2-epi-GA_{29}. By comparison, extracts from l plants contained only two major metabolites, and these were identified by GC–SIM as GA_{29} and 2-epi-GA_{29} (Fig. 3). Estimation of endogenous GA levels by the dilution of the ^{13}C compounds indicates that the l gene is a leaky mutant, a fact that is not surprising given that it is not an extreme dwarf (ca. 40% of the length of the wild-type plants).

In addition to determining the site of action of gene l and providing further evidence as to the generality of the view that GA_1 levels determine elongation in plants possessing a dominant early-13-hydroxylation pathway, these results make two other general points. Firstly, *Lathyrus odoratus* is the fourth species in which a mutant gene appears to control the conversion of GA_{20} to GA_1, the others being peas,[6] maize,[19] and rice.[14] Given the limited number of mutants that have been examined in detail, this reinforces the view that this conversion is an important regulatory step, particularly in limiting the production of GA_1. Secondly, a distinct difference exists between sweet peas and peas or maize. A significant amount (about 15%) of 2-epi-GA_{29} appears to be produced after feeds with [^{3}H, ^{13}C]GA_{20} (Fig. 3). It is also present endogenously at approximately half the level of GA_{29} although internal standards were not available for accurate quantification by GC–SIM.[17] Whether 2-epi-GA_{29} occurs in sweet peas because of the presence of a separate enzyme or because of the nonspecificity of the 2β-hydroxylase has not been examined.

Fig. 2. The response of wild-type (L) and dwarf (l) sweet peas to the application of 1 µg of GA_1 or GA_{20} to the fourth leaf. All plants were dwarfed by the application of 10 µg of paclobutrazol to the seed

4 Gibberellin-Sensitivity Mutants in Peas and Sweet Peas

Genes influencing the response to applied GA_1 are common and may be grouped into a broad sensitivity category.[20] In peas, eleven mutants that alter the response to applied GA_1 have been characterized[1] while in sweet peas one such mutant has been characterized.[16,17]

These mutants presumably either directly or indirectly influence some

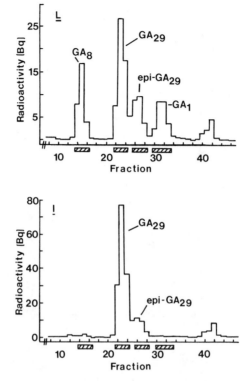

Fig. 3. HPLC of the methylated metabolites from [^{13}C,^{3}H]GA$_{20}$ feeds to tall (*L*) or dwarf (*l*) sweet peas. The derivatives of [^{13}C,^{3}H]GA$_1$, [^{13}C,^{3}H]GA$_8$, [^{13}C,^{3}H]GA$_{29}$ and 2-*epi*-[^{13}C,^{3}H]GA$_{29}$ were identified by GC–MS from the fractions indicated. Gibberellin A$_{20}$ was separated from the metabolites by prior HPLC of the free acids. J.J. Ross et al., unpublished data

step between reception of the GA signal and elongation.[1] They may result in either increased (e.g., *la cry*[s], *lv*[21,22]) or decreased (e.g., *lk, lka, lkb, lw, lb*[3,16,23,24]) stem elongation and either an enhanced (*lv*[22]) or depressed (e.g., *lk, lka, lkb, la cry*[s], *lb*[3,17,21,24]) response to GA$_1$. The steps that these genes influence have not been fully determined, but it seems likely that cell wall properties are intimately involved with certain of these mutants. Consequently, in collaboration with other workers, we are examining the properties and composition of the cell walls.

5 Slender Types

Slender plants behave as if treated with saturating quantities of GA$_1$ regardless of the endogenous GA level.[21,25] The original slender phenotype results from the action of the duplicate gene combination *la cry*[s] and is

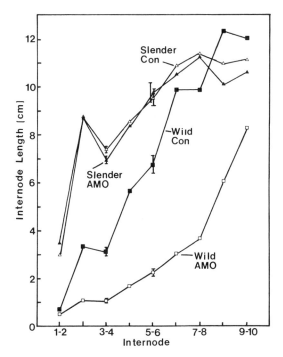

Fig. 4. Mean internode length versus internode number for plants of either the wild-type cv. Torsdag or the new slender mutant. The plants were either treated on the dry testa with 100 μg of AMO-1618 in 5 μg of ethanol or ethanol alone. Some representative SE are indicated. $n \geq 9$. J.B. Reid, unpublished data

thought to influence GA reception or some step close to this point since all GA-influenced responses are affected (e.g., internode elongation, rate of leaf expansion, flowering node, senescence, and parthenocarpic pod production[21]) and the dwarf GA-insensitivity gene *lk* is epistatic to *la cry*[s].[4]

A second mutant conferring this phenotype has recently been identified (J.B. Reid, unpublished), suggesting that more than one step may be involved with the slender phenotype. The new mutant, as with the *la cry*[s] slender types, is not influenced by GA-synthesis inhibitors that markedly reduce elongation in wild-type plants (Fig. 4). Consequently, even when mutants are isolated that possess the phenotype expected if their action involved the reception of the GA signal (either by an alteration in the level of the receptor or in its affinity for GA_1), the likelihood is that there is still a chain of events involved, and consequently receptor mutants will be difficult to identify and may indeed be rare. Consistent with this suggestion is the fact that none of the eight dwarf mutants with reduced GA-sensitivity in peas appears to involve GA reception since none is an exact phenocopy of GA-deficient dwarfs.

6 The GA-Hypersensitive Mutant *lv* and the End-of-Day FR Response

The mutant *lv* is substantially longer and shows an enhanced response to applied GA_1 compared with *Lv* peas when grown in white light.[22] However, *lv* plants show similar elongation to *Lv* plants when grown in darkness or FR light.[26] The mutant *lv* therefore behaves as if it does not fully de-etiolate and is similar to certain phytochrome photomorphogenic mutants isolated by Koornneef and co-workers.[27–29] However, the level of spectrophotometrically detectable soluble phytochrome appears to be similar in *Lv* and *lv* plants grown in the light or the dark.[26] In immunoassays with monoclonal antibodies raised against pea phytochrome I and II, the immunologically detectable levels of phytochrome I and II also appear to be similar in both genotypes in the light and the dark, suggesting that gene *lv* may influence the transduction chain for the phytochrome signal.[26]

Plants possessing gene *lv* lack the normal end-of-day FR response.[26] Normally, plants exposed to FR light briefly or to a photoperiod extension with FR-rich light (e.g., incandescent bulbs) following a white-light photoperiod exhibit considerable elongation compared with plants transferred directly to darkness.* On the basis of work with phytochrome-deficient mutants, this response has been suggested to be under the control of physiologically light-stable phytochrome.[27,28]

The GAs have been implicated with this response in peas since GA-deficient mutants show a reduced response to FR-rich photoperiod extensions compared to wild-type tall plants.[23,30] In cowpeas (*Vigna sinensis* L.) it has been concluded from application experiments with GA_1 and GA_{20} that FR may increase 3β-hydroxylation of GA_{20} to GA_1,[31] and a similar role is a possibility in peas.[30] However, this suggestion is not in accord with the action of gene *lv*, which appears to raise tissue sensitivity to GA_1 (and also GA_{20})[22] but eliminates the end-of-day FR response.[26] In order to clarify the role of GA_1 in the end-of-day FR response, the levels of GA_1 and GA_{20} were determined by GC–SIM using deuterated internal standards in wild-type plants exposed to an 8-h photoperiod followed by dark or a photoperiod extension with incandescent light. The levels of GA_{29} are estimates only, due to the lack of a suitable internal standard. No major differences were apparent in GA levels, even though a substantial difference in elongation was present (Table 1).

Further, when the end-of-day FR response was examined in a *le* dwarf mutant (*le^5839*,[23]), the change in elongation again could not be explained by differences in GA_1 levels (Table 1). However, the effect of the *Le/le^5839* gene difference in causing a similar proportional change in internode length is readily apparent since *le^5839* plants possessed only one-sixth the

---

*References 22, 23, 26, and 30.

Table 1. The levels of GA_1, GA_{20} and GA_{29} in isogenic tall (Le) and dwarf (le^{5839}) peas exposed to either an 8-h photoperiod of natural light or an 8-h photoperiod extended with 16 h of low-intensity incandescent light [3 μmol/(m$^2 \cdot$s) at pot top].

Genotype	Photoperiod	GA levels (ng/g fr wt)			Length between nodes 6 and 9 (cm)
		GA_1	GA_{20}	GA_{29}	
Le	8	12.4	23.6	32.5	16.6 ± 0.4
	8 + incandescent	10.1	21.7	33.6	30.9 ± 1.0
le^{5839}	8	2.3	84.7	79.8	5.2 ± 0.2
	8 + incandescent	1.8	59.3	118.4	7.8 ± 0.3

O. Hasan et al., unpublished data
Gibberellin A_1 and GA_{20} levels were determined by GC–SIM using deuterated internal standards. Internal standards were not available for GA_{29} determinations. The mean ± SE stem length between nodes 6 and 9 is also indicated.

level of GA_1 compared with isogenic Le plants (Table 1). GA_{20} and GA_{29} levels were elevated in the le^{5839} plants (Table 1), a result similar to that found in the analogous maize dwarf (d_1) by Fujioka et al.[11]

These results suggest that the end-of-day FR response is not mediated by changes in the level of GA_1 due to enhanced 3β-hydroxylation of GA_{20}. An alternative explanation for the end-of-day FR response is that the photoperiod extension acts in a similar manner to the lv gene (i.e., it enhances tissue sensitivity to GA_1). This could explain why the end-of-day FR response is absent in lv plants.[26] This hypothesis was examined in GA-deficient na plants. Untreated na plants are very short[9] and show no response to photoperiod extensions with incandescent light (Table 2). However, if na plants are treated with 1 μg of GA_1, a substantial response is restored, similar in magnitude to that found in essentially isogenic wild-type plants (Table 2). This is consistent with the suggestion that the end-of-day FR response is due to enhanced tissue sensitivity rather than to altered levels of GA_1 since the metabolism of GA_1 appeared to be unaffected by the photoperiod extension (Reid et al., unpublished). Preliminary results suggest that changes in cell division rather than cell elongation may be the major cause of the response.

7 Conclusions

These results point to both the level of GA_1 and the tissue sensitivity to GA_1 limiting elongation in certain circumstances. End-of-day FR appears to act by enhancing tissue sensitivity to GA_1, and preliminary results suggest that continuous darkness may also increase sensitivity to GA_1 compared with white-light.[32] The biochemical nature of this "sensitivity" is not

Table 2. The response of GA deficient *na* plants or isogenic wild-type (*Na*) plants to an extension of an 8-h photoperiod of natural light with 16 h of weak incandescent light [3 μmol/(m$^2 \cdot$s) at pot top]

Treatment	Response in Internode					
	4–5	5–6	6–7	7–8	8–9	9–10
Untreated, *na*	1.07	0.91	1.00	1.03	1.03	1.00
1 μg GA$_1$, *na*	1.04	1.04	1.47	1.50	1.44	1.29
Untreated, wild-type	1.70	1.56	1.60	1.52	1.67	1.66

J.B. Reid et al., unpublished data
The plants were given either 1 μg of GA$_1$ on leaf 4 or left untreated. The response is the ratio of the length in the extended photoperiod to the length under the 8-h photoperiod.

known, but the results with *lv* plants suggest that it may involve the transduction chain for light-stable phytochrome.

The view that GA$_1$ may be the only major endogenous GA active per se in plants possessing a predominant early-13-hydroxylation pathway for GA biosynthesis can now be expanded to include *Lathyrus odoratus* (sweet peas). Gene *l* in sweet peas partially blocks the 3β-hydroxylation of GA$_{20}$ to GA$_1$ and is consequently analogous to the genes *le* in peas[2], d_1 in maize[19] and *dy* in rice.[14]

Acknowledgments. We wish to thank Drs. I.C. Murfet and A. Nagatani and Professors J. MacMillan, L. Mander, M. Furuya, P. Davies, and Mr. O. Hasan for the provision of labeled compounds or assistance and advice during this work. We also thank the Australian Research Grants Scheme for financial support.

References

1. Reid JB. Gibberellin synthesis and sensitivity mutants in *Pisum*. In: Pharis RP, Rood SB, eds. Plant growth substances 1988. Berlin: Springer-Verlag, 1990: in press.
2. Ingram TJ, Reid JB, Murfet IC, et al. Internode length in *Pisum*. The *Le* gene controls the 3β-hydroxylation of gibberellin A$_{20}$ to gibberellin A$_1$. Planta. 1984; 160:455–463.
3. Reid JB, Potts WC. Internode length in *Pisum*. Two further mutants, *lh* and *ls*, with reduced gibberellin synthesis, and a gibberellin insensitive mutant, *lk*. Physiol Plant. 1986; 66:417–426.
4. Reid JB. Internode length in *Pisum*. Three further loci, *lh*, *ls*, and *lk*. Ann Bot. 1986; 57:577–592.
5. Ross JJ, Reid JB, Gaskin P, et al. Internode length in *Pisum*. Estimation of GA$_1$ levels in genotypes *Le*, *le* and *le^d*. Physiol Plant. 1989; 76:173–176.
6. Ingram TJ, Reid JB, MacMillan J. The quantitative relationship between

gibberellin A_1 and internode elongation in *Pisum sativum* L. Planta. 1986; 168: 414–420.
7. Ingram TJ, Reid JB. Internode length in *Pisum*. Gene *na* may block gibberellin biosynthesis between *ent*-7α-hydroxykaurenoic acid and gibberellin A_{12} aldehyde. Plant Physiol. 1987; 83:1048–1053.
8. Gaskin P, Gilmour SJ, MacMillan J, et al. Gibberellins in immature seeds and dark grown shoots of *Pisum sativum*. Gibberellins identified in the tall cultivar Alaska in comparison with those in the dwarf Progress No. 9. Planta. 1985; 163:283–289.
9. Potts WC, Reid JB. Internode length in *Pisum* III. The effect and interaction of the *Na/na* and *Le/le* gene differences on endogenous gibberellin-like substances. Physiol Plant. 1983; 57:448–454.
10. Ingram TJ, Reid JB, MacMillan J. Internode length in *Pisum sativum* L. The kinetics of growth and [^{3}H] gibberellin A_{20} metabolism in genotype *na Le*. Planta. 1985; 164:429–438.
11. Fujioka S, Yamane H, Spray CR, et al. Qualitative and quantitative analyses of gibberellins in vegetative shoots of normal, dwarf-1, dwarf-2, dwarf-3 and dwarf-5 seedlings of *Zea mays* L. Plant Physiol. 1988; 88:1367–1372.
12. Phinney BO. Gibberellin A_1, dwarfism and the control of shoot elongation in higher plants. In: Crozier A, Hillman JR, eds. The biosynthesis and metabolism of plant hormones. Soc Exp Biol Seminar Series 23. London: Cambridge University Press, 1984: p. 17–41.
13. Lenton JR, Hedden P, Gale MD. Gibberellin insensitivity and depletion in wheat—consequences for development. In: Hoad GV, Lenton JR, Jackson MB, Atkin BK, eds. Hormone action in plant development—A critical appraisal. London: Butterworths, 1987: pp. 145–160.
14. Kobayashi M, Sakurai A, Saka H, et al. Quantitative analysis of endogenous gibberellins in normal and dwarf cultivars of rice. Plant Cell Physiol. 1989; 30:963–969.
15. Rood SB, Pearce D, Williams PN, et al. A gibberellin-deficient *Brassica* mutant. Plant Physiol. 1989; 89:482–487.
16. Ross JJ, Murfet IC, Reid JB. Internode length in *Lathyrus odoratus* L. The expression and interaction of genes *L* and *Lb*. J. Heredity. 1990: in press.
17. Ross JJ, Reid JB, Davies NW, et al. Internode length in *Lathyrus odoratus* L. The involvement of gibberellins. Physiol Plant. 1990: in press.
18. Graebe JE. Gibberellin biosynthesis and control. Ann Rev Plant Physiol. 1987; 38:419–465.
19. Spray C, Phinney BO, Gaskin P, et al. Internode length in *Zea mays* L. The dwarf-1 mutant controls the 3β-hydroxylation of gibberellin A_{20} to gibberellin A_1. Planta. 1984; 160:464–468.
20. Reid JB. Phytohormone mutants in plant research. J Plant Growth Regul. 1990; 9:97–111.
21. Potts WC, Reid JB, Murfet IC. Internode length in *Pisum*. Gibberellins and the slender phenotype. Physiol Plant. 1985; 63:357–364.
22. Reid JB, Ross JJ. Internode length in *Pisum*. A new gene, *lv*, conferring an enhanced response to gibberellin A_1. Physiol Plant. 1988; 72:595–604.
23. Jolly CJ, Reid JB, Ross JJ. Internode length in *Pisum*. Action of gene *lw*. Physiol Plant. 1987; 69:489–498.

24. Reid JB, Ross JJ. Internode length in *Pisum*. Two further gibberellin-insensitivity genes, *lka* and *lkb*. Physiol Plant. 1989; 75:81–88.
25. Ingram TJ, Reid JB. Internode length in *Pisum*: Biochemical expression of the *le* and *na* mutations in the slender phenotype. J Plant Growth Regul. 1987; 5:235–243.
26. Nagatani A, Reid JB, Ross JJ, et al. Internode length in *Pisum*. The response to light quality, and phytochrome I and II levels in *lv* plants. J Plant Physiol. 1990; 135:667–674.
27. Adamse P, Jaspers PAPM, Bakker JA, et al. Photophysiology and phytochrome content of long-hypocotyl mutant and wide-type cucumber seedlings. Plant Physiol. 1988; 87:264–268.
28. Adamse P, Jaspers PAPM, Bakker JA, et al. Photophysiology of a tomato mutant deficient in labile phytochrome. J Plant Physiol. 1988; 133:436–440.
29. Koornneef M, Cone JW, Dekens EG, et al. Photomorphogenic responses of long-hypocotyl mutants of tomato. J Plant Physiol. 1985; 120:153–165.
30. Murfet IC. Internode length in *Pisum*. Variation in response to a daylength extension with incandescent light. Ann Bot. 1988; 61:331–345.
31. Garcia-Martinez J, Keith B, Bonner BA, et al. Phytochrome regulation of the response to exogenous gibberellins by epicotyls of *Vigna sinensis*. Plant Physiol. 1987; 85:212–216.
32. Ross JJ, Reid JB. Internode length in *Pisum*. Biochemical expression of the *le* gene in darkness. Physiol Plant. 1989; 76:164–172.

CHAPTER 6

The Relationship of Different Gibberellin Biosynthetic Pathways in *Cucurbita maxima* Endosperm and Embryos and the Purification of a C-20 Oxidase from the Endosperm

J.E. Graebe, T. Lange, S. Pertsch, and D. Stöckl

1 Introduction

Cell-free preparations of *Cucurbita maxima* endosperm have been used for two decades to investigate gibberellin (GA) biosynthesis. The pathway as originally defined in this system has been described in several recent reviews.[1-4] This paper reports new properties of the system that have been uncovered only recently by using the conditions optimal for a partially purified GA C-20 hydroxylase from pea cotyledons, and also by using GA_{53} instead of GA_{12} as a substrate. We feel that we are keeping up a tradition in reporting these results here, since it was in Tokyo that a paper with the title "Gibberellin Biosynthesis: New Intermediates in the *Cucurbita* System" was presented sixteen years ago.[5] We will also report on the inhibition of the 2-oxoglutarate-dependent enzymes by a new plant growth retardant and on the properties of a purified C-20 hydroxylase.

2 Material and Methods

Details of the methods used were as described by Lange and Graebe.[6] [^{14}C]Gibberellin A_{12} (6.7×10^{-12} Bq/mol) was made with the pumpkin system, and some of it was 13-hydroxylated with the pea system to yield [^{14}C]GA_{53} (5.4×10^{-12} Bq/mol). The data shown in Figs. 2, 3, 5, 6, and 7 and Table 1 were obtained by incubation with Sephadex G-25 filtered $200,000 \times g$ supernatants of *Cucurbita maxima* endosperm or embryo extracts, the products being separated by high-performance liquid chromatography (HPLC). The incubation mixtures (0.1 ml) consisted of 0.1 M Tris hydrochloride buffer, pH 7.0 (at 30°C) containing 4 mM ascorbic acid, 4 mM 2-oxoglutaric acid, 0.5 mM $FeSO_4$, 0.6 mg bovine serum albumin, 0.01 mg catalase, and substrate as noted, with the following exceptions: the experiments described in Figs. 4, 6, and 7 were done with bis-Tris buffer, and the incubation mixture in Fig. 7 contained 0.2 mg bovine serum albumin and no Fe^{2+}. The protein concentration was varied as shown on the ordinates of the figures or otherwise noted. The incubation temperature was 30°C.

The products were identified by full mass spectra obtained by combined gas chromatography–mass spectrometry (GC–MS). The identifications of 12α-hydroxy-GA_{12}, 12α-hydroxy-GA_{25}, 12α-hydroxy-GA_{43} (= 2β-hydroxy-GA_{39}), 2β-hydroxy-GA_{44}, and 2β-hydroxy-GA_{28} are preliminary, being based on the mass spectra and considerations of the structures of the precursors. So, for instance, the presumed 12α-hydroxy-GA_{43} is a trihydroxylated tricarboxylic GA obtained both from GA_{43} and GA_{39}, whereas the putative 12α-hydroxy-GA_{25} is obtained from GA_{25} and converted to GA_{39} (unpublished data with Dr. P. Hedden, Long Ashton Research Station, Bristol, U.K.). The other three presumed structures are more speculative.

The purification of the enzyme preparations used in obtaining the data shown in Table 2 consisted of gel filtering a 20,000 × g supernatant through a PD10 (Sephadex G-25) column, ammonium sulfate precipitation (50–70%), Sephadex G-100 gel filtration, and DEAE ion exchange chromatography.

3 The Gibberellin Biosynthesis Pathway

3.1 The Old Work

Gibberellin biosynthesis was originally investigated in endosperm preparations by using GA_{12}-aldehyde as a substrate. The main pathway proceeded via GA_{12}, GA_{15}-hydroxy acid, GA_{24}, GA_{36}, and GA_{13} to GA_{43}. The latter was the main end product. In addition, small amounts of GA_{25} were formed from GA_{24}, and small amounts of GA_4 were formed from GA_{36}. Gibberellin A_4 was the only C_{19} GA obtained in the system. This pathway was described by Hedden and Graebe[7] and in the reviews mentioned above.

A thorough investigation of the endogenous GAs in the endosperm showed that large amounts of GA_{43} were present,[8] which was in agreement with this GA being a major end product in the cell-free system. We therefore concluded that the cell-free system was a fair representation of GA biosynthesis in vivo. However, there were discrepancies. The endosperm contained large amounts of GA_{58} and GA_{49},[8,9] which are both 12α-hydroxylated C_{19} GAs, whereas the in vitro system originally showed neither 12α-hydroxylation nor a very active conversion to C_{19} GAs. Gibberellin A_4 accounted for no more than 2% of the total products. The first discrepancy seemed to be resolved when Hedden et al.[10] discovered a microsomal 12α-hydroxylating system. However, some doubt remained

6. GA Biosynthetic Pathways in *C. maxima*

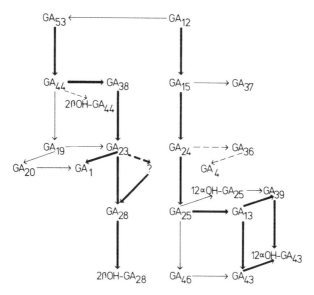

Fig. 1. Gibberellin A_{12} and GA_{53} metabolism in *Cucurbita maxima* endosperm preparations. The identifications of 2β-hydroxy-GA_{28}, 12α-hydroxy-GA_{43}, and 2β-hydroxy-GA_{44} are preliminary

as to whether this was a natural feature of the system, since it only worked at acid pH and only GA_{12}-aldehyde was 12α-hydroxylated, whereas GA_{12} was 13-hydroxylated to GA_{53}.

3.2 Conversion of GA_{12} by Endosperm Preparations

After a C-20 hydroxylase had been partially purified from pea cotyledons,[6] we tried incubating the soluble (200,000 × g supernatant) fraction of *C. maxima* endosperm with GA_{12} under the conditions found to be optimal for the pea enzyme. These conditions included the use of Tris–HCl buffer (pH 7.0) instead of phosphate buffer (pH 8.0) and other concentrations of 2-oxoglutarate, ascorbate, and Fe^{2+} than had been used in the previous work. Mg^{2+}, which had been used routinely in preparing the endosperm, was left out. These changes resulted in the pathway shown in the right-hand part of Fig. 1, which differs from the original pathway in two aspects. Firstly, the oxidation of C-20 is favored over the 3β-hydroxylation, so that the latter occurs at the oxidation stage of GA_{25} rather than at that of GA_{24}. Secondly, there is a very strong 12α-hydroxylation resulting in GA_{39} and 12α-hydroxy-GA_{43} as the main products. The quantitative relationships of the products at different protein concentrations are shown in Fig. 2. We conclude that there is a soluble 12α-hydroxylase in the endo-

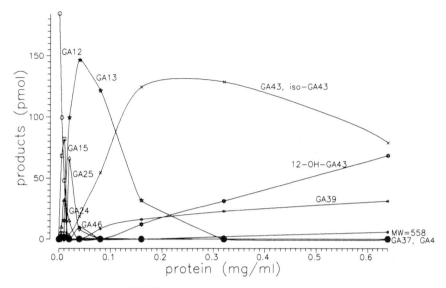

Fig. 2. Metabolism of $[^{14}C]GA_{12}$ in endosperm preparations at different protein concentrations (this gives a picture equivalent to a time curve). Substrate concentration, 1.9 μM; incubation time, 5 min

sperm and that incubation under optimal conditions for C-20 oxidation allows for 3β-hydroxylation at the end of the pathway only, when C-20 is fully oxidized.

One might ask why the soluble 12α-hydroxylase was not discovered before. There are several reasons for this. Data presented at this symposium by us show that the soluble 12α-hydroxylating activity is inhibited by phosphate (slide only) and that it has a sharp optimum at pH 7.0 (slide only) as well as a strong requirement for added Fe^{2+} (Table 1). The conditions used in the original experiments (phosphate buffer, a pH of 7.6–8.0, and no added Fe^{2+}) all worked against the 12α-hydroxylation. The use of thin-layer chromatography (TLC) for the identification of products in routine incubations also would not have separated GA_{39} and 12α-hydroxy-GA_{43} from GA_{43}. It is likely that the conversion of deuterated GA_9 to deuterated GA_{58} observed by Hedden et al.[10] involved the activity of the soluble 12α-hydroxylase, since a microsomal 12α-hydroxylation at this oxidation state of C-20 is less likely. Although phosphate buffer was used in that incubation, Fe^{2+} was present and the identification was by GC–MS.

3.3 Conversion of GA_{53} by Endosperm Preparations

In addition to the microsomal 12α-hydroxylation, Hedden et al.[10] also observed a microsomal 13-hydroxylation of GA_{12}, resulting in the formation of GA_{53}. This was surprising, because no 13-hydroxylated GAs had

Table 1. The dependency of 12α-hydroxylation on added Fe^{2+}. *Cucurbita maxima* endosperm preparation (0.7 mg protein/ml) was incubated with [^{14}C]GA_{12} (1.1 μM) for 3 h

Incubation mixture	Products (pmol)		
	GA_{13}	GA_{43}	12α-Hydroxy-GA_{43}
Complete	0	22	89
Fe^{2+} omitted	4	101	6

been identified in the endosperm.[8] We have confirmed the presence of a microsomal 13-hydroxylase with activities comparable to those of a microsomal 13-hydroxylase in pea cotyledons (unpublished work) and now report on the metabolism of GA_{53} and the identification of GA_1 in the endosperm.

The pathway of GA_{53} conversion is seen in the left-hand part of Fig. 1. It differs from the conversion of GA_{12} in three ways. Firstly, 3β-hydroxylation occurs earlier in the pathway (the conversion of GA_{44} to GA_{38}); secondly, there is a much greater production of the main C_{19} GA, GA_1; and thirdly, there is no 12α-hydroxylation. We also note that the tricarboxylic C_{20} GA, GA_{28}, gets 2β-hydroxylated (preliminary identification), whereas there is no 2β-hydroxylation of the C_{19} GAs, GA_1 and GA_{20}. This corresponds to the results obtained when the endosperm is

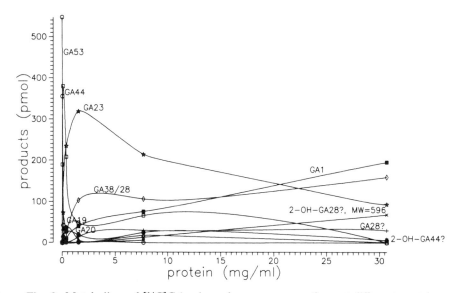

Fig. 3. Metabolism of [^{14}C]GA_{53} in endosperm preparations at different protein concentrations. Substrate concentration 5.5 μM; incubation time, 5 min

incubated with GA_{12} (Fig. 1) but is opposite to the results obtained with the embryo, as will be shown later. The quantitative data for incubation with GA_{53} are shown in Fig. 3. It is seen that GA_1 and, probably, GA_{28} are the main end products. Gibberellin A_{28} was not separated from GA_{38} by the HPLC system used, and the question of whether the one or the other accumulated is here a minor one.

The formation of the C_{19} GA GA_1 is particularly interesting. We believe that the presence of a 13-hydroxyl group slows down the oxidation at C-20 to a carboxyl function, allowing the decarboxylation to take place instead. Endogenous GA_1 was identified in the endosperm by a full mass spectrum, thus confirming that the 13-hydroxylation pathway exists in vivo as well.

3.4 Conversion of GA_{12} by Embryo Preparations

The conversion of GA_{12} by the soluble fraction of cell-free embryo preparations is shown in Fig. 4. The core of the pathway from GA_{12} to GA_{25} is identical with that in the endosperm, but then there are differences. In the embryo preparation, the 12α-hydroxylation dominates over the 3β-hydroxylation and there is no 2β-hydroxylation of C_{20} GAs. The main end product is therefore 12α-hydroxy-GA_{25} rather than GA_{43} and 12α-hydroxy-GA_{43} as was the case in the endosperm. Instead, there is a very active 2β-hydroxylation of C_{19} GAs. The quantitative data for conversion of GA_{12} in the embryo preparations are shown in Fig. 5.

Interestingly, less than optimal conditions for the C-20 hydroxylase, such as high substrate concentrations or the presence of Mn^{2+} (Fig. 6), result in

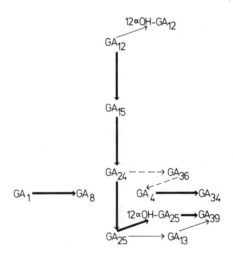

Fig. 4. Gibberellin A_{12} metabolism in *C. maxima* embryo preparations. The identifications of 12α-hydroxy-GA_{12} and 12α-hydroxy-GA_{25} are preliminary

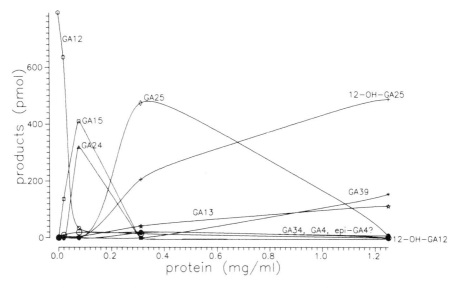

Fig. 5. Metabolism of [^{14}C]GA$_{12}$ in embryo preparations of different protein concentrations. Substrate concentration, 7.9 μM; incubation time, 2 h

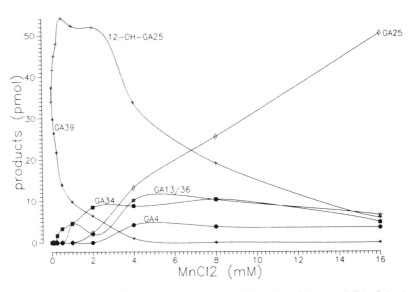

Fig. 6. The effect of Mn^{2+} on the proportions of C$_{19}$ GAs (GA$_4$ and GA$_{34}$) in the embryo system. Substrate concentration, 0.71 μM; incubation time, 2 h; protein concentration, 1.25 mg/ml

increased proportions of C_{19} GAs among the products. Thus, in Fig. 6, the formation of GA_4 and GA_{34} increases strongly when Mn^{2+} is added.

4 The Effects of Growth Retardants

Several growth retardants inhibit the formation of *ent*-kaurene from geranylgeranyl pyrophosphate while others inhibit the further oxidation of *ent*-kaurene, both resulting in lower GA contents and reduced internode elongation. The latter type of retardants, including the well-known compounds ancymidol, tetcyclacis, BAS 111...W, and triapenthenol, block the three first steps of *ent*-kaurene oxidation but have no effect on the conversion of *ent*-kaurenoic acid or later precursors to C_{20} and C_{19} GAs.[3,4,11] In checking several plant growth retardants, which had been supplied by Dr. W. Rademacher at BASF, Limburger Hof, Federal Republic of Germany, we found that 4-propylcyclohexane-3,5-dione-1-carboxylic acid diethyl ester (cyclohexanetrione), a new type of plant growth retardant, did change the pattern of products obtained by the in-

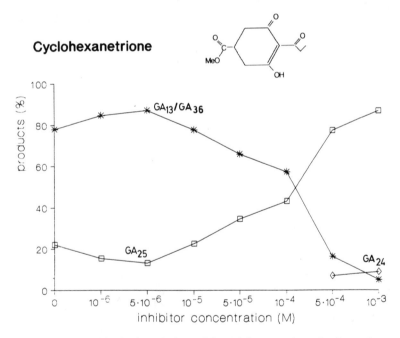

Fig. 7. Inhibition of 3β-hydroxylation with cyclohexanetrione in the endosperm system. The fraction labeled GA_{13}/GA_{36} contained only GA_{13} in the control, then increasing proportions of GA_{36} with increasing inhibitor concentrations (identification by HPLC/TLC). Substrate [^{14}C]GA_{12} concentration, 0.8 μM; protein concentration, 0.06 mg/ml; incubation time, 5 min

cubation of GA_{12} with *C. maxima* endosperm preparation. As Fig. 7 shows, 3β-hydroxylation seems to be particularly sensitive to increasing concentration of cyclohexanetrione, since the 3β-hydroxylated GAs, GA_{13} and GA_{36}, decrease whereas the 3-deoxy GA, GA_{25}, accumulates. A concentration slightly higher than 10^{-4} M inhibits 3β-hydroxylation to 50% whereas ca. 10^{-3} M (the compound did not dissolve completely at this concentration) inhibits this activity completely. The fact that the proportions of GA_{36} increase (not visible from the diagram) and that GA_{24} starts accumulating at the highest concentration used indicate that C-20 oxidation may to some degree be affected by the retardant as well. The concentrations needed to inhibit 3β-hydroxylation are ca. three orders of magnitude greater than those required for the inhibition of *ent*-kaurene oxidation by ancymidol, tetcyclacis, and triapenthenol in the pumpkin system. Chapters 8 and 10 in this volume are concerned with further aspects of this inhibition.

5 Purification of a GA C-20 Hydroxylase

Volger and Graebe[12] reported partial purification of a GA_{12}-aldehyde C-7 oxidase and a 3β-hydroxylase from the *C. maxima* endosperm. The preparations were obtained reasonably pure, but the activities were very low. After having partially purified a C-20 hydroxylase from pea cotyledons and studied its properties,[6,13] we have now returned to the pumpkin system to purify the C-20 hydroxylase, taking advantage of the experience gained with the pea system. The pumpkin system is very much more active, but the enzymes are also more prone to become inactivated.

A very active preparation of the pumpkin C-20 hydroxylase was obtained after ammonium sulfate precipitation, filtration over Sephadex G-100, and DEAE ion exchange chromatography. Table 2 compares the

Table 2. Properties of partially purified C-20 hydroxylases from pea cotyledons and pumpkin endosperm

Property	*Pisum sativum*	*Cucurbita maxima*
M_r (kDa)	44	48
pI	5.8	5.4
K_m GA_{53} (μM)	0.7	—
K_m GA_{12} (μM)	—	0.11
Spec. act. [μmol/(mg·h)]	0.12	32
K_m 2-oxoglutarate (mM)	0.1	0.06
K_m ascorbate (mM)	1.6	2.4
Fe^{2+} optimum (μM)	8	8
Yield (%)	15	26
Purification factor	270	52

properties of this preparation with those of the corresponding enzyme from pea cotyledons. It is seen that molecular weights, pIs, and Fe^{2+} optima are very similar for the two preparations, whereas the K_m value for the best substrate is considerably lower in the *C. maxima* preparation and the specific activity is 267 times higher with a 26% yield of the activity. Further purification of this preparation by hydrophobic interaction and gel filtration HPLC has yielded a preparation giving only two distinct bands on sodium dodecyl sulfate–polyacrylamide gel electrophoresis (silver staining).

6 Conclusions

The experiments reported here show that the *Cucurbita maxima* endosperm contains most of the known pathways of GA biosynthesis. The high yield of GA_1 in the incubations with GA_{53} shows that the production of C_{19} GAs is dependent on the substrate. We believe that the presence of a 13-hydroxyl group slows down the oxidation of C-20 to a carboxylic acid, favoring the decarboxylation to form a C_{19} GA. Presumably, the presence of a 12α-hydroxyl group would have the same effect. Until this has been tested and although we have come a little closer to the solution, we may state almost with the exact words of the paper by Hedden et al.[10] that the route of formation of GA_{58} in the endosperm of *C. maxima* is still unknown.

Acknowledgment. Our work was supported by the Deutsche Forschungsgemeinschaft.

References

1. Hedden P. In vitro metabolism of gibberellins. In: Crozier A, ed. The biochemistry and physiology of gibberellins, Vol. 1. New York: Praeger Press, 1983: p. 99–149.
2. Takahashi N, Yamaguchi I, Yamane H. Gibberellins. In: Takahashi N, ed. Chemistry of plant hormones. Boca Raton, Florida: CRC Press, 1986: p. 57–151.
3. Graebe JE. Gibberellin biosynthesis and control. Ann Rev Plant Physiol. 1987; 38:419–465.
4. Sponsel VM. Gibberellin biosynthesis and metabolism. In: Davies PJ, ed. Plant hormones and their role in plant growth and development. Dordrecht: Martinus Nijho Publishers, 1987: p. 43–75.
5. Graebe JE, Hedden P, MacMillan J. Gibberellin biosynthesis: new intermediates in the *Cucurbita* system. In: Sumiki Y, ed. Plant growth substances 1973. Tokyo: Hirokawa Publishing Co., 1974: p. 260–266.
6. Lange T, Graebe JE. The partial purification and characterization of a gibberellin C-20 hydroxylase from immature *Pisum sativum* seeds. Planta. 1989; 179:211–221.

7. Hedden P, Graebe JE. Cofactor requirements for the soluble oxidases in the metabolism of the C_{20}-gibberellins. J Plant Growth Regul. 1982; 1:105–116.
8. Blechschmidt S, Castel U, Gaskin P, et al. GC/MS analysis of the plant hormones in seeds of *Cucurbita maxima*. Phytochemistry. 1984; 23:553–558.
9. Beale MH, Bearder JR, Hedden P, et al. Gibberellin A_{53} and *ent*-6α, 7α, 12α-trihydroxykaur-16-en-19-oic acid from seeds of *Cucurbita maxima*. Phytochemistry. 1984; 23:565–567.
10. Hedden P, Graebe JE, Beale MH, et al. The biosynthesis of 12α-hydroxylated gibberellins in a cell-free system from *Cucurbita maxima* endosperm. Phytochemistry. 1984; 23:569–574.
11. Rademacher W. Gibberellins: metabolic pathways and inhibitors of biosynthesis. In: Böger P, Sandmann G, eds. Target sites of herbicide action. Boca Raton, Florida: CRC Press, 1989; p. 128–145.
12. Volger H, Graebe JE. Isolation and characterization of soluble enzymes involved in gibberellin metabolism. In: Bopp M, Knoop B, Rademacher W, ed. Abstracts, 12th Int Conf Plant Growth Substances. Heidelberg: Botanisches Institut der Universität. 1985; p. 10.
13. Graebe JE, Lange T. The dioxygenases in gibberellin biosynthesis after gibberellin A_{12}-aldehyde. In: Pharis RP, Rood SB, eds. Plant growth substances 1988. Berlin: Springer-Verlag. 1990; in press.

CHAPTER 7

Enzymatic 3β-Hydroxylation of Gibberellins A_{20} and A_5

V.A. Smith, K.S. Albone, and J. MacMillan

1 Introduction

Following the elucidation of the gibberellin (GA) metabolic pathways from GA_{12}-aldehyde, there is now considerable interest in the enzymes that catalyze the steps in these pathways. The enzymes that catalyze the post-GA_{20} steps (Fig. 1) in the early-13-hydroxylation pathway are of particular interest in view of the evidence that GA_1 and possibly GA_3 are the endogenous GAs active per se for stem elongation in maize seedlings.[1-5] No studies of the enzymes for these post-GA_{20} conversions have been published for vegetative tissue (but see Chapter 3 by Phinney et al. in this volume). Current information on the characterization of these enzymes comes from developing seeds.

In a series of papers, Kamiya et al.,[6] Takahashi et al.,[7] and Kwak et al.[8] have shown that GA_{20} is metabolized to GA_1 and GA_5 in immature seeds of *Phaseolus vulgaris* and that both activities require Fe^{2+}, ascorbate, 2-oxoglutarate, and O_2. More recently, Kwak et al.[9] claim to have purified the GA_{20} 3β-hydroxylase to homogeneity after a 300-fold purification. However, the latter authors used an assay procedure[10] that does not discriminate between 2β-hydroxylation, 3β-hydroxylation, and 2,3-dehydrogenation of [2β,3β-$^3H_2$]GA_{20} and did not identify and quantify the products of the enzyme reaction or establish the stoichiometric ratio between tritiated water (THO) and [^{3}H]GA_1 formation.

In our published work,[11] relating enzyme activities for GA_{20} metabolism with seed development, we showed that the activities for the 3β-hydroxylation of GA_{20} to GA_1 and for the dehydrogenation of GA_{20} to GA_5 were confined to the cotyledons. Both activites were maximal at 21 days after anthesis, and both co-occurred with a low activity for the 2β-hydroxylation of GA_{20} to GA_{29}. Using GA_{20} substrates stereospecifically labeled with deuterium at the 2- and 3-positions, we also showed (Fig. 1) that 3β-hydroxylation of GA_{20} to GA_1 occurs with loss of the 3β-hydrogen and 2,3-desaturation of GA_{20} to GA_5 occurs with loss of the 2β- and 3β-

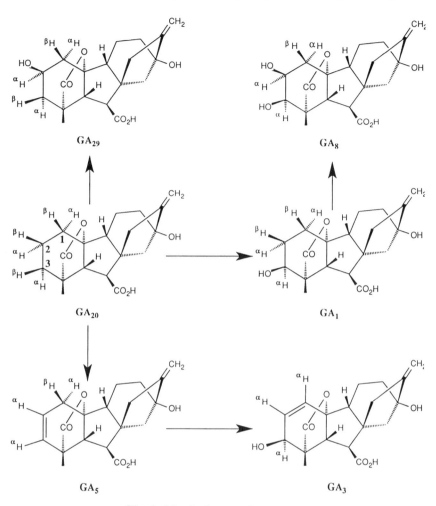

Fig. 1. Metabolic steps from GA_{20}

hydrogens. It has subsequently been shown that 2β-hydroxylation of GA_{20} to GA_{29} occurs with loss of the 2β-hydrogen. Under a range of incubation conditions, the relative rates of synthesis of GA_1, GA_5, and GA_{29} are constant, suggesting that all three products may be formed from GA_{20} by the same enzyme. This possibility is pursued in the present paper, together with details of the cofactor requirements for the GA_{20}-metabolizing activities.

We[12,13] have also described an enzyme preparation from the cotyledons/endosperm of seeds of *Marah macrocarpus* that establishes the biosynthetic origin of GA_7 from GA_9 and not from GA_4. This enzyme preparation also catalyzes the conversion of GA_5 to GA_3. In each case, the 1β-hydrogen of the substrate is lost. In seeds of *P. vulgaris*, GA_5 is converted[7,8] into GA_6. However, in maize seedlings, GA_5 is metabolized[5] to GA_3, a conversion of potential significance for stem elongation. This paper provides further information on the cofactor requirements for the formation of GA_3 from GA_5 by the *M. macrocarpus* enzyme preparation.

2 The GA_{20}-Metabolizing Enzyme Activity from Cotyledons of *Phaseolus vulgaris*

2.1 Enzyme Preparation

Unless stated otherwise, the enzyme preparations were obtained by ammonium sulfate precipitation of the soluble proteins from cotyledon homogenates as previously described,[11] followed by elution through DEAE-cellulose and gel filtration on Sephadex G-100.

2.2 Assay Procedures and Substrates

[1β,2β,3β-$^3H_3$]Gibberellin A_{20}[14] was used to assay enzyme activity by measuring the release of THO as previously described by Smith and MacMillan[10]. [17-^{13}C,$^3H_2$]Gibberellin A_{20}[15] and [2β-$^2H_2$,17-^{13}C,$^3H_2$]GA_{20} (L. Birch and C.L. Willis, unpublished) were used for product analyses and determination of kinetic parameters using reverse-phase high-performance liquid chromatography (HPLC), followed by radiocounting and combined gas chromatography–mass spectrometry (GC–MS) of the appropriate fractions.

2.3 Product Analysis

The partially purified enzyme preparation catalyzed the conversion of [17-^{13}C,$^3H_2$]GA_{20} (10 μM) to the appropriately labeled GA_1, GA_5, and GA_{29} in the ratio of 28:8:1 with the cofactors Fe^{2+}, 2-oxoglutarate, and ascorbate. By contrast, when [1β,2β,3β-$^3H_3$]Ga_{20} was used as substrate at high (10 μM) or low (0.01 μM) concentration under the same conditions, only [^{3}H]GA_1 and [^{3}H]GA_{29} were detected by radio-HPLC in the ratio of 3.5:1. The possibility that GA_5 is also formed but with loss of the 1β-^{3}H label as well as the 2β- and 3β-^{3}H labels has been excluded on two grounds. Firstly, no ^{2}H-labeled reaction products were detected after incubation of GA_{20} with the enzyme in deuterated buffer, ruling out exchange at the 1β-position. Secondly, the total amount of THO formed (0.106 pmol) was equal to the sum of the amounts of GA_{29} (0.022 pmol) and GA_1 (0.075 pmol) formed. Clearly, the synthesis of GA_5 is subject to a primary kinetic isotope effect and, with [1β, 2β,3β-^{3}H$_3$]GA$_{20}$ as substrate, the synthesis of GA$_{29}$ is favored over the synthesis of GA$_5$.

2.4 Enzyme Cofactors

a Ferrous Ion

Ferrous ion is an essential cofactor for the formation of GA$_1$, GA$_5$, and GA$_{29}$. Maximal activity is obtained with an Fe^{2+} concentration of ca. 10 μM, but significantly higher concentrations (10 mM) are not inhibitory. The absolute requirement for Fe^{2+} is demonstrated without the addition of chelators (2,2'-bipyridyl, EDTA) and contrasts with most of the 2-oxoglutarate-dependent oxidases, which appear to bind Fe^{2+} tightly.

b Ascorbate

The enzyme activity shows an absolute requirement for ascorbate, unlike many of the 2-oxoglutarate-dependent oxidases for which ascorbate is found to be stimulatory only. Ascorbate cannot be replaced by dithioerythritol, suggesting a specific role for ascorbate in the catalytic mechanism.

c 2-Oxoglutarate

After, but not before, purification on Sephadex G-100, the enzyme activity shows total dependence for 2-oxoglutarate under the standard assay conditions with [^{3}H$_3$]GA$_{20}$ at 0.01 μM. However, the partially purified enzyme does show some activity when assayed in the absence of 2-oxoglutarate but at high concentration (1.0 μM) of [^{3}H$_3$]GA$_{20}$.

Double reciprocal plots of reaction rate versus 2-oxoglutarate concentration are parallel for the syntheses of GA$_1$, GA$_5$, and GA$_{29}$ from [17-^{13}C,^{3}H$_2$]GA$_{20}$. This result supports the hypothesis that all three activities reside in the same catalytic protein. However, these plots do not conform to Michaelis–Menten kinetics. At high concentrations, 2-oxoglutarate is inhibitory.

2.5 ^{18}O-Labeling Experiments

When [17-^{13}C,^{3}H$_2$]GA$_{20}$ was incubated in a complete medium, product analyses showed that no products were formed under N$_2$ and that the yields of GA$_1$, GA$_5$, and GA$_{29}$ were equally increased by 60–70% under pure ^{18}O$_2$ compared with air. This result shows that oxygen is required and that its availability is rate limiting. Analyses of the GA$_1$ and GA$_{29}$ from incubations under ^{18}O$_2$ showed the incorporation of 0.76 atoms of ^{18}O in each case.

2.6 Inhibitor Studies

Using the release of THO from $[^3H_3]GA_{20}$ (0.01 μM) as an assay, GA_1, GA_3, GA_6, and GA_{29} did not inhibit enzyme activity within the concentration range of 0.001–1.0 μM. However, GA_5 and GA_9 are both inhibitors; at a 0.01-μM concentration of $[^3H_3]GA_{20}$, GA_5 and GA_9 showed 50% inhibition at 1.0 and 0.1 μM, respectively. Product analysis, using [17-$^{13}C,^3H_2]GA_{20}$ as substrate, showed that GA_5 equally inhibited the formation of GA_1, GA_5, and GA_{29}.

Gibberellin A_9 is a competitive substrate. The relative rate of the metabolism of $[1\beta,2\beta,3\beta-^3H_3]GA_9$ was ca. eight times less than that of $[1\beta,2\beta,3\beta-^3H_3]GA_{20}$. The main products were $[^3H]GA_4$ and THO in a 1:1 stoichiometric ratio.

A monoclonal antibody (McAb), prepared from the partially purified enzyme (F.M. Semenenko and V.A. Smith, unpublished), inhibits the conversion of $[17-^{13}C,^3H_2]GA_{20}$ to $[^3H]GA_1$, $[^3H]GA_5$, and $[^3H]GA_{29}$ to the same extent with maximal inhibition of 85%. The purified McAb inhibits the release of THO from $[^3H_3]GA_{20}$ in a linear manner with respect to the concentration of the antibody protein.

2.7 Kinetic Studies

In the course of the work described in the previous sections, it was found that the ratio of GA_1, GA_5, and GA_{29} (ca. 28:8:1) formed from $[17-^{13}C,^3H_2]GA_{20}$ was constant under a variety of conditions. In contrast, the product ratio was altered when the substrate, GA_{20}, was labeled with tritium or deuterium at the 2β-position. For example, with $[1\beta,2\beta,3\beta-^3H_3]GA_{20}$ as substrate, GA_5 was not formed and the ratio of GA_{29} to GA_1 increased from ca. 1:20 to 1:3.5. Also, with $[2\beta-^2H,17-^3H_2]GA_{20}$ as substrate, the ratio of GA_1 to GA_5 increased from 3.5:1 to greater than 10:1. These results show that there is a primary kinetic isotope effect for the formation of GA_5 from $[2\beta-^3H]$- and $[2\beta-^2H]GA_{20}$. They also suggest that GA_1, GA_5, and GA_{29} may be formed from GA_{20} by the same catalytic site on a single enzyme. To obtain further information on the latter point, a comparison was made of the kinetic parameters for the formation of GA_1, GA_5, and GA_{29} from $[17-^{13}C,^3H_2]GA_{20}$ and from $[2\beta-^2H, 17-^{13}C,^3H_2]GA_{20}$.

It was found that the rates of formation of GA_1, GA_5 and GA_{29} were altered by the presence of a 2β-deuterium atom. The D_V (V_H/V_D) values for the formation of GA_1, GA_5, and GA_{29} were 0.6, 2.5, and 0.26, respectively, and the corresponding D_V (V_{max}/K_m) values were 0.76, 9.0, and 2.3. These data support the hypothesis that GA_1, GA_5, and GA_{29} are formed from GA_{20} at the same catalytic site. For three separate enzymes, a primary isotope effect at C-2 would have been expected to result in no change in the V_{max} for GA_1 formation and a decrease in V_{max} for GA_5 and GA_{29} formation.

3 The Enzyme from Seeds of *Marah macrocarpus*; Conversion of GA_5 to GA_3

3.1 Enzyme Preparation and Assay Procedure

The enzyme preparation was obtained by ammonium sulfate precipitation of the solube proteins from a homogenate of cotyledons/endosperm as described by Albone et al.[13] The preparation was then desalted by elution through a column of Sephadex G-25. The GA_3 formed from [17-$^{13}C,^3H_2$]GA_5 was determined by reverse-phase HPLC and radiocounting; the conversion is expressed as a percentage of the total radioactivity of the substrate.

3.2 Cofactor Requirements

The results are shown in Tables 1 and 2.

a Ferrous Ion

No requirement for added Fe^{2+} was observed (Table 1), and the enzyme activity was not reduced by adding 2,2'-bipyridyl (Table 1) or EDTA (data not shown). Metal ions inhibited the conversion of GA_5 to GA_3 (Table 2). The inhibition by Mn^{2+} was reversed by Fe^{2+} although high concentrations of Fe^{2+} caused some inhibition of enzyme activity.

Table 1. *Marah* enzyme: Cofactor requirements for the conversion of [17-$^{13}C,^3H_2$]GA_5 to [17-$^{13}C,^3H_2$]GA_3[a]

Fe^{2+} (0.5 mM)	Ascorbate (0.5 mM)	2-Oxoglutarate (0.5 mM)	2,2'-Bipyridyl (0.5 mM)	Conversion (%)
−	−	−	−	0.8
+	+	+	−	88.8
−	+	+	−	88.2
+	−	+	−	85.6
+	+	−	−	6.4
−	−	+	−	63.2
−	+	−	−	5.9
+	−	−	−	8.6
−	+	+	+	90.9
+	+	+	+	90.3

[a] Determined by RC–RP–HPLC radio-counting, reverse phase—HPLC

Table 2. *Marah* Enzyme: Effect of metal ions on the percentage conversion of [17-^{13}C,^{3}H$_2$]GA$_5$ to [17-^{13}C,^{3}H$_2$]GA$_3$[a]

Incubation Conditions	Conversion (%)
Control (complete medium)	80
2,2′-Bipyridyl (0.2 mM)	90
Mn^{2+} (5 mM)	16
Cu^{2+} (5 mM)	4
Zn^{2+} (5 mM)	13
Mn^{2+} + 1 mM Fe^{2+}	93
Mn^{2+} + 5 mM Fe^{2+}	91
Mn^{2+} + 10 mM Fe^{2+}	80

[a] Determined by RC–RP–HPLC

b Ascorbate

Ascorbate is stimulatory but not obligatory (Table 1).

c 2-Oxoglutarate

2-Oxoglutarate is required (Table 1). [^{14}C]Succinic acid is produced from [^{14}C(U)]2-oxoglutarate during the enzymatic conversion of GA$_5$ to GA$_3$, but it was impossible to determine the stoichiometric ratio between the [^{14}C]succinate and GA$_3$ because of the large and variable amount of succinate formed in the absence of substrate by the crude enzyme preparation.

4 Discussion

The enzyme activity from immature cotyledons of *P. vulgaris* that catalyzes the formation of GA$_1$, GA$_5$, and GA$_{29}$ from GA$_{20}$ is a soluble oxidase requiring Fe^{2+}, ascorbate, 2-oxoglutarate, and O$_2$ as cofactors. However, it differs in several respects from the 2-oxoglutarate-dependent dioxygenases, described in the literature. Thus, Fe^{2+} is not tightly bound to the protein and ascorbate is an absolute requirement. Further evidence that the same enzyme catalyzes the conversion of GA$_{20}$ to GA$_1$, GA$_5$, and GA$_{29}$ has been obtained. Inhibition studies show that the rates of formation of GA$_1$, GA$_5$, and GA$_{29}$ are equally affected, and kinetic studies show that the presence of a 2β-deuterium atom in GA$_{20}$ alters the relative rates of formation of GA$_1$, GA$_5$, and GA$_{29}$.

The GA$_1$ 2β-hydroxylase activity, detected[11] in cotyledons from about 37 days after anthesis, has not been directly studied. However, this activity is presumed to be the same as that from imbibed mature seeds,[10] which shows no GA$_{20}$ 2β-hydroxylase or GA$_{20}$ 2,3-desaturase activity.

Speculative intermediates and reaction pathways for the formation of GA$_1$, GA$_5$, and GA$_{29}$ from GA$_{20}$ by the enzyme preparation from imma-

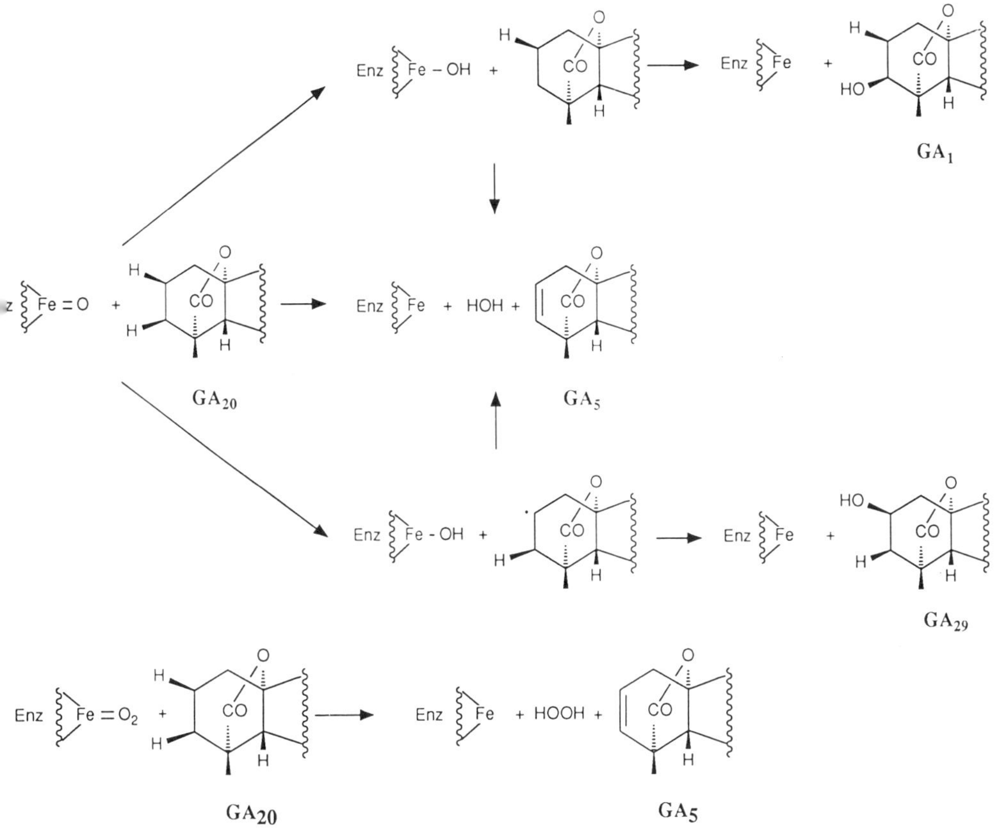

Fig. 2. Speculative enzymatic pathways from GA_{20}

ture seeds of *P. vulgaris* are shown in Fig. 2. However, further kinetic and inhibition studies are required before a mechanism of enzyme activity can be formulated.

The seeds of *M. macrocarpus* are a rich source of enzyme that catalyzes the conversion of GA_5 to GA_3. The protein appears to bind Fe^{2+} tightly and requires. 2-oxoglutarate as a cofactor, possibly as a cosubstrate. These properties are characteristic of the 2-oxoglutarate dependent dioxygenases, but the conversion of an olefin to an allylic alcohol with abstraction of an allylic hydrogen and double-bond migration by these dioxygenases has not been reported previously. A speculative stepwise process is shown in Fig. 3 and compared with the chemical oxygenation and lipoxygenase-type enzymatic conversion of an olefin to the allylic hydroperoxide.

The properties of the enzyme activities from the seeds of *P. vulgaris* and *M. macrocarpus* are different. Thus, for these systems, at least, the conver-

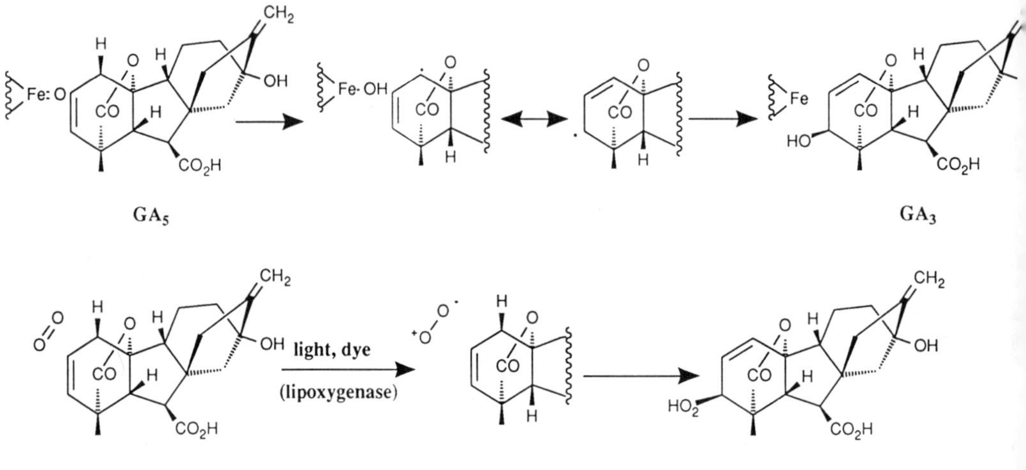

Fig. 3. Speculative pathways from GA_5 to GA_3

sion of GA_{20} to GA_3 via GA_5 (Fig. 1) is catalyzed in two steps by two different enzymes.

Acknowledgments. We gratefully acknowledge funding through an SERC-ICI plc Co-operative Research Grant and an AFRC Grant. We also thank Paul Gaskin for GC–MS analyses and Dr. C.L. Willis for providing the substrates.

References

1. Phinney BO, Spray CR. Chemical genetics and the gibberellin pathway in *Zea mays* L. In: Wareing PF, ed. Plant growth substances 1982. London: Academic Press, 1982: p. 101–110.
2. Spray CR, Phinney BO, Gaskin P, et al. Internode length in *Zea mays* L. The *dwarf-1* mutation controls the 3β-hydroxylation of gibberellin A_{20} to gibberellin A_1. Planta 1984; 160:464–468.
3. Fujioka S, Yamane H, Spray CR, et al. Qualitative and quantitative analyses of gibberellins in vegetative shoots of normal *dwarf-1*, *dwarf-2*, *dwarf-3*, and *dwarf-5* seedlings of *Zea mays* L. Plant Physiol. 1988; 88:1367–1372.
4. Fujioka S, Yamane H, Spray CR, et al. The dominant non-gibberellin responding dwarf mutant (*D8*) of maize accumulates native gibberellins. Proc Natl Acad Sci USA. 1988; 85:9031–9035.
5. Fujioka S, Yamane H, Spray CR, et al. Metabolism of gibberellins A_{20}, A_5 and A_1 in normal and dominant *Dwarf-8* seedlings of maize. Biogenesis of gibberellin A_3. Plant Physiol. 1990; submitted.
6. Kamiya Y, Takahashi M, Takahashi N, et al. Conversion of gibberellin A_{20} to gibberellins A_1 and A_5 in a cell-free system from *Phaseolus vulgaris*. Planta. 1984; 162:154–158.
7. Takahashi M, Kamiya Y, Takahashi N, et al. Metabolism of gibberellins in a

cell-free system from immature seeds of *Phaseolus vulgaris* L. Planta. 1986; 168:190–199.
8. Kwak S, Kamiya Y, Takahashi M, et al. Metabolism of [^{14}C]GA$_{20}$ in a cell-free system from developing seeds of *Phaseolus vulgaris* L. Plant Cell Physiol. 1988; 29:707–711.
9. Kwak S, Kamiya Y, Sakurai A. et al. Partial purification and characterization of gibberellin 3β-hydroxylase from immature seeds of *Phaseolus vulgaris* L. Plant Cell Physiol. 1988; 29:935–943.
10. Smith VA, MacMillan J. Purification and partial characterization of a gibberellin 2β-hydroxylase from *Phaseolus vulgaris*. J Plant Growth Regul. 1984; 2:251–264.
11. Albone KS, Gaskin P, MacMillan J, et al. Enzymes from seeds of *Phaseolus vulgaris* L.: Hydroxylation of gibberellins A$_{20}$ and A$_1$ and 2,3-dehydrogenation of gibberellin A$_{20}$. Planta. 1989; 177:108–115.
12. MacMillan J. Metabolism of gibberellins A$_{20}$ and A$_9$ in plants: Pathways and enzymology. In: Pharis RP, Rood, SB, eds. Plant growth substances 1988. Berlin: Springer-Verlag, 1990: in press.
13. Albone KS, Gaskin P, MacMillan, J. et al. The biosynthetic origin of gibberellins A$_3$ and A$_7$ in cell-free extracts from seeds of *Marah macrocarpus* and *Malus domestica*. Plant Physiol. 1990: in press.
14. Willis CL, Gaskin P, MacMillan J. [1β,2β,3β-^{3}H$_3$]Gibberellin A$_{20}$: Confirmation of structure by ^{3}H NMR and by mass spectrometry. Phytochemistry. 1988; 27:3970–3972.
15. Ingram TJ, Reid JB, Murfet IC. et al. Internode length in *Pisum*. The *le* gene controls the 3β-hydroxylation of gibberellin A$_{20}$ to gibberellin A$_1$. Planta. 1984; 160:455–463.

CHAPTER 8

Partial Characterization of the Gibberellin 3β-Hydroxylase from Immature Seeds of *Phaseolus vulgaris*

Y. Kamiya and S.-S. Kwak

1 Introduction

Much information about gibberellin (GA) biosynthesis in higher plants has come from studies using cell-free systems prepared from immature seeds (for a recent review, see Graebe[1]). Cell-free systems from immature seeds of *Phaseolus vulgaris* catalyze the conversion of GA_{20} to GA_1, GA_5, and GA_6 by 3β-hydroxylation, 2,3-dehydrogenation, and 2,3-epoxidation of GA_{20}, respectively.[2-5] Of these reactions, 3β-hydroxylation is particularly interesting because it has been suggested that GA_1 is the active GA in the regulation of internode elongation in maize,[6] pea,[7] and bean.[8] Although the physiological function of GAs in immature seeds is unknown, immature seed is a convenient material for the isolation of the enzyme, 3β-hydroxylase.

In this paper, we report (1) partial purification of the 3β-hydroxylase from immature seeds of *P. vulgaris*, (2) characterization of the 3β-hydroxylase, (3) effects of GAs and deoxygibberellin C on the 3β-hydroxylase, and (4) epoxidation of GA_5 to GA_6 by the partially purified enzyme preparation.

2 Partial Purification of the 3β-Hydroxylase from the Embryo of *P. vulgaris*

During the purification of the 3β-hydroxylase, the enzyme activity was measured by the release of tritiated water with the use of $[2\beta,3\beta\text{-}^3H_2]GA_{20}$ as a substrate. This method is convenient and speedy but has some disadvantages because the conversion of GA_{20} to GA_5 and GA_{29} also produces tritiated water and thus confuses the results. Recently, Albone et al. have reported that the formation of GA_5 from GA_{20} occurs with loss of 2β- and 3β-hydrogens in a *cis*-2β,3β-dehydrogenation.[5] To study the ratio of conversion of GA_{20} to GA_1, to GA_5, and to GA_6, $[17\text{-}^{13}C,^3H_2]GA_{20}$ was used as a substrate to eliminate the isotope effects of tritium at 2β- and 3β-positions. Figure 1 shows the products of metabolism of $[17\text{-}^{13}C,^3H_2]GA_{20}$

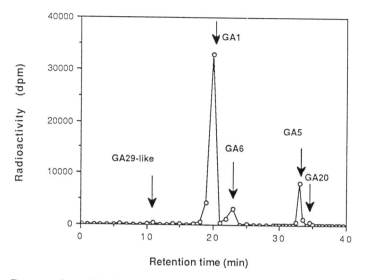

Fig. 1. Reverse-phase HPLC of products formed from [17-^{13}C,^{3}H$_2$]GA$_{20}$ (833 Bq, 1,75 GBq/mmol) in a soluble enzyme preparation from embryos (13DAF) of *Phaseolus vulgaris*. Products were injected onto a Nucleosil 5C$_{18}$ column (4.6 × 100 mm) and eluted with a 32-min linear gradient of 19% methanol to 46% methanol in 0.05% acetic acid at a flow rate of 1 ml/min. The eluate was collected in 1 ml (1–29 min) and 0.5 ml (29.5–40 min) and analyzed by a liquid scintillation counter. For the separation of GA$_5$ and GA$_{20}$, a Nucleosil 5N(CH$_3$)$_2$ column (4.6 × 100 mm) was used

by soluble enzyme extracts of immature seeds. [17-^{13}C,^{3}H$_2$]Gibberellin A$_1$, GA$_5$, and GA$_6$ were identified by full-scan gas chromatography–mass spectrometry (GC–MS). Gibberellin A$_{29}$ was detected by high-performance liquid chromatography (HPLC) only.

Prior to the large-scale purification of the 3β-hydroxylase, the time course of enzyme activity during seed maturation was studied using two cultivars, Kentucky Wonder (normal) and Masterpiece (dwarf).[8] Very young seeds of both cultivars contain high levels of 3β-hydroxylase activity (Fig. 2A). Similar results were obtained using cv. Canadian Wonder (normal).[5] A plot of the enzyme specific activity per seed shows that the maximal enzyme activity was reached 20 days after flowering (DAF) (Fig. 2B). The enzyme activity decreased with seed maturation.

Immature seeds of Masterpiece were used for the large-scale purification of the 3β-hydroxylase because enzyme preparations from Masterpiece were more active than those from Kentucky Wonder, as shown in Fig. 2A. Immature seeds were harvested twice, approximately 12 and 16 DAF. Seeds were separated into testae and embryos. Embryos were extracted with a phosphate buffer, and the extracts were centrifuged at 200,000 × g to give a crude soluble enzyme fraction. Details of the purification procedure have

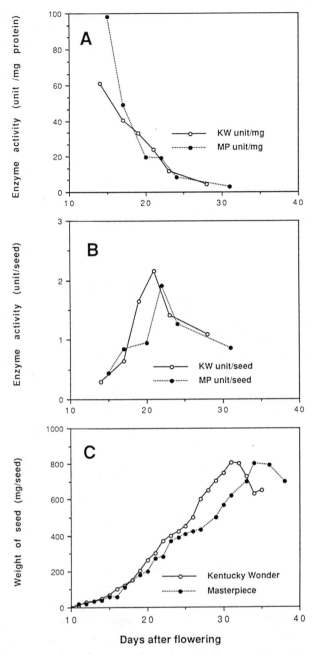

Fig. 2. (A) and (B). Time course of enzyme activity for [2, 3-^{3}H$_2$]GA$_{20}$ from embryos at different ages. Butyl-Toyopearl-purified preparations of two cultivars, Kentucky Wonder (normal) and Masterpiece (dwarf), were used for enzyme assay. The 3β-hydroxylase activity per mg protein (A) and the 3β-hydroxylase activity per seed (B) are shown. (C) Growth curves of seeds of *P. vulgaris* cv. Kentucky Wonder and cv. Masterpiece. Plants were grown in the field near the institute in the summer of 1986. Seeds were harvested in July. One unit of enzyme activity is defined in Table 1

Table 1. Purification of the GA 3β-hydroxylase from immature seeds of *Phaseolus vulgaris* L.

Purification step	Total protein (mg)	Total activity units[a]	Specific activity units/mg	Yield (%)	Purification (fold)
Crude extract	1586.2	21,750	13.7	100.0	1.0
Methanol precipitation	1206.0	19,575	16.2	90.0	1.2
Butyl-Toyopearl	49.14	4,989	101.5	22.9	7.4
DEAE-Toyopearl	4.18	1,979	473.5	9.	34.5
Gel filtration HPLC	0.086	369	4290	1.7	313

[a] 1 unit is 3×10^{-13} mol/h when [^{3}H]GA$_{20}$ (333 Bq) was used as a substrate

been reported.[9] Crude soluble extracts were purified by methanol precipitation, hydrophobic interaction fast-flow liquid chromatography (FFLC) using a Butyl-Toyopearl column, and ion exchange FFLC using a DEAE-Toyopearl column. The 3β-hydroxylase activity was separated into two peaks by DEAE-FFLC, suggesting the presence of isoenzymes. The main active fraction was purified by gel filtration HPLC (TSKgel G3000SW). A summary of the enzyme purification is shown in Table 1. The enzyme preparation was labile and rapidly lost activity during purification. The addition of 30% glycerol, 0.2M sucrose, and 2 mM dithiothreitol (DTT) to the elution buffer increased enzyme stability. The overall purification of the 3β-hydroxylase obtained after gel filtration HPLC was estimated from very young seed to be 313-fold in a 1.7% yield. The enzyme from older seed (20–30 DAF) could not be purified more than 50-fold.

3 Characterization of the 3β-Hydroxylase

3.1 Molecular Weight Estimation

The molecular weight of the enzyme was estimated to be 42,000 by gel filtration HPLC and sodium dodecyl sulfate (SDS)–polyacrylamide gel electrophoresis.[9]

3.2 Enzyme Kinetics

The optimum pH of the 3β-hydroxylase was 7.7 in the phosphate buffer. The DEAE-purified enzyme preparation was used for a preliminary kinetics study. The K_m values for [2,3-^{3}H$_2$]GA$_{20}$, [2,3-^{3}H$_2$]GA$_9$, and 2-oxoglutarate were 0.29 μM, 0.33 μM, and 250 μM, respectively.[9] The K_m value of 2-oxoglutarate was thus approximately 1000 times larger than that of the GA substrates.

3.3 Cofactor Requirement

The 3β-hydroxylase requires Fe^{2+} as a cofactor; at least 0.2 mM Fe^{2+} was necessary for full activity. About 12% of the maximum activity was observed in the absence of added Fe^{2+}. Addition of the Fe^{2+} ion chelator 2,2'-bipyridine (10 μM) to the enzyme mixture completely abolished this residual enzyme activity. The enzyme activity in the absence of added Fe^{2+} was probably due to Fe^{2+} remaining associated with the enzyme preparation during purification.[10] Fe^{2+} could not be replaced by other metallic ions except Fe^{3+}. The enzyme activity was strongly inhibited by 0.5 mM Mn^{2+}, Co^{2+}, Ni^{2+}, Cu^{2+}, Zn^{2+}, Cd^{2+}, and Hg^{2+}. The level of enzyme activity was stimulated 10-fold by the addition of 2 mM ascorbate. The activity level without ascorbate was very low.

4 Effects of Nonlabeled GAs and Deoxygibberellin C on the 3β-Hydroxylase

The effects of nonlabeled GAs on the 3β-hydroxylase activity were studied using [^{3}H]GA$_{20}$ and [^{3}H]GA$_9$ as substrates. [2β,3β-^{3}H$_2$]Gibberellin A$_9$ is a better substrate than GA$_{20}$ for the enzyme assay because GA$_9$ is converted to GA$_4$ only, and neither 13-deoxy-GA$_5$ nor 13-deoxy-GA$_6$ was produced.[3] As shown in Fig. 3, GA$_5$, GA$_9$, GA$_{15}$, GA$_{20}$, GA$_{44}$, and deoxygibberellin C inhibited the 3β-hydroxylase activity. Gibberellins A$_1$, A$_3$, A$_4$, A$_8$, A$_{12}$, A$_{19}$, A$_{22}$, A$_{24}$, A$_{29}$, and A$_{53}$ did not inhibit the 3β-hydroxylase. Methyl esters of GA$_9$, GA$_{20}$, and GA$_{24}$ also did not inhibit the activity. We have reported[3] that GA$_9$, GA$_{15}$, GA$_{20}$, and GA$_{44}$ are converted to GA$_4$, GA$_{37}$, GA$_1$, and GA$_{38}$, respectively by the crude soluble 3β-hydroxylase preparation. The reaction products were confirmed by full-scan GC–MS identification.[3] The DEAE-purified preparation also catalyzes the same reactions. [17-^{2}H$_2$]Gibberellin A$_9$, [20-^{2}H]GA$_{15}$, [17-^{13}C,^{3}H$_2$]GA$_{20}$, and [^{18}O]GA$_{44}$ were used as substrates, and the products were identified by full-scan GC–MS. When the DEAE-purified preparation was used, GA$_{15}$ and GA$_{44}$ were converted to GA$_{37}$ and GA$_{38}$, respectively, without opening of the lactone. [17-^{13}C,^{3}H$_2$]Gibberellin A$_5$, which was not metabolized by the crude enzyme preparation, was epoxidized to GA$_6$ by the DEAE-purified preparation as described below, but GA$_{12}$, GA$_{19}$, and GA$_{24}$ were not 3β-hydroxylated by the DEAE-purified preparation.

Deoxygibberellin C, an isomer of GA$_{20}$, inhibited the growth of normal rice and maize seedlings. Interaction of deoxygibberellin C and other GAs was studied by Hashimoto using the dwarf mutants of maize, d-2 and d-5. Deoxygibberellin C inhibited the growth-promoting activity of GAs such as GA$_9$, GA$_{19}$, and GA$_{20}$ but did not affect the activity of GA$_1$.[11] This suggests that deoxygibberellin C might be an inhibitor of the 3β-hydroxylase. Deoxygibberellin C inhibited the 3β-hydroxylase at lower

8. GA 3β-Hydroxylase from *P. vulgaris* 77

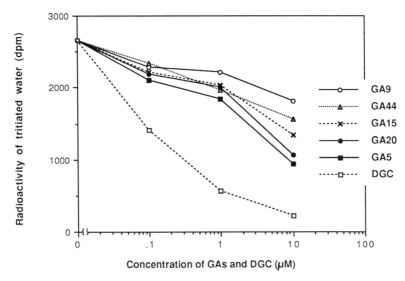

Fig. 3. Effects nonlabeled GAs, methyl esters of GAs, and deoxygibberellin C on the 3β-hydroxylase using [2, 3-^{3}H$_2$]GA$_9$ as a substrate. The DEAE-purified enzyme preparation (2.4 μg protein) was incubated with [2, 3-^{3}H$_2$]GA$_9$ (666 Bq, 0.4 pmol), 2-oxoglutarate (5 mM), FeSO$_4$ (0.5 mM), and ascorbate (5 mM) for 1.5 h at 30°C in the phosphate buffer containing 30% glycerol, 0.2 M sucrose, and 2 mM DTT. Noninhibitory GAs and methyl esters of GAs are not shown in the figure

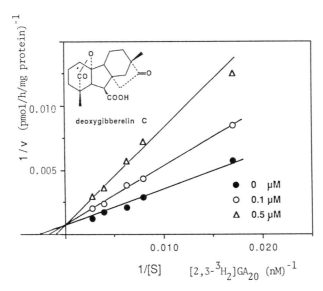

Fig. 4. Lineweaver–Burk plots of the 3β-hydroxylase at different concentrations of [2, 3-^{3}H$_2$]GA$_{20}$ and deoxygibberellin C. The DEAE-purified enzyme preparation was incubated under the conditions described in the legend to Fig. 3. ●, Control; ○, 0.1 μM deoxygibberellin C; △, 0.5 μM deoxygibberellin C

concentrations than GA_5, GA_9, GA_{15}, GA_{20}, and GA_{44} (Fig. 3). These GAs were metabolized by the 3β-hydroxylase during the incubation. Lineweaver–Burk plots of the 3β-hydroxylase activity at different concentrations of deoxygibberellin C and $[2,3\text{-}^3H_2]GA_{20}$ are shown in Fig. 4. The figure shows clearly that deoxygibberellin C is a competitive inhibitor of the 3β-hydroxylase. Although deoxygibberellin C inhibited the 3β-hydroxylase from *Phaseolus vulgaris*, it did not inhibit the conversion of GA_{15} to GA_{37} by a crude 3β-hydroxylase prepared from the endosperm of *Cucurbita maxima* (S.-S. Kwak, T. Saito, Y. Kamiya, et al., unpublished results).

Deoxygibberellin C and its analogues could be used for affinity chromatography of the 3β-hydroxylase if deoxygibberellin C is a competitive inhibitor. Other candidates for affinity chromatography are calcium 3,5-dioxo-4-propionylcyclohexane carboxylate (BX-112) and its analogues. These are the inhibitors of 2-oxoglutarate-dependent GA hydroxylases. Details of these inhibitors are described in Chapter 30 of this volume by Nakayama et al.

5 Epoxidation of GA_5 to GA_6 by the Partially Purified 3β-Hydroxylase Preparation

When $[17\text{-}^{13}C,^3H_2]GA_5$ was incubated with the DEAE-purified enzyme preparation, it was converted to a GA_6-like compound as shown in Fig. 5. The GA_6-like fraction was collected and identified as $[17\text{-}^{13}C,^3H_2]GA_6$ by full-scan GC–MS.

Since epoxidation of GA_5 to GA_6 was catalyzed by the DEAE-purified preparation, the cofactor requirements of the enzyme reaction were studied. Epoxidation requires oxygen, Fe^{2+}, 2-oxoglutarate, and ascorbate, like the 3β-hydroxylase. When 2-oxoglutarate was eliminated from the complete cofactor mixtures, production of GA_6 was reduced to 6.3% of that in the control incubation. Under anaerobic conditions, GA_5 was converted to GA_6, but only to 25% of the control value. The enzyme catalyzing the epoxidation of GA_5 to GA_6 has the same characteristics as the 3β-hydroxylase. Effects of nonlabeled GAs on the epoxidase activity were studied using $[17\text{-}^{13}C,^3H_2]GA_5$ as a substrate. Gibberellins A_5, A_9, A_{15}, A_{20}, and A_{44} inhibited the epoxidation reaction. Since these GAs are substrates of the 3β-hydroxylase, the catalytic site of the 3β-hydroxylase and the epoxidase will be quite similar. Figure 6 shows the conversion of $[17\text{-}^{13}C,^3H_2]GA_{20}$ to GA_1, GA_5, and GA_6 by different enzyme preparations prepared according to the seed maturation. Butyl-Toyopearl-purified enzyme preparation was used for this study. When the same levels of the 3β-hydroxylase activity were used for incubation, the ratios of products obtained with different enzyme preparations were almost identical. At present, it is unknown whether the same enzyme catalyzes 3β-hydroxy-

8. GA 3β-Hydroxylase from *P. vulgaris*   79

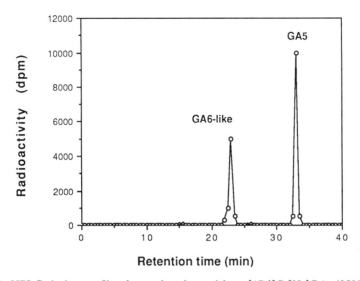

Fig. 5. HPLC elution profile of a product formed from [17-^{13}C,^{3}H$_2$]GA$_5$ (2500 Bq, 1.51 GBq/mmol) in the DEAE-purified preparation (8.5 μg protein). The enzyme was incubated under the conditions described in Fig. 3. One-fifth of the incubation extracts were analyzed by HPLC under the same conditions as described in the legend to Fig. 1

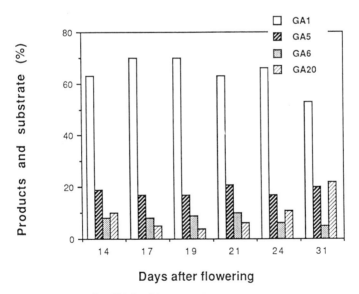

Fig. 6. Conversion of [17-^{13}C,^{3}H$_2$]GA$_{20}$ with Butyl-Toyopearl-purified preparations from seeds of different ages. Each enzyme preparation containing 1 pmol/h of the 3β-hydroxylase activity was incubated with [17-^{13}C,^{3}H$_2$]GA$_{20}$ (833 Bq, 1.75 GBq/mmol), and products were analyzed by ODS and Nucleosil N(CH$_3$)$_2$ column under the conditions described in the legend to Fig. 1

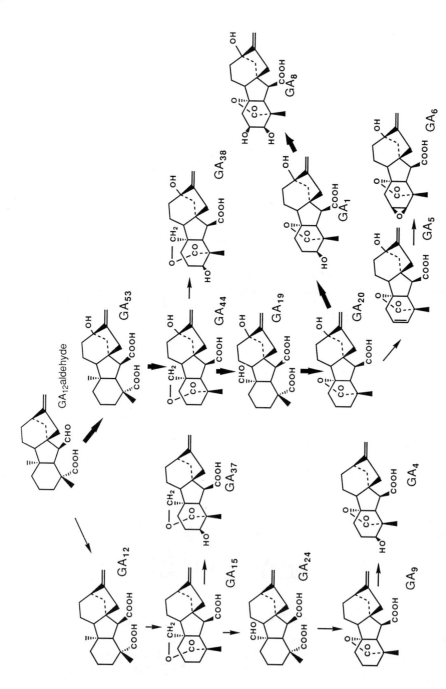

Fig. 7. Biosynthetic pathways of GAs in immature seeds of *Phaseolus vulgaris*. The pathways were refined by studies on the substrate specificity of the 3β-hydroxylase. Thick arrows show the main pathway and light arrows show the minor pathway. All the intermediates were incubated in the cell-free system and products were identified by full-scan GC–MS. Conversion of GA₄ to GA₁, see ref. 5.

lation and epoxidation or different enzymes that comigrate during the purification catalyze each specific reaction.

6 Conclusion

The GA 3β-hydroxylase was purified 313-fold in four purification steps from very immature embryos of *Phaseolus vulgaris* L. cv. Masterpiece. The 3β-hydroxylase requires Fe^{2+}, ascorbate, and 2-oxoglutarate. The K_m values for [2,3-$^3H_2$]GA$_{20}$ and [2,3-$^3H_2$]GA$_9$ were 0.29 μM and 0.33 μM, respectively. The K_m value for 2-oxoglutarate was 250 μM. Among many endogenous GAs of the immature embryos, GA$_9$, GA$_{15}$, GA$_{20}$, and GA$_{44}$ were 3β-hydroxylated by the partially purified enzyme preparation. Gibberellin A$_5$ was epoxidized to GA$_6$ by the same preparation; epoxidation requires the same cofactors as the 3β-hydroxylase. From the substrate specificity of the 3β-hydroxylase, biosynthetic pathways of GAs in immature embryos are now refined as shown in Fig. 7. An isomer of GA$_{20}$, deoxygibberellin C is a competitive inhibitor of the 3β-hydroxylase of *P. vulgaris*.

Acknowledgments. We thank Professor N. Takahashi and Professor J.E. Graebe for their useful suggestions and Professor N. Murofushi, Dr. T. Yokota, Professor B. O. Phinney, and Dr. H. Yamane for providing [2,3-$^3H_2$]GA$_9$, [2,3-$^3H_2$]GA$_{20}$, [17-^{13}C,$^3H_2$]GA$_5$, [17-^{13}C,$^3H_2$]GA$_{20}$, and deoxygibberellin C.

References

1. Graebe JE. Gibberellin biosynthesis and control. Ann Rev Plant Physiol. 1987; 38:419–465.
2. Kamiya Y, Takahashi M, Takahashi N, et al. Conversion of gibberellin A$_{20}$ to giberellins A$_1$ and A$_5$ in a cell-free system from *Phaseolus vulgaris* Planta. 1984; 162:154–158.
3. Takahashi M, Kamiya Y, Takahashi N, et al. Metabolism of gibberellins in a cell-free system from immature seeds of *Phaseolus vulgaris* L. Planta. 1986; 168:190–199.
4. Kwak S-S, Kamiya Y, Takahashi M, et al. Metabolism of [^{14}C]GA$_{20}$ in a cell-free system from developing seeds of *Phaseolus vulgaris* L. Plant Cell Physiol. 1988; 29:707–711.
5. Albone K, Gaskin P, MacMillan J, et al. Enzymes from seeds of *Phaseolus vulgaris* L.:Hydroxylation of gibberellin A$_{20}$ and A$_1$ and 2,3-dehydrogenation of gibberellin A$_{20}$. Planta. 1989; 177:108–115.
6. Spray C, Phinney BO, Gaskin P, et al. Internode length in *Zea mays* L. The dwarf-1mutant controls the 3β-hydroxylation of gibberellin A$_{20}$ to gibberellin A$_1$. Planta. 1984; 160:464–468.
7. Ingram TJ, Reid JB, Gaskin P, et al. Internode length in *Pisum*. The *Le* gene controls the 3β-hydroxylation of gibberellin A$_{20}$ to gibberellin A$_1$. Planta. 1984; 160:454–463.

8. Endo K, Yamane H, Nakayama M, et al. Endogenous gibberellins in the vegetative shoots of tall and dwarf cultivars of *Phaseolus vulgaris* L. Plant Cell Physiol. 1989; 30:137–142.
9. Kwak S-S, Kamiya Y, Sakurai A, et al. Partial purification and characterization of gibberellin 3β-hydroxylase from immature seeds of *Phaseolus vulgaris* L. Plant Cell Physiol. 1988; 29:935–943.
10. Hedden P, Graebe JE. Cofactor requirements for the soluble oxidases in the metabolism of the C_{20}-gibberellins. J Plant Growth Regul. 1982; 1:105–116.
11. Hashimoto T. Gibberellin structure-dependent interaction between gibberellins and deoxygibberellin C in the growth of dwarf maize seedlings. Plant Physiol. 1987; 83:910–914.

CHAPTER 9

Gibberellin Metabolism in Cell-Free Preparations from *Phaseolus coccineus*

A. Crozier, C.G.N. Turnbull, J.M. Malcolm, and J.E. Graebe

1 Introduction

Many studies on the metabolism of gibberellins (GAs) have been made with cell-free preparations following the elucidation of in vitro GA-biosynthesis pathways in preparations derived from the liquid endosperm of immature *Cucurbita maxima* seed[1-3] and cotyledons of developing pea seed.[4] The endogenous GAs in the seeds of several species of the Leguminosae are well characterized,[5] and the corresponding metabolic pathways fall into distinct categories. Non-and 13-hydroxylated routes leading to the C_{19} GAs are features of *Pisum sativum*[4] whereas in *Phaseolus vulgaris* 3β- and 13-hydroxylations occur.[6] There is information on the properties of the enzymes in *P. vulgaris* seed responsible for 3β-hydroxylation of GA_{20} to GA_1 as well as 2,3-desaturation of GA_{20} to GA_5 and 2β-hydroxylation of GA_{20} to GA_{29} and GA_1 to GA_8.[7] There are also reports on the partial purification and characterization of GA 2β-hydroxylases from *P. vulgaris* and *Pisum sativum* seed[8,9] and a 3β-hydroxylase catalyzing the conversion of GA_{20} to GA_1 from immature seed of *P. vulgaris*.[10]

Although many GAs have been shown to be endogenous constituents of immature *Phaseolus coccineus* seed,[5,11] there are relatively few reports on GA biosynthesis and metabolism in this tissue. Cell-free preparations from suspensors of immature *P. coccineus* cv. Prizewinner convert [^{14}C]mevalonic acid to *ent*-kaurene and *ent*-kaurenol[12] and [^{14}C]*ent*-kaurene to *ent*-kaurenol, *ent*-kaurenal, *ent*-kaurenoic acid, and *ent*-7α-hydroxykaurenoic acid.[13] Incubation with [^{14}C]*ent*-7α-hydroxykaurenoic acid results in label being incorporated into three C_{19} GAs, GA_1, GA_5, and GA_8.[14]

This report will summarize studies on GA biosynthesis in cell-free preparations derived from immature *P. coccineus* seed per se rather than their suspensors. In addition, the in vitro metabolism of [^{3}H]GA_4 by germinating seed of *P. coccineus* will be discussed.

2 In Vitro GA Biosynthesis in Preparations from Immature Seed

2.1 Mevalonic Acid to *ent*-Kaurene, GA_{12}-Aldehyde, and GA_{12}

When incubated in phosphate buffer (pH 7.4, 50 mM) containing ATP (10 mM), phosphoenolpyruvate (PEP., 10 mM), Mg^{2+} (1 mM), NADPH (1 mM), and bovine serum albumin (10 mg/ml), cell-free preparations obtained from the 1000 g supernatant (S-1) of immature seed of *P. coccineus* cv. Prizewinner convert [2-^{14}C]mevalonic acid to *ent*-kaurene, *ent*-kaurenoic acid, *ent*-kauradienoic acid, *ent*-7α-hydroxykaurenoic acid, *ent*-6α, 7α-dihydroxykaurenoic acid, GA_{12}-aldehyde, and GA_{12}. Similar incubations with [^{14}C]*ent*-kaurene, in the absence of ATP, PEP, and Mg^{2+}, yield *ent*-kaurenol, *ent*-kaurenal, *ent*-kaurenoic acid, and *ent*-7α-hydroxykaurenoic acid.[15] The products obtained in these incubations are identical to the intermediates in the pathway from *ent*-kaurene to GA_{12} in cell-free systems from other seeds[16,17] and the fungus *Gibberella fujikuroi*.[18] *ent*-Kauradienoic acid is known to be the precursor of hydroxykaurenolide by-products in both *Cucurbita maxima*[19] and *G. fujikuroi*.[20] *ent*-6α, 7α-Dihydroxykaurenoic acid is the first member of another branch from the GA pathway.[16,17]

2.2 GA_{12}-Aldehyde to C_{19} GAs

Incubation of [^{14}C]GA_{12}-aldehyde with the *P. coccineus* cv. Prizewinner S-1 preparation in 50 mM phosphate buffer (pH 7.4) containing Fe^{2+} (0.5 mM), ascorbate (5 mM), 2-oxoglutarate (5 mM), NADPH (1 mM), and bovine serum albumin (20 mg/ml) leads to the production of isotopically labeled GA_1, GA_4, GA_5, GA_6, GA_{15}, GA_{17}, GA_{19}, GA_{20}, GA_{24}, GA_{37}, GA_{38}, GA_{44}, and GA_{53}-aldehyde.[21] All of these metabolites, with the exception of GA_{15}, GA_{24}, and GA_{53}-aldehyde, are known endogenous GAs of *P. coccineus* seed. In order to elucidate the exact pathways involved in the production of C_{19} GAs by the *P. coccineus* S-1 cell-free system, further feeds were carried out using ^{14}C-, ^{2}H- and ^{3}H-labeled substrates.[22] The data obtained are summarized in Table 1, and the likely biosynthetic sequence is illustrated in Fig. 1.

The cell-free preparation converted [^{14}C]GA_{12} to GA_{53}, GA_{44}, and GA_{15}, indicating the operation of early-13-hydroxylation and early-nonhydroxylation pathways. The latter route was investigated incubating [^{3}H]GA_{15}, which underwent 3β-hydroxylation to yield GA_{37} (Table 1). When [^{3}H]GA_{15} was treated with KOH, prior to incubation, to yield GA_{15}-open lactone, GA_{37} was again the sole metabolite, and conversion to the C-20 aldehyde, GA_{24}, was not observed.

When used as a substrate, [^{3}H]GA_{37} was not further metabolized by the cell-free preparation. Treatment of GA_{37} with KOH does not yield GA_{37}-

Table 1. Summary of GA metabolism in cell-free preparations from immature seed of *Phaseolus coccineus* cv. Prizewinner

Substrate	Metabolites
[^{14}C]GA$_{12}$	GA$_{53}$ (55%), GA$_{44}$ (7%), GA$_{15}$ (15%)
[^{3}H]GA$_{15}$	GA$_{37}$ (52%)
[^{3}H]GA$_{15}$-open lactone	GA$_{37}$ (30%)
[^{3}H]GA$_{37}$	No metabolites detected
[17-^{3}H]GA$_{14}$	GA$_{37}$ (56%), GA$_{36}$ (7%), GA$_{4}$ (29%)
[^{3}H]GA$_{24}$	GA$_{4}$ (11%)
[^{3}H]GA$_{36}$	GA$_{4}$ (49%)
[1,2-^{3}H]GA$_{4}$	No metabolites detected
[^{14}C]GA$_{53}$	GA$_{44}$ (51%)
[^{14}C]GA$_{44}$	No metabolites detected
[^{14}C]GA$_{44}$-open lactone	No metabolites detected
[^{2}H$_{3}$]GA$_{19}$	GA$_{20}$, GA$_{1}$, GA$_{5}$
[2,3-^{3}H]GA$_{20}$	GA$_{1}$ (24%), GA$_{5}$ (15%)
[17-^{13}C, ^{3}H$_{2}$]GA$_{20}$	GA$_{1}$ (18%), GA$_{5}$ (16%)
[1-^{3}H]GA$_{5}$	GA$_{6}$ (31%)
[^{3}H]GA$_{6}$	No metabolites detected
[1,2-^{3}H]GA$_{1}$	No metabolites detected

Incubation conditions as outlined in the text for [^{14}C]GA$_{12}$-aldehyde. ^{14}C- and ^{2}H-labelled metabolites identified by combined gas chromatography-mass spectrometry. ^{3}H-labelled metabolites identified by reversed- and normal phase high performance liquid chromatography-radio counting (HPLC-RC) of both underivatized samples and their methoxycoumaryl esters.[22]

open lactone because the 3β-hydroxylated A ring undergoes a reversible retroaldol rearrangement.[23] In an attempt to overcome this problem, incubations were carried out with [^{3}H]GA$_{14}$ as a substrate. Although not an endogenous constituent of *P. coccineus* seed, it was thought that oxidation of GA$_{14}$ at C-20 might yield GA$_{37}$-open lactone, in situ, which would then be metabolized in a similar manner to the endogenous constituent. In practice, this approach proved successful as the cell-free preparation metabolized [^{3}H]GA$_{14}$ to GA$_{4}$, GA$_{36}$, and GA$_{37}$. This suggests the existence of a GA$_{15}$ → GA$_{37}$ → GA$_{36}$ → GA$_{4}$ pathway, which was confirmed by incubation of [^{3}H]GA$_{36}$, which was converted to GA$_{4}$. Although GA$_{24}$ was not detected when [^{3}H]GA$_{15}$ was used as a substrate, its presence as one of the many metabolites of GA$_{12}$-aldehyde implies that it could be an alternative intermediate to GA$_{37}$ in the conversion of GA$_{15}$ to GA$_{36}$ (see Fig. 1). Support for this possibility was obtained when GA$_{4}$ was detected as a metabolite of [^{3}H]GA$_{24}$. [^{3}H]Gibberellin A$_{4}$ was an end product in the metabolic sequence as it was not further metabolized (Table 1).

The early-13-hydroxylation pathway began with the conversions of [^{14}C]GA$_{12}$-aldehyde and [^{14}C]GA$_{12}$ to GA$_{53}$-aldehyde and GA$_{53}$, respectively.[21] Limited substrate availability prevented confirmation of the likely conversion of GA$_{53}$-aldehyde to GA$_{53}$ (see Fig. 1). Incubation of

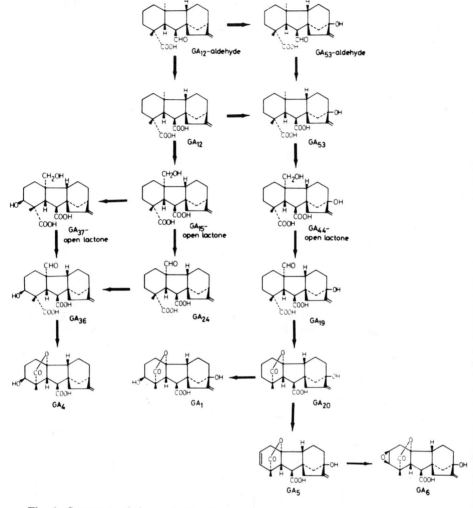

Fig. 1. Summary of the probable GA metabolic steps occurring in S-1 cell-free preparations from immature seed of *P. coccineus* cv. Prizewinner

[^{14}C]GA$_{53}$ resulted in metabolism to GA$_{44}$, which was not further metabolized as either the δ-lactone or the open lactone (Table 1). [^{2}H$_{3}$]Gibberellin A$_{19}$ was, however, converted to the C$_{19}$ GAs GA$_{1}$, GA$_{5}$, and GA$_{20}$. The use of either [2,3-^{3}H$_{2}$]- or [17-^{13}C,^{3}H$_{2}$]GA$_{20}$ as a substrate resulted in the accumulation of GA$_{1}$ and GA$_{5}$. [^{3}H]Gibberellin A$_{5}$ was converted to GA$_{6}$ while [^{3}H]GA$_{1}$ and [^{3}H]GA$_{6}$ were end products and were not further metabolized (Table 1).

The failure of the *P. coccineus* S-1 cell-free system to convert GA$_{1}$ to GA$_{8}$ is perhaps unexpected as the preparations do contain endogenous

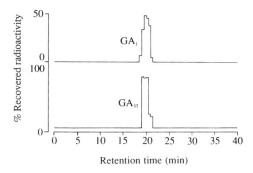

Fig. 2. Post-FPLC incubation of an S-1 cell-free preparation from immature seed of *P. coccineus* cv. Prizewinner with [2, 3-^{3}H]GA$_{20}$ and [17-^{3}H]GA$_{14}$. FPLC column: 50 × 5 mm i.d. Mono Q. Mobile phase: (A) 20 mM Tris hydrochloride, pH 7.5; (B) 20 mM Tris hydrochloride, pH 7.5, containing 1.0 M NaCl. Program: 0–22 min 0% B, 22–24 min 0–4.5% B, 24–26 min 4.5% B, 26–40 min 4.5–35% B, 40–50 min 35–100% B. Flow rate: 1 ml/min. Successive 0.5-ml fractions were collected and aliquots incubated at 30°C for 2 h with (i) [2, 3-^{3}H]GA$_{20}$ and (ii) [17-^{3}H$_2$]GA$_{14}$ in the presence of the cofactors outlined in Section 2.2. At the end of the incubation period, samples were analyzed by reversed-phase HPLC–RC to assess post-FPLC enzyme activity. Data expressed as the percentage of recovered radioactivity in each fraction associated with GA$_1$ in [2, 3-^{3}H$_2$]GA$_{20}$ incubations and with GA$_{37}$ in the [17-^{3}H]GA$_{14}$ incubations

GA$_8$.[21] It should be noted, however, that, in contrast to in vivo feeds, in vitro preparations from *P. vulgaris* seed are similarly unable to produce the 2β-hydroxylated metabolites of GA$_1$ and GA$_{20}$ although GA$_8$ and GA$_{29}$ are both endogenous constituents of the immature seed.[6,24,25] Unlike other test systems,[5,6,19,26] the *P. coccineus* preparations did not convert either GA$_{15}$, GA$_{44}$, or their open lactones to the corresponding C-20 aldehyde. This is probably due to experimental conditions rather than intrinsic differences in the C-20 oxidation pattern in *P. coccineus*. The GA-biosynthesis pathways illustrated in Fig. 1 are broadly similar to those operating in cell-free preparations from *P. vulgaris* seed.[6] The main differences relate to the occurrence of 13-hydroxylation. In *P. coccineus,* it is restricted to the conversion of GA$_{12}$-aldehyde to GA$_{53}$-aldehyde and GA$_{12}$ to GA$_{53}$. In *P. vulgaris* preparations, 13-hydroxylation can also occur at later stages in the biosynthesis sequence as indicated by the conversion of GA$_{15}$ to GA$_{44}$ and that of GA$_{24}$ to GA$_{20}$ via GA$_9$.[6]

2.3 Fast Protein Liquid Chromatography

Following fast protein liquid chromatography (FPLC) of the *P. coccineus* S-1 cell-free preparation on a Mono-Q ion exchange column, successive fractions were collected and their ability to metabolize GAs determined by

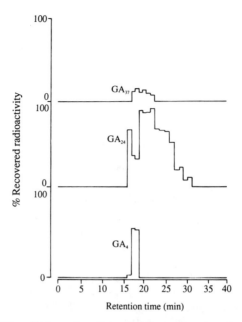

Fig. 3. Post-FPLC incubation of an S-1 cell-free preparation from immature seed of *P. coccineus* cv. Prizewinner with [^{3}H]GA$_{12}$. FPLC conditions as described in legend to Fig. 2. Successive 1.0-ml fractions collected and incubated with [^{3}H]GA$_{12}$ prior to analysis by reverse-phase HPLC–RC. Data expressed as percentage of radioactivity in each fraction associated with GA$_{37}$, GA$_{24}$, and GA$_4$

incubating aliquots with [^{3}H]GA$_{20}$ and [^{3}H]GA$_{14}$.[22] The 3β-hydroxylase converting GA$_{20}$ to GA$_1$ eluted as a discrete band of activity (Fig. 2). The crude *P. coccineus* homogenate converted [^{3}H]GA$_{14}$ to GA$_4$, GA$_{36}$, and GA$_{37}$ (see Table 1). In contrast, GA$_{37}$ was the only metabolite to be detected when the FPLC fractions were assayed using [^{3}H]GA$_{14}$ as a substrate (Fig. 2). This implies that FPLC separates the C-20 CH$_2$OH oxidase, which converts GA$_{37}$ to GA$_{36}$, from the C-20 CH$_3$ oxidase responsible for GA$_{37}$ production.

When [^{14}C]GA$_{12}$ was incubated with the crude cell-free preparation, both 13-and non-13-hydroxylated metabolism occurred, yielding GA$_{15}$, GA$_{44}$, and GA$_{53}$ (Table 1). After FPLC, [^{3}H]GA$_{12}$ incubation resulted in the accumulation of the non-13-hydroxylated products, GA$_4$, GA$_{24}$, and GA$_{37}$ (Fig. 3). It is likely that the 13-hydroxylase, which is particulate in *Pisum sativum*,[4] was removed by ultrafiltration prior to FPLC; hence the conversion of GA$_{12}$ to GA$_{53}$ could not take place.

The production of both GA$_{37}$ and GA$_4$, after incubation of FPLC fractions with [^{3}H]GA$_{12}$, was relatively restricted whereas that of the major metabolite GA$_{24}$ was much more widespread (Fig. 3). The GA$_4$ that accumulated would have been produced from GA$_{36}$, which originated from either GA$_{24}$ or GA$_{37}$. It seems probable that GA$_{24}$ is the major precursor as (1) the peak in GA_4 production corresponds with a trough in the level of GA_{24}, and (2) there was a complete absence of metabolism beyond GA_{37} when FPLC fractions were incubated with [^{3}H]GA_{14} (Fig. 2).

The lack of oxidation of GA_{37} to GA_{36} in the [^{3}H]GA_{14} FPLC incubations (Fig. 2) contrasts with the seemingly ready conversion of GA_{15} to GA_{24}, by the same C-20 oxidation step, in the [^{3}H]GA_{12} incubations (Fig. 3). This suggests that one enzyme performs both C-20 oxidations and than GA_{24} is more successful at competing for active sites than GA_{37}. Alternatively, two separate C-20 oxidation enzymes exist and the protein responsible for the conversion of GA_{37} to GA_{36} is separated by FPLC from the fractions in which GA_{37} biosynthesis occurs.

3 In Vitro [^{3}H]GA_4 Metabolism in Preparations from Germinating Seed

Although the in vitro studies with immature seed of *P. coccineus*, described in Section 2.2, indicate that GA_4 is an end product, S-1 cell-free preparations from the cotyledons of 24-h-germinated *P. coccineus* seed, incubated in Tris hydrochloride (50 mM, pH 7.8) containing Fe^{2+} (0.5 mM), ascorbate (5 mM), 2-oxoglutarate (5 mM), NADPH (1 mM), and uridine 5'-diphosphate glucose (1 mM), metabolize [^{3}H]GA_4 to GA_4-glucosyl ester, GA_{34}, a glucoside of GA_{34}, presumably GA_{34}-2-*O*-glucoside, and GA_1.[27] In addition, the preparations also convert [^{3}H]GA_1 to GA_8. The cell-free system, thus, contains at least three types of GA-metabolizing enzyme, namely, (1) 2β-hydroxylase(s) converting GA_4 to GA_{34} and GA_1 to GA_8, (2) 13-hydroxylase metabolizing GA_4 to GA_1, and (3) glucosyl transferase(s) catalyzing the conversion of GA_4 to GA_4-glucosyl ester and GA_{34} to GA_{34}-glucoside. The probable metabolic relationships involved in these conversions are illustrated in Fig. 4.

High-speed centrifugation (100,000 × g) showed that the glucosyl transferase activity was associated with the soluble fraction (S-100) (Table 2). In contrast, 2β-hydroxylase activity was detected in both the supernatant and the membrane pellet (P-100). It is possible that this was caused by incomplete washing of the pellet although this appears to be unlikely as there was no detectable glucosyl transferase activity in the P-100 pellet. An entirely soluble 2β-hydroxylase catalyzing the conversion of GA_1 to GA_8 has been found in a cell-free system from germinating *P. vulgaris* seed.[29]

Cell-free preparations were also made from shoots, roots, and cotyledons of *P. coccineus*, 0–7 days after imbibition of the seed. Incubations with [^{3}H]GA_4, and in some instances [^{3}H]GA_1, indicated the presence of glucosyl transferase(s), 2β-hydroxylase(s), and a 13-hydroxylase in almost all preparations from cotyledons obtained 3–24 h after imbibition. Preparations from the embryo obtained 3–24 h after the start of imbibition contained only 2β-hydroxylase activity which converted [^{3}H]GA_4 to GA_{34} and [^{3}H]GA_1 to GA_8 (Table 3). S-1 preparations from dry seed and seed-

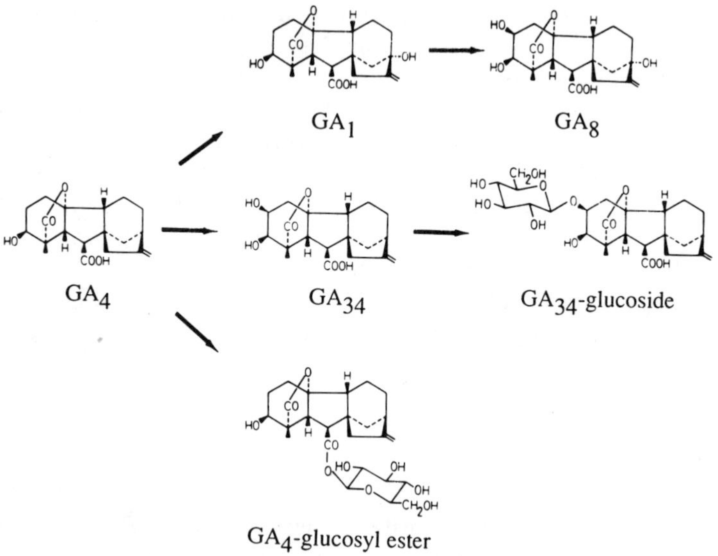

Fig. 4. Probable metabolic relationships of products formed from [1, 2-^{3}H$_2$]GA$_4$ in S-1 cell-free preparations from cotyledons of 24-h-germinated seed of *P. coccineus* cv. Prizewinner

Table 2. Summary of [^{3}H]GA$_4$ metabolites produced by S-1 preparation, S-100 supernatant and P-100 pellet obtained from the cotyledons of 24 h-germinated *Phaseolus coccineus* seed

Compound	S-1	S-100	P-100
[^{3}H]GA$_4$	30 ± 3	19 ± 2	38 ± 4
[^{3}H]H$_2$O	19 ± 1	23 ± 3	16 ± 1
[^{3}H]GA$_1$	1 ± 1	1 ± 1	1 ± 1
[^{3}H]GA$_{34}$	43 ± 3	47 ± 2	45 ± 5
[^{3}H]GA$_4$-glucosyl ester/ [^{3}H]GA$_{34}$-glucoside	7 ± 1	10 ± 1	n.d.

Data expressed as % radioactivity detected at the end of 2h incubation period ± standard error. n.d. = not detected (<0.5%). Metabolites identified by HPLC procedures described by Turnbull et al.[28]

lings older than 24 h failed to metabolize [^{3}H]GA$_4$, despite considerable experimentation with isolation and incubation conditions.[27]

The degree of metabolism of [^{3}H]GA$_4$ to GA$_{34}$ and that of [^{3}H]GA$_1$ to GA$_8$ were broadly equivalent in the various S-1 preparations (Table 3). In this context, it is of interest to note that a purified GA 2β-hydroxylase from cotyledons of *P. vulgaris* seed catalyzes the conversion of both GA$_4$ to GA$_{34}$ and GA$_1$ to GA$_8$.[8]

Table 3. Summary of conversions of [³H]GA$_1$ and [³H]GA$_4$ detected in S-1 cell-free preparations from germinating seed of *Phaseolus coccineus*

Tissue	Time after start of imbibition (h)	Reactions observed (max. % conversion)				
		2β-Hydroxylation		13-Hydroxylation	Glucosylation	
		GA$_1$ → GA$_8$	GA$_4$ → GA$_{34}$	GA$_4$ → GA$_1$	GA$_4$ → GA$_{34}$-OG[1]	GA$_4$ → GA$_4$-GE[2]
Cotyledons	0(dry)	no metabolism observed				
	3	n.a.	68	5	9	n.d.
	6	73	82	n.d.	2	n.d.
	24	79	60	7	22	2
Shoot	3	n.a.	50	n.d.	n.d.	n.d.
	24	n.a.	8	n.d.	n.d.	n.d.
Root	3	n.a.	46	n.d.	n.d.	n.d.
Whole embryo	6	35	54	n.d.	n.d.	n.d.

n.a., not assayed; n.d., not detected (<1% of total radioactivity). Metabolites identified by HPLC-RC procedures described by Turnbull et al.[28]
[1] GA$_{34}$-*O*glucoside; [2] GA$_4$-glucosyl ester

Acknowledgments. The investigations reported in this article were supported by a Science and Engineering Research Council grant to A.C., by a grant from the Deutsche Forschungsgemeinschaft to J.E.G., and by a Postgraduate Fellowship from The Carnegie Trust for Scottish Universities to J.M.M. The authors would like to thank Dr. Y. Kamiya (Institute of Physical and Chemical Research, Saitama), Professor J. MacMillan F.R.S. (University of Bristol), and Professor R.P. Pharis (University of Calgary) for kindly supplying some of the isotopically labeled GAs used in the investigations. We are also grateful to Ludger Schwenen (University of Göttingen) and Einar Jensen (University of Tromsö) for help with GC–MS analyses and Dr. P. Hedden (Long Ashton) for advice on the use of FPLC.

References

1. Graebe JE, Hedden P. Gaskin P, et al. Biosynthesis of gibberellins A_{12}, A_{15}, A_{24}, A_{36} and A_{37} by a cell-free system from *Cucurbita maxima*. Phytochemistry. 1974; 13:1433–1400.
2. Graebe JE, Hedden P, Gaskin et al. The biosynthesis of a C_{19}-gibberellin from mevalonic acid in a cell-free system from a higher plant. Planta. 1974; 120:307–309.
3. Hedden P, Graebe JE, Beale MH, et al. The biosynthesis of 12α-hydroxylated gibberellins in a cell-free system from *Cucurbita maxima* endosperm. Phytochemistry. 1984; 23:569–574.
4. Kamiya Y, Graebe JE. The biosynthesis of all major pea gibberellins in a cell-free system from *Pisum sativum*. Phytochemistry. 1983; 22:681–689.
5. Sponsel VM. *In vitro* gibberellin metabolism in higher plants. In: Crozier A, ed. The biochemistry and physiology of gibberellins, Vol. 1. New York: Praeger, 1983: p. 151–250.
6. Takahashi M, Kamiya Y, Takahashi N, et al. Metabolism of gibberellins in a cell-free system from immature seeds of *Phaseolus vulgaris*. Planta. 1986; 168:190–199.
7. Albone K, Gaskin P, MacMillan J. et al. Enzymes from seeds of *Phaseolus vulgaris* L.: Hydroxylation of gibberellins A_{20} and A_1 and 2,3-dehydrogenation of gibberellin A_{20}. Planta. 1989; 177:108–115.
8. Smith VA, MacMillan J. Purification and partial characterisation of gibberellin 2β-hydroxylase from *Phaseolus vulgaris* seed. J Plant Growth Regul. 1984; 2:251–264.
9. Smith VA, MacMillan J. Purification and partial characterisation of gibberellin 2β-hydroxylases from seeds of *Pisum sativum*. Planta. 1986; 167:9–18.
10. Kwak S-S, Kamiya Y, Sakurai A, et al. Partial purification of gibberellin 3β-hydroxylase from immature seed of *Phaseolus vulgaris* L. Plant Cell Physiol. 1988; 29:935–943.
11. Albone KS, Sponsel VM, Gaskin P, et al. Identification and localisation of gibberellins in maturing seeds of the cucurbit *Sechium edule* and a comparison between this cucurbit and the legume *Phaseolus coccineus*. Planta. 1984; 162:560–565.

12. Ceccarelli N, Lorenzi R, Alpi A. Kaurene and kaurenol biosynthesis in cell-free system of *Phaseolus coccineus* suspensor. Phytochemistry. 1979; 18:1657–1658.
13. Ceccarelli N, Lorenzi R, Alpi A. Kaurene metabolism in cell-free extracts of *Phaseolus coccineus* suspensors. Plant Sci Lett. 1981; 21:325–332.
14. Ceccarelli N, Lorenzi R, Alpi A. Gibberellin biosynthesis in *Phaseolus coccineus* suspensor. Z. Pflanzenphysiol. 1981; 102:37–44.
15. Turnbull CGN, Crozier A, Schwenen L. et al. Biosynthesis of gibberellin A_{12}-aldehyde, gibberellin A_{12} and their kaurenoid precursors from [^{14}C]mevalonic acid in a cell-free system from immature seed of *Phaseolus coccineus*. Phytochemistry. 1986;25:97–101.
16. Graebe JE. Gibberellin biosynthesis in cell-free systems from higher plants. In: Wareing PF, ed. Plant growth substances 1982. London: Academic Press, 1982: p. 71–80.
17. Hedden P. *In vitro* metabolism of gibberellins. In: Crozier A., ed. The biochemistry and physiology of gibberellins, Vol. 1. New York: Praeger, 1983: p. 99–149.
18. Bearder JR. *In vivo* diterpenoid biosynthesis in *Gibberella fujikuroi*: the pathway after *ent*-kaurene. In: Crozier A., ed. The biochemistry and physiology of gibberellins, Vol. 1. New York: Praeger, 1983: p. 251–387.
19. Hedden P, Graebe JE. Cofactor requirements for the soluble oxidases in the metabolism of C_{20}-gibberellins. J Plant Growth Regul. 1982; 1:105–116.
20. Beale MH, Bearder JR, Down GH, et al. The biosynthesis of kaurenolide diterpenoids by *Gibberella fujikuroi*. Phytochemistry. 1982; 21:1279–1287.
21. Turnbull CGN, Crozier A, Schwenen L, et al. Conversion of [^{14}C]gibberellin A_{12}-aldehyde to C_{19}- and C_{20}-gibberellins in a cell-free system from immature seed of *Phaseolus coccineus* L. Planta. 1985; 165:108–113.
22. Malcolm JM. Gibberellin metabolism in seeds and seedlings of the runner bean, *Phaseolus coccineus* L. Ph.D. thesis, University of Glasgow, 1988
23. MacMillan J, Pryce RJ. The gibberellins. In: Miller LP, ed. Phytochemistry. New York: Van Nostrand-Reinhold, 1973: p. 283–286.
24. Yamane H, Murofushi N, Takahashi N. Metabolism of gibberellins in maturing and germinating bean seed. Phytochemistry. 1975; 14:1195–1200.
25. Yamane H, Murofushi N, Osada H, Takahashi N. Metabolism of gibberellins in early immature bean seeds. Phytochemistry. 1977; 16:831–835.
26. Gilmour SJ, Zeevaart JAD, Schwenen L, Graebe JE. Gibberellin metabolism in cell-free extracts from spinach leaves in relation to photoperiod. Plant Physiology. 1986; 82:190–195.
27. Turnbull CGN, Crozier A. Metabolism of [1, 2-^{3}H]gibberellin A_4 by epicotyls and cell-free preparations from *Phaseolus coccineus* L. seedlings. Planta. 1989; 178:267–274.
28. Turnbull CGN, Crozier A, Schneider G. HPLC-based methods for the identification of gibberellin conjugates: metabolism of [^{3}H]gibberellin A_4 in seedlings of *Phaseolus coccineus*. Phytochemistry. 1986; 25:1823–1828.
29. Patterson RJ, Rappaport L. The conversion of gibberellin A_1 to gibberellin A_8 by a cell-free enzyme system. Planta. 1974; 119:183–191.

CHAPTER 10

Gibberellin Biosynthetic Enzymes and the Regulation of Gibberellin Concentration

P. Hedden

1 Introduction

The numerous gibberellin (GA)-deficient dwarf mutants that have been characterized are spectacular demonstrations of the importance of GAs for shoot elongation. However, the growth rates of many tall genotypes are also increased by application of GAs, suggesting that the endogenous GA concentration may be limiting for growth in "normal" varieties. It has been shown for maize that the increased vigor of F1 hybrids, a phenomenon known as heterosis, is associated with higher GA concentrations in the hybrids.[1] More dramatically, the rapid stem extension that precedes flowering in rosette plants, such as spinach, is accompanied by, and dependent on, increased GA biosynthesis.[2,3] There is, therefore, considerable evidence that GA concentration is a determinant of plant height, and any consideration of how plant height is regulated must include an understanding of the control of GA concentration.

The metabolic relationship of the GAs present in vegetative tissues is well understood. It is generally accepted that of the compounds in the metabolic pathway from GA_{53} to GA_8, only GA_1 is physiologically active for shoot growth.[4] However, it was demonstrated recently[5] that GA_3 also occurs naturally in shoot tissue and is formed from GA_{20} via GA_5. The importance of GA_3 as a growth regulator has still to be established.

The concentration of GA_1 at its site of action will be determined by its rates of biosynthesis and catabolism and, if the sites of action and synthesis are remote, by the rates of transport of GA_1 or its precursors to, and from, the active site. There is, as yet, little information on which, if any, of the GA biosynthetic steps are rate limiting. Such steps would be candidates for points of natural metabolic control as well as prime targets for the genetic or chemical manipulation of GA concentration and therefore also of plant stature.

In the following discussion, aspects of work at Long Ashton Research Station (part of the AFRC Institute of Arable Crops Research) on the regulation of GA metabolism will be described. Firstly, the changes in GA metabolism apparent in GA-response mutants will be considered. Secondly, the partial purification and properties of 2β-hydroxylases from cotyledons of *Phaseolus vulgaris* will be described. Finally, the mode of action of a new class of growth retardants will be discussed. Since these compounds inhibit the later stages of GA biosynthesis, they offer an opportunity to determine the consequences for plant growth of altering the relative levels of GAs and precursors.

2 GA Metabolism in Stature Mutants of Cereals

It is only within the last ten years, with the increased sensitivity of combined gas chromatography–mass spectrometry (GC–MS) and particularly with the routine use of selected ion monitoring (SIM) for quantitative analysis, that the GA content of vegetative tissues has been determined accurately. There is now detailed information on GA concentrations in vegetative shoots of several monocotyledons. A common feature is the relatively high concentration of GA_{19}, as has been reported for rice,[6] maize,[7] and *Sorghum*,[8] indicating that the conversion of GA_{19} to GA_{20} may be a rate-limiting step for GA_1 biosynthesis in these species.

Dwarf mutants of cereals that are insensitive to applied GA have been shown to have highly elevated levels of GA_1. Examples include the *Rht3* mutant of wheat[9] and the *D8* mutant of maize.[7] In Table 1 the concentrations of 13-hydroxy-GAs in shoots of *Rht3* wheat and *D8* maize are compared with those in the corresponding isogenic tall lines. As well as containing much higher levels of GA_1 than the normal lines, the dwarf genotypes have increased GA_8 and GA_{20} levels. However, the concentration of GA_{19} is lower in the dwarf mutants. In contrast, shoots of the *slender* mutant of barley, an overgrowth mutant which has no requirement for GA,[10–12] contains substantially lower levels of GA_{20}, GA_1, GA_3, and GA_8 than the normal genotype (Table 1) (S.J. Croker, P. Hedden, J.R. Lenton, and J.L. Stoddart, unpublished). The concentration of GA_{19} in *slender* is higher than that in the parent line.

One interpretation of the data in Table 1 is that the rate of conversion of GA_{19} to GA_{20} is altered in the stature mutants: increased in the dwarfs and decreased in the overgrowth mutant. We have tested this by comparing the metabolism of $[^3H]GA_{19}$ in *rht3* (tall) and *Rht3* (dwarf) wheat and in normal and *slender* barley. The substrate, labeled by catalytic exchange, was applied in water to 5-day-old seedlings by injection into the cavity within the curled first leaf. Plants were harvested after 24 h, and the lower expanding section of the first leaf together with the younger leaves was extracted. Products were purified and separated by high-performance liquid chromatography (HPLC) as described elsewhere.[13] The radioactivity profiles of the extracted free acids on C_{18} reverse-phase HPLC are shown in Fig. 1. The degree of conversion into free acid products is greater in the

Table 1. Concentrations of GAs in shoots of GA-response mutants of cereals and in those of their corresponding wild-type

Species/genotype	Concentration (ng/g fr wt)						
	GA_{44}	GA_{19}	GA_{20}	GA_{29}	GA_1	GA_3	GA_8
Triticum aestivum[a]							
Rht3 (dwarf)	—[b]	2.18	0.93	—	10.93	1.64	5.61
rht3 (normal)	—	4.01	0.50	—	0.44	0.34	4.64
Zea mays[c]							
D8 (dwarf)	1.20	0.38	2.22	0.22	13.10	—	3.13
normal	0.79	5.09	0.62	0.12	0.22	—	0.26
Hordeum vulgare[d]							
slender	—	6.32	0.35	—	0.02	0.04	0.78
normal	—	3.58	0.72	—	0.23	0.25	5.51

The GA quantifications were by GC–SIM using deuterated or ^{13}C-labeled internal standards
[a] First leaf sheath and enclosed younger leaves from 12-day-old seedlings (N.E.J. Appleford and J.R. Lenton, unpublished data)
[b] not determined
[c] Whole shoots from 3-week-old seedlings[7]
[d] First leaf sheath and enclosed younger leaves from 7-day-old seedlings (S.J. Croker, P. Hedden, J.R. Lenton, and J.L. Stoddart, unpublished data).

Rht3 (dwarf) wheat seedlings (63%) than in the tall (rht3) genotype (37%) (Fig. 1A). On the other hand, *slender* barley produced only 7% free acid products compared with 27% by the normal genotype (Fig. 1B). Application of [17-^{2}H$_2$]GA$_{19}$ to the wheat seedlings and GC–MS analysis of the products confirmed the assignments in Fig. 1A. It could be argued that the apparent genotypic differences in metabolism of applied GA$_{19}$ are due to the different GA$_{19}$ pool sizes, which, for the wheat tissue, were measured as 510 pg per plant in the tall genotype and 35 pg in the dwarf. However, when the amount of GA$_{19}$ applied to each wheat seedling was increased to 20 ng, that is, well in excess of the endogenous pool size, differences in metabolism still occurred.

Stoddart[14] found that the major difference in GA$_1$ metabolism between Rht3 and rht3 wheat seedlings was that the tall genotype produced more conjugate. We have confirmed this observation after applications of ^{3}H-labeled GA$_1$, GA$_{20}$, and GA$_{19}$ and are presently investigating its significance for the pool sizes of the free acids. However, reduced conjugate formation cannot explain readily the low levels of GA$_{19}$ in the slower-growing genotypes, including the GA$_1$-deficient maize dwarf, *d1*,[15] in which the lesion is after GA$_{20}$.[16] Whatever the explanation, it appears that GA metabolism, and consequently the pool size of GA$_1$, is affected by growth rate in monocotyledons. The implications of this for the regulation of GA concentrations warrant further investigation.

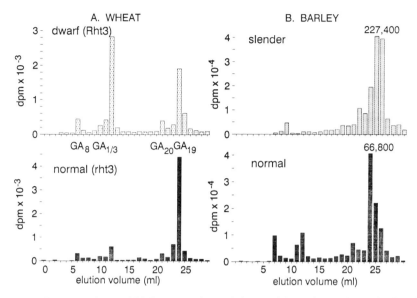

Fig. 1. Reverse-phase HPLC separation of free acid products from feeds of [^{3}H]GA$_{19}$ to 5-day-old seedlings of *Rht3* and normal wheat cv. Maris Huntsman (A) (0.66×10^6 dpm applied to six seedlings of each genotype) and *slender* and normal barley cv. Herta (B). (1.38×10^6 dpm applied to six seedlings)

3 2β-Hydroxylases from *Phaseolus vulgaris*

The 2β-hydroxylation of GA$_1$ is, as the first catabolic step, an important factor determining the concentration of this GA. We have been purifying GA$_1$ 2β-hydroxylase from cotyledons of imbibed *Phaseolus vulgaris* seeds. Enzyme activity for the conversion of GA$_1$ to GA$_8$ was first detected in this tissue by Patterson and Rappaport.[17] The enzyme was partially purified subsequently by Smith and MacMillan,[18] who showed it to be a dioxygenase requiring 2-oxoglutarate, Fe^{2+}, and ascorbate and to be stimulated by catalase.

The assay for enzyme activity involves measuring the release of ^{3}H from [1,2-^{3}H$_2$]GA$_1$ as ^{3}HOH according to Scheme 1. After the incubation, pro-

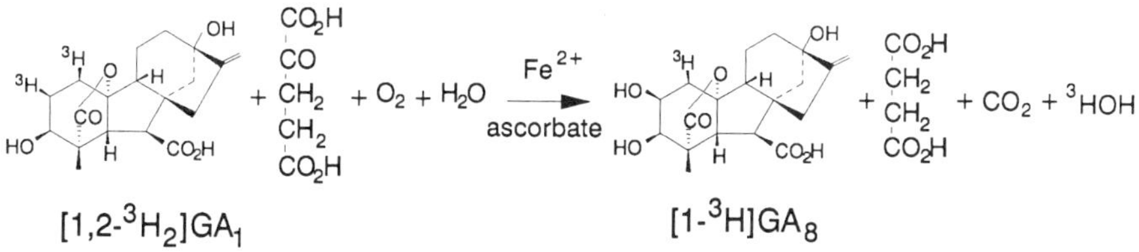

Scheme 1. 2β-Hydroxylation of [1,2-^{3}H$_2$]GA$_1$ to [1-^{3}H$_2$]GA$_8$ catalyzed by a 2-oxoglutarate-dependent dioxygenase. ^{3}H is released in the reaction as ^{3}HOH

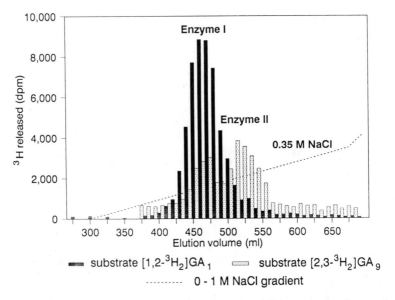

Fig. 2. Separation of *P. vulgaris* 2β-hydroxylases I and II by cation exchange chromatography on S-Sepharose. After partial purification by batch anion exchange with DE-52 and ammonium sulfate precipitation, the sample was applied in 0.1 *M* sodium acetate buffer at pH 6 and enzyme activity eluted at 5 ml/min in 5-ml fractions with a gradient of NaCl. Alternate fractions were assayed for release of ^{3}H from [1,2-^{3}H$_2$]GA$_1$ or [2,3-^{3}H$_2$]GA$_9$

Table 2. Purification of 2β-hydroxylases (Enzymes I and II) from mature *Phaseolus vulgaris* seeds

Purification step	Total protein	Specific activity (fold purification) (pmol h^{-1}mg^{-1})		Yield (%)	
		[^{3}H]GA$_1$	[^{3}H]GA$_9$	[^{3}H]GA$_1$	[^{3}H]GA$_9$
Homogenate	28.7 g	0.57(1)	0.07(1)	100	100
Batch anion exchange (DE-52)	10.9 g	2.0(3.5)	0.34(4.9)	133	184
Cation exchange, S-Sepharose (×2)					
Enzyme I	77.8 mg	7.4(13)		3.5	
Enzyme II	19.5 mg		3.2(46)		3.1
Size exclusion, Superose-12					
Enzyme I	3.1 mg	31.2(55)		0.6	
Enzyme II	7.8 mg		5.5(79)		2.0
Cation exchange, Mono-S					
Enzyme I	0.6 mg	53.2(93)		0.2	
Enzyme II	0.6 mg		5.9(84)		0.2

Table 3. Kinetic parameters for P. vulgaris 2β-hydroxylases with [³H]GA substrates and 2-oxoglutarate

Substrate	Enzyme I			Enzyme II		
	K_m (nM)	V_{max} (nmol h^{-1}mg^{-1})	V_{max}/K_m (ml h^{-1}mg^{-1})	K_m (nM)	V_{max} (nmol h^{-1}mg^{-1})	V_{max}/K_m (ml h^{-1}mg^{-1})
[1,2-³H$_2$]GA$_1$	103	0.206	2.52	52	0.004	0.08
[2,3-³H$_2$]GA$_9$	538	0.083	0.15	27	0.007	0.26
[1,2-³H$_2$]GA$_4$	60	0.182	3.03	135	0.014	0.10
[2,3-³H$_2$]GA$_{20}$	302	0.008	0.02	12,440	0.300	0.02
2-Oxoglutarate	58,500[a]	—	—	2,600[b]	—	—

[a] Measured with respect to 2β-hydroxylation of [1,2-³H$_2$]GA$_1$
[b] Measured with respect to 2β-hydroxylation of [2,3-³H$_2$]GA$_9$

ducts and remaining substrate are precipitated by the addition of activated charcoal and centrifugation. Enzyme activity is determined by the amount of ^{3}HOH in the supernatant, measured by liquid scintillation counting.

Smith and MacMillan[18] showed that the *P. vulgaris* preparation 2β-hydroxylates GA_4, GA_9, and GA_{20} in addition to GA_1, with GA_1 and GA_4 the preferred substrates. After cation exchange chromatography, we separated two enzyme activities (Fig. 2) with different substrate specificities. These activities, designated Enzymes I and II, have been partially purified (80–90-fold) as outlined in Table 2. The enzymes are separated after the cation exchange step on S-Sepharose. The molecular masses of the two enzymes, as estimated after size exclusion chromatography on TSK-125, are ca. 26 and 42 kDa, respectively. This compares with a value of 36 kDa reported by Smith and MacMillan. Both enzymes have a pH optimum of 6.5.

Table 3 compares the kinetics of 2β-hydroxylation of four [^{3}H]GAs by Enzymes I and II as determined by a nonlinear regression analysis of the variation of reaction rate with substrate concentration. The ratio V_{max}/K_m is used for comparison of substrate specificities. Enzyme I has a similar substrate specificity to the enzyme reported by Smith and MacMillan with the 3β-hydroxy-GAs, GA_1, and GA_4, the preferred substrates. [2,3-^{3}H$_2$]Gibberellin A_9 is the favored substrate for Enzyme II. The low V_{max} values obtained for Enzyme II might reflect a low abundance of this enzyme. The K_m values for 2-oxoglutarate were determined from the effect of the concentration of this compound on the release of ^{3}H from [1,2-^{3}H$_2$]GA$_1$ and [2,3-^{3}H$_2$]GA$_9$ by Enzymes I and II, respectively. The value obtained for Enzyme II is about an order of magnitude less than that for Enzyme I. High concentrations of 2-oxoglutarate are inhibitory, with the maximum activity of Enzyme I reduced to 70% at 5 mM 2-oxoglutarate and that of Enzyme II to 80% at 1 mM 2-oxoglutarate. This inhibition may be due to 2-oxoglutarate interfering with the access of Fe^{2+} and/or ascorbate to the active site, as has been proposed for prolyl 4-hydroxylase, a related enzyme.[19]

4 Inhibitors of 2-Oxoglutarate-Dependent Dioxygenases

In a collaborative project with Dr W. Rademacher, BASF AG, we are investigating the mode of action of a new class of growth retardants. These compounds are cyclohexanedione carboxylic acid derivatives based on structure I and are structurally related to herbicides that inhibit acetyl coenzyme A carboxylase.[20]

The effects of one such derivative (BASF code no. LAB 236735; I: R_1 = C_2H_5) on the GA levels in wheat and barley shoots are presented in Table 4. The compound was applied, at the time of sowing, at 6 mg per pot containing ten seeds, and the plants were harvested after 16 days. Gib-

[Structure I: cyclohexane ring with OH, R₁, R₂, and carbonyl substituents]

Table 4. Concentrations of GAs in whole shoots of 16-day-old wheat and barley seedlings with and without soil treatment with LAB 236735

Sample	Concentration (ng/g dry weight)							
	GA_{44}	GA_{19}	GA_{17}	GA_{20}	GA_{29}	GA_1	GA_3	GA_8
Wheat cv. Ralle								
Untreated	100%[a]	2.8	100%[a]	3.3	0.4	2.5	4.8	12.8
Treated	69%[a]	7.4	166%[a]	12.1	4.0	1.3	4.5	0.0
Barley cv. Aramir								
Untreated	100%[a]	22.5%	100%	8.5	3.2	4.0	7.3	24.3
Treated	31%[a]	13.0%	31%	18.9	18.3	1.3	3.7	2.3

[a] Relative amount as there was no internal standard available

berellin concentrations were measured by GC–SIM using ^{2}H-labeled GAs as internal standards.[13] The levels of some GAs for which no labeled analogues were available as internal standards are given in Table 4 only as relative values. Plant height in wheat and barley was reduced by 62% and 51%, respectively, by the treatment, relative to controls, with corresponding reductions of 47% and 67%, respectively, in the concentrations of GA_1. Gibberellin A_3, which is present in these cereals at similar levels to GA_1, was reduced in barley by 50% after treatment with the retardant. However, there was very little reduction of the level of GA_3 in wheat (5%), and the importance of this GA to shoot elongation must be considered questionable. Since treatment with LAB 236735 resulted in large increases in the levels of GA_{20} and GA_{29} in both species, the primary site of action appears to be the 3β-hydroxylation of GA_{20} to GA_1. This specificity is interesting considering that the same types of enzymes, that is, 2-oxoglutarate-dependent dioxygenases, are involved in each step from GA_{53} to GA_8. In barley, but not in wheat, the concentrations of the C_{20} GAs, GA_{44}, GA_{19}, and GA_{17}, are reduced by the retardant, suggesting that oxidation at C-20 is also inhibited. These results provide further support for the hypothesis that GA_1, but not GA_{20} or earlier precursors, is active in promoting growth per se[4].

Table 5. Effect of 3,5-dioxo-4-butyrylcyclohexanecarboxylic acid ethyl ester on the kinetics of 2β-hydroxylation with respect to 2-oxoglutarate

Inhibitor concentration (μM)	Enzyme I[a]		Enzyme II[b]	
	K_m (μM)	V_{max} (pmol h^{-1} mg^{-1})	K_m (μM)	V_{max} (pmol h^{-1} mg^{-1})
0	58.5	31.8	2.6	9.9
5	1355.0	32.4	—[c]	—
10	1012.0	15.8	51.8	8.6
15	655.8	9.3	65.1	6.1

[a] Relationship between 2-oxoglutarate concentration and release of ^{3}H from [1,2-^{3}H$_2$]GA$_1$
[b] Relationship between 2-oxoglutarate concentration and release of ^{3}H from [2,3-^{3}H$_2$]GA$_9$
[c] Not determined

Although the 3β-hydroxylase appears to be the most sensitive enzyme to the cyclohexanedione retardants, preliminary experiments indicated that the 2β-hydroxylase was also affected. Thus, the mechanism of enzyme inhibition was investigated using the partially purified 2β-hydroxylases from mature *P. vulgaris* seeds and 3,5-dioxo-4-butyrylcyclohexanecarboxylic acid ethyl ester (I: $R_1 = C_3H_7$, $R_2 = OC_2H_5$) as the inhibitor. Since the cyclohexanedione derivatives bear some structural similarity to 2-oxoglutarate, the effect of 2-oxoglutarate concentration on the rate of 2β-hydroxylation was

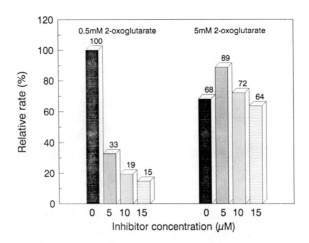

Fig. 3. Inhibition of *P. vulgaris* 2β-hydroxylase I by 3,5-dioxo-4-butyrylcyclohexanecarboxylic acid ethyl ester at two concentrations of 2-oxoglutarate. Activity (release of ^{3}HOH from [1,2-^{3}H$_2$]GA$_1$) was expressed relative to the highest value, which was obtained with 0.5 mM 2-oxoglutarate in the absence of inhibitor

investigated in the presence of different concentrations of the inhibitor. The influence of inhibitor concentration on the apparent K_m for 2-oxoglutarate and V_{max} for 2β-hydroxylation by Enzymes I and II was determined by nonlinear regression analysis as described above (Table 5). For Enzyme I, the lowest inhibitor concentration (5 μM) caused a >20-fold increase in the apparent K_m with little change in V_{max}, indicating competitive inhibition with respect to 2-oxoglutarate; the calculated K_i is 0.22 μM. At higher inhibitor concentrations, both K_m and V_{max} change such that their ratio remains almost constant. Inhibition appears, therefore, to be uncompetitive with respect to 2-oxoglutarate at these higher inhibitor concentrations. For Enzyme II, the apparent K_m for 2-oxoglutarate is increased ca. 25-fold in the presence of 10 μM inhibitor with only a slight reduction in V_{max}. Assuming purely competitive inhibition, the calculated K_i is 0.5 μM. Again, at higher inhibitor concentrations, the inhibition is not purely competitive.

Thus, the K_i values for the cyclohexanedione inhibitor are similar for Enzymes I and II. However, since the K_m for 2-oxoglutarate is an order of magnitude higher for Enzyme I than for Enzyme II, the former enzyme is more sensitive to the inhibitor. At low concentrations, the retardant acts as a competitive inhibitor with respect to 2-oxoglutarate, but at higher inhibitor concentrations the mechanism is more complicated. A similar situation exists for inhibition of prolyl 4-hydroxylase by certain pyridine dicarboxylic acids,[21] high concentrations of which are thought to remove Fe^{2+} through chelation. This mechanism may also pertain to the cyclohexanediones. It is of interest that 2,5-dicarboxypyridine, the most potent prolyl hydroxylase inhibitor, is several orders of magnitude less effective than the cyclohexanedione as an inhibitor of the 2β-hydroxylases (D.L. Griggs, unpublished data).

The interaction between 2-oxoglutarate and the cyclohexanedione inhibitor was examined further by comparing the extent of inhibition of Enzyme I at two concentrations of 2-oxoglutarate (Fig. 3). At 0.5 mM 2-oxoglutarate, there is considerable inhibition of enzyme activity at each inhibitor concentration (5–15 μM). However, at 5 mM 2-oxoglutarate, a concentration which is slightly inhibitory (see above), 5 μM inhibitor increases enzyme activity; the activity decreases with increasing inhibitor concentration, but only to just below the level without inhibitor present. The explanation for this observation is unclear. If, as suggested above, high 2-oxoglutarate concentrations interfere with the access of Fe^{2+}, or ascorbate, to the enzyme active site, the inhibitor may antagonize this effect.

There is clearly much to learn about these new and exciting growth retardants. They will undoubtedly prove of great benefit to agriculture, and at present they are a valuable tool in our work to understand the metabolism and mode of action of the GAs.

Acknowledgments. The work described in this article results from the efforts of a number of Long Ashton staff and students. I wish to acknowledge the valuable contributions of Dr. John Lenton, Mr. Stephen Croker, and Miss Karen McCrea to the work on GA metabolism in cereals and of Mr. David Griggs and Miss Kay Temple-Smith to the enzyme characterization and inhibitor studies. I thank Professors Lewis Mander, Australian National University, Canberra, and Bernard Phinney, University of California, Los Angeles, for, respectively, ^{2}H- and ^{13}C-labeled GAs, Dr. Alan Crozier, University of Glasgow, for the [^{3}H]GA$_9$, and Dr. Yuji Kamiya, The Institute of Physical and Chemical Research, Wako, Japan, for help in obtaining the [^{3}H]GA$_{19}$.

References

1. Rood SB, Buzzell RI, Mander LN, et al. Gibberellins: A phytohormonal basis for heterosis in maize. Science. 1988; 241:1216–1218.
2. Zeevaart JAD. Effects of photoperiod on growth rate and endogenous gibberellins in the long-day rosette plant spinach. Plant Physiol. 1971; 47:821–827.
3. Metzger JD, Zeevaart JAD. Photoperiodic control of gibberellin metabolism in spinach. Plant Physiol. 1982; 69:287–291.
4. Phinney BO. Gibberellin A$_1$, dwarfism and the control of shoot elongation in higher plants. In: Crozier A, Hillman, JR, eds. The biosynthesis and metabolism of plant hormones. Cambridge: Cambridge University Press, 1984: p. 17–41.
5. MacMillan J. Metabolism of gibberellins A$_{20}$ and A$_9$ in plants: Pathways and enzymology. In: Pharis RP, Rood SB, eds. Plant growth substances 1988. Berlin: Springer-Verlag, 1990: In press.
6. Kobayashi M, Yamaguchi I, Murofushi N, et al. Fluctuation and localization of endogenous gibberellins in rice. Agric Biol Chem. 1988; 52:1189–1194.
7. Fujioka S, Yamane H, Spray CR, et al. The dominant non-gibberellin-responding dwarf mutant (*D8*) of maize accumulates native gibberellins. Proc Natl Acad Sci USA. 1988; 85:9031–9035.
8. Rood SB, Larsen KM, Mander LN, et al. Identification of endogenous gibberellins from *Sorghum*. Plant Physiol. 1986; 82:330–332.
9. Lenton JR, Hedden P, Gale MD. Gibberellin insensitivity and depletion in wheat—consequences for development. In:Hoad GV, Lenton JR, Jackson MB, Atkin RK, eds. Hormone action in plant development—A critical appraisal. London: Butterworths, 1987: p. 145–160.
10. Foster CA. Slender: An accelerated extension growth mutant of barley. Barley Genet Newslett. 1977; 7:24–27.
11. Lanahan MB, Ho T-HD. Slender barley: a constitutive gibberellin-response mutant. Planta. 1988; 175:107–114.
12. Chandler PM. Hormonal regulation of gene expression in the "slender" mutant of barley (*Hordeum vulgare* L.). Planta. 1988; 175:115–120.
13. Hedden P, Croker SJ. GC–MS analysis of gibberellins in plant tissues. In: Kutáček M, Elliott MC, Macháčková I, eds. Molecular aspects of hormonal regulation of plant development. The Hague: SPB Academic Publishing. 1990; p. 19–30.

14. Stoddart JL. Growth and gibberellin-A_1 metabolism in normal and gibberellin-insensitive (*Rht3*) wheat (*Triticum aestivum* L.) seedlings. Planta. 1984; 161:432–438.
15. Fujioka S, Yamane H, Spray CR, et al. Qualitative and quantitative analyses of gibberellins in vegetative shoots of normal, *dwarf*-1, *dwarf*-2, *dwarf*-3, and *dwarf*-5 seedlings of *Zea mays* L. Plant Physiol. 1988; 88:1367–1372.
16. Spray C, Phinney BO, Gaskin P, et al. Internode length in *Zea mays* L. The dwarf-1 mutation controls the 3β-hydroxylation of gibberellin A_{20} to gibberellin A_1. Planta. 1984; 160:464–468.
17. Patterson RJ, Rappaport L. The conversion of gibberellin A_1 to gibberellin A_8 by a cell-free enzyme system. Planta. 1974; 119:183–191.
18. Smith VA, MacMillan J. Purification and partial characterization of a gibberellin 2β-hydroxylase from *Phaseolus vulgaris*. J Plant Growth Regul. 1984; 2:251–264.
19. Nietfeld JJ, De Long L, Kemp A. The influence of 2-oxoglutarate on the activity of prolyl 4-hydroxylase. Biochim Biophys Acta. 1982; 704:321–325.
20. Harwood JL. The site of action of some selective gramineous herbicides is identified as acetyl CoA carboxylase. Trends Biochem Sci. 1988; 13:330–331.
21. Majamaa K, Hanauske-Abel M, Günzler V, et al. The 2-oxoglutarate binding site of prolyl 4-hydroxylase. Identification of distinct subsites and evidence for 2-oxoglutarate decarboxylation in a ligand reaction at the enzyme-bound ferrous ion. Eur J Biochem. 1984; 138:239–245.

CHAPTER 11

Gibberellin A_3-Regulated α-Amylase Synthesis and Calcium Transport in the Endoplasmic Reticulum of Barley Aleurone Cells

D.S. Bush, L. Sticher, and R.L. Jones

1 Introduction

The cereal aleurone layer has emerged as an excellent model system for studies of the mode of action of gibberellins (GAs).[1] Experiments with protoplasts from the aleurone layer of wild oat have been used to probe the nature of the putative GA receptor,[2] and protoplasts from the aleurone layer of barley have been used for the isolation of transcriptionally active nuclei for studies of the molecular basis of GA action.[3]

The synthesis and secretion of acid hydrolases from the cereal aleurone cell are dependent on the presence of Ca^{2+} as well as GA. Chrispeels and Varner[4] were the first to show that isolated barley aleurone layers required high levels (10–20 mM) of Ca^{2+} for the synthesis of α-amylase. This observation has now been expanded to include a requirement for Ca^{2+} in the synthesis of other proteins such as β-glucanases[5] and acid phosphatase[6] in the barley aleurone and in the synthesis of α-amylase and other proteins in aleurone and scutellum tissue isolated from the grain of other cereals.[2,7,8]

Until recently, little has been done to understand the role of Ca^{2+} in the regulation of protein synthesis and secretion in aleurone tissue. Experiments from our laboratory have shown that millimolar levels of extracellular Ca^{2+} are not required for the expression of barley aleurone α-amylase genes.[9] We have shown that, in the absence of added Ca^{2+}, GA_3 can promote the accumulation of α-amylase mRNA.[9] More recent work from our laboratory indicates a role for Ca^{2+} in the synthesis of the α-amylase holoenzyme. Barley α-amylase (EC 3.2.1.1), like bacterial, animal, and fungal α-amylases, is a Ca^{2+}-containing metalloprotein.[10] The holoenzyme binds one atom of Ca^{2+} per molecule of protein, and when Ca^{2+} is removed from α-amylase, the apoenzyme exhibits no or low enzymatic activity and is sensitive to proteolytic degradation.[11] We speculated that Ca^{2+} was required by the aleurone cell to sustain the synthesis of active, stable holoenzyme.[11]

In this paper we describe experiments that demonstrate the occurrence of a flux of Ca^{2+} through the cytoplasm into the endoplasmic reticulum

(ER), where α-amylase and other secreted enzymes are synthesized. We show that Ca^{2+} is transported into the ER of the aleurone cell by an ATP-dependent pump whose activity is stimulated manyfold by GA_3. This pump maintains the Ca^{2+} concentration of the ER at, at least, 4 μM, while cytosolic Ca^{2+} is maintained at around 350 nM. Our experiments show that elevated levels of ER Ca^{2+} are required to maintain the activity of α-amylase holoenzyme.

2 Calcium Homeostasis in the Cytosol

There is a paucity of information on cytosolic Ca^{2+} homeostasis in plants (for reviews, see refs. 12 and 13). This derives in part from the technical problems that are posed by the cell wall and vacuole. The use of Ca^{2+}-sensitive microelectrodes is hampered by the rigid wall, which resists impalement and gives rise to high turgor pressures that tend to occlude microelectrodes, and the vacuole, which occupies most of the volume of the cell.[14] The cell wall also hinders measurement of Ca^{2+} using fluorescent dyes because of the pronounced autofluorescence of phenolic groups in the wall. To overcome these difficulties, we[15-17] and others[18,19] have used Ca^{2+}-sensitive fluorescent dyes to measure cytosolic Ca^{2+} in protoplasts isolated from plant tissues.

We have developed a nonintrusive method for loading the Ca^{2+}-sensitive dye indo-1 into cells of higher plants. The method relies on the fact that indo-1 is a weak acid that at pH values below 5 exists in the protonated, uncharged form.[15] The uncharged dye crosses the plasma membrane readily and accumulates in the cytosol because the dye dissociates at the slightly basic pH (pH 7.4)[16] of the cytosol. This strategy of dye loading is well suited to the barley aleurone cell since this cell maintains the pH of the external medium at about 4. At this pH, indo-1 loads almost exclusively into the cytosol of aleurone protoplasts suspended in 15 μM indo-1 for 2 h.[15] This method has now been used to load indo-1 into protoplasts from corn roots[18] and guard cells (A.J. Trewavas, personal communication), walled cells from barley coleoptiles [A. Wrona, personal communication), and yeast cells (J. Gao and R.Y. Tsien, personal communication).

Using the acid-loading method and a ratio technique for determining Ca^{2+} concentration, we have monitored the concentration of Ca^{2+} in barley aleurone protoplasts under various conditions of incubation. Protoplasts incubated in 5 μM GA_3 in the absence of added Ca^{2+} (external Ca^{2+}, ca. 5 μM) contain cytosolic Ca^{2+} in the range of 200–250 nM (Fig. 1). Incubation of protoplasts in GA_3 in the presence of 20 mM added Ca^{2+} elevates cytosolic Ca^{2+} levels. These cells contain cytosolic Ca^{2+} in the range of 350 nM (Fig. 1). Cytosolic Ca^{2+} levels in GA_3-treated protoplasts are strongly correlated with α-amylase synthesis and secretion (Fig. 1). Thus, α-amylase levels are low in protoplasts incubated in GA_3 without

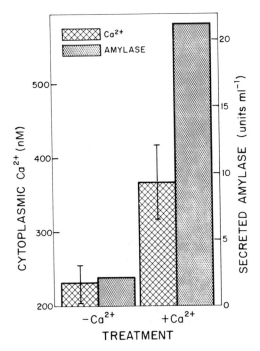

Fig. 1. The effect of extracellular Ca^{2+} on the concentration of cytoplasmic Ca^{2+} and α-amylase secretion from barley aleurone protoplasts. Protoplasts were incubated in GA_3 (5 μM) in the absence of added Ca^{2+} ($-Ca^{2+}$) or with 20 mM Ca^{2+} ($+Ca^{2+}$). Cytoplasmic Ca^{2+} was measured with indo-1. Reproduced from Bush and Jones[17]

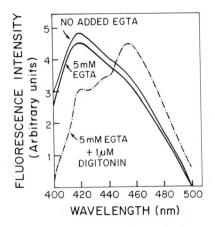

Fig. 2. Fluorescence spectra of ER preparations isolated from aleurone protoplasts incubated in GA_3 and 10 mM Ca^{2+} and loaded with indo-1. Fluorescence emission of membrane preparations in the presence of added EGTA, EGTA plus digitonin, and controls with no additions is shown. Reproduced from Bush et al.[21]

added Ca^{2+} (2 units/10^5 protoplasts), and enzyme levels are increased 10-fold by the addition of 20 mM Ca^{2+} to the incubation medium (Fig. 1).

The levels of cytosolic Ca^{2+} are homeostatically maintained in barley aleurone cells during incubation in GA_3 and Ca^{2+}. Aleurone cells undergo progressive vacuolation during incubation, and small protein body vacuoles coalesce to form a large central vacuole.[20] This pronounced change in cytoplasmic organization is not accompanied by large changes in cytosolic Ca^{2+} levels.[15] Cells at early stages of vacuolation have cytosolic Ca^{2+} in the range of 357 ± 103 nM, whereas cells that posses one large central vacuole have cytosolic Ca^{2+} levels of 281 ± 86 nM.[15]

3 Endoplasmic Reticulum Calcium Levels in GA_3-Treated Tissue

The ER of aleurone protoplasts loaded with indo-1 using the acid loading technique also accumulates the Ca^{2+}-sensitive form of the dye. Approximately 2% of the indo-1 in the cytoplasm of GA_3-treated aleurone protoplasts could be recovered with the ER fraction when protoplasts were osmotically lysed and the microsomal fraction was isolated on a discontinuous sucrose gradient.[15] The accumulation of indo-1 in the ER allowed us to measure the Ca^{2+} concentration of this compartment of the endomembrane system.

Microsomes isolated from protoplasts loaded with indo-1 were found to contain indo-1 in the high-Ca^{2+} form (Fig. 2). To establish that this high-Ca^{2+} signal was reporting Ca^{2+} in the lumen of the ER and not Ca^{2+} bound to the surface of microsomal membranes, we incubated microsomes in a medium containing 5 mM EGTA in the presence or absence of 1 μM digitonin. In the presence of both EGTA and digitonin, the high-Ca^{2+} signal from indo-1 was completely quenched, whereas in the presence of EGTA alone, indo-1 reported high levels of Ca^{2+} in the microsomes (Fig. 2). This result demonstrates that the detergent treatment is necessary for EGTA to chelate Ca^{2+} in isolated barley microsomes, indicating that the Ca^{2+} is sequestered within the lumen of microsomal vesicles.

Using the ratio technique for determining Ca^{2+} levels, we estimate the Ca^{2+} concentration of the ER to be at least 4 μM. This represents a minimum estimate of ER Ca^{2+} concentration because this concentration is at the upper limit of the sensitivity of indo-1 to Ca^{2+}.

4 Gibberellin-Stimulated Calcium Transport into the ER

Our measurements of cytosolic (Fig. 1) and ER (Fig. 2) Ca^{2+} levels show at least ten times more Ca^{2+} in the ER than in the cytosol. To understand the mechanisms that permit the accumulation of Ca^{2+} in the lumen of the

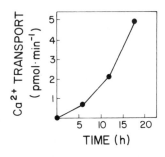

Fig. 3. The effect of GA_3 on the transport of $^{45}Ca^{2+}$ into microsomal membranes isolated from barley aleurone layers. The data show the difference in uptake between GA_3-treated and control aleurone layers. Reproduced from Bush et al.[21]

ER, we studied Ca^{2+} transport into ER membranes isolated from GA_3-treated cells. Calcium transport was studied by following the uptake of $^{45}Ca^{2+}$ into membrane fractions isolated from GA_3-treated barley aleurone layers and protoplasts. The accumulation of $^{45}Ca^{2+}$ in microsomal membranes is dependent on the presence of added ATP and Mg^{2+}. When membranes isolated from aleurone layers incubated with GA_3 and Ca^{2+} for 16 h are purified by isopycnic sucrose density gradient centrifugation, $^{45}Ca^{2+}$ transport is confined to a peak at 28–32% (w/w) sucrose, corresponding to the position of the ER marker enzyme cytochrome c reductase (CCR).[21] Whereas this peak of $^{45}Ca^{2+}$ uptake does not contain membranes derived from the tonoplast or plasma membrane, membranes having inosine diphosphatase (IDPase) activity characteristic of the Golgi apparatus accumulate in this region of the density gradient. We showed that transport of $^{45}Ca^{2+}$ was primarily into the ER and not Golgi apparatus using the sodium ionophore monensin.[21] In the presence of monensin, only membranes with IDPase activity, not membranes with CCR and $^{45}Ca^{2+}$ transport activity, are displaced to a more dense region of the sucrose density gradient.[21]

Endoplasmic reticulum isolated from barley aleurone layers preincubated in GA_3 and Ca^{2+} for at least 6 h transport $^{45}Ca^{2+}$ more rapidly than do membranes isolated from layers incubated in the absence of GA_3 (Fig. 3). After 16-h incubation in GA_3, membranes transport Ca^{2+} up to five times more rapidly than controls (Fig. 3). The effect of GA_3 cannot be observed by its addition to membrane fractions in vitro. Thus, enhanced Ca^{2+} transport is found only in ER membranes isolated from aleurone layers preincubated in GA_3 for at least 6 h.

The high affinity of the Ca^{2+} transporter indicates that this pump could participate in the transport of Ca^{2+} from the cytosol into the ER lumen. Based on the rate of $^{45}Ca^{2+}$ uptake into microsomal membranes from solutions of increasing Ca^{2+} concentration, we calculated the K_m for Ca^{2+}

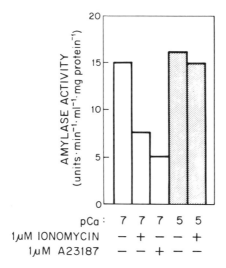

Fig. 4. The effects of ionomycin and A23187 on the activity of α-amylase sequestered in the lumen of microsomal membranes isolated from barley aleurone layers. Membranes were incubated in pCa 7 (0.1 μM) or pCa 5 (10 μM) for 30 min before measuring α-amylase activity. Reproduced from Bush et al.[11]

at 0.5 μM.[21] Since cytosolic Ca^{2+} is around 0.35 μM in GA_3-treated aleurone cells, this Ca^{2+} pump could function to accumulate Ca^{2+} in the ER from the cytosol.

5 The Role of Elevated ER Calcium

Because α-amylase is a Ca^{2+}-containing metalloenzyme,[10,11] we speculated that one role of elevated ER Ca^{2+} was to maintain the activity and stability of newly synthesized α-amylase. Our data show that affinity-purified barley α-amylase is rapidly denatured if the Ca^{2+} concentration of solutions falls below 1 μM,[11] indicating that if the enzyme is exposed to cytosolic concentrations of Ca^{2+}, it will be inactivated and possibly proteolytically degraded.[11]

We have shown that newly synthesized α-amylase within the lumen of the ER is also sensitive to low Ca^{2+} concentrations, indicating that the level of ER Ca^{2+} must be high enough to maintain the activity and stability of this enzyme. Microsomal membranes containing high levels of newly synthesized α-amylase were isolated from GA_3-treated barley aleurone layers and incubated in solutions of differing Ca^{2+} concentrations in the presence and absence of the Ca^{2+} ionophores ionomycin or A23187 (Fig. 4). When microsomal vesicles were incubated in 100 nM Ca^{2+} with either ionomycin or A23187, α-amylase activity was markedly reduced relative to that in vesicles incubated in 100 nM Ca^{2+} alone (Fig. 4). Addition of ionomycin to microsomal vesicles incubated in 10 μM Ca^{2+} had no effect on the activity of α-amylase (Fig. 4).

These data show that newly synthesized α-amylase in the ER lumen is like purified α-amylase in its response to low Ca^{2+} concentrations. This result validates our finding that ER Ca^{2+} levels are in the low micromolar range (Fig. 2) and helps to confirm our speculation that one of the roles of the GA$_3$-stimulated Ca^{2+} pump is to maintain elevated ER Ca^{2+} levels.

6 Conclusions and Speculations

We conclude that one of the roles of Ca^{2+} in the cereal aleurone layer is to support the synthesis of active and stable α-amylase molecules. The synthesis of active α-amylase is achieved by the pumping of Ca^{2+} from the cytosol into the lumen of the ER by a transporter whose activity is stimulated by GA$_3$. We conclude that, in addition to regulating the expression of the genes encoding α-amylase and other secreted proteins, GA$_3$ also regulates the transport of Ca^{2+} from cytosol to ER. Gibberellin A$_3$ can control the synthesis of α-amylase proteins by limiting the supply of mRNA, or the activity and stability of the proteins by regulating the supply of Ca^{2+}.

Another role for GA$_3$ in regulating Ca^{2+} levels in the aleurone cell should not be overlooked. By stimulating the ER Ca^{2+} pump, GA$_3$ may play an important regulatory role in setting cytosolic Ca^{2+} levels. In this way, Ca^{2+} may serve as a second messenger for GA$_3$. Elevated levels of GA$_3$ would stimulate uptake of Ca^{2+} into the ER, lowering cytosolic Ca^{2+} levels; conversely, lowered levels of GA$_3$ would result in elevated cytosolic Ca^{2+}. Since barley aleurone layers also contain calmodulin (R.L. Jones and P. Simon, unpublished data), the elements necessary for control of cellular functions via altered cytosolic Ca^{2+} are present in the aleurone cell.

Acknowledgments. This work was supported by grants from the U.S. Department of Energy and the National Science Foundation to R.L. Jones. The assistance of Eleanor Crump in the preparation of this manuscript is gratefully acknowledged.

References

1. Fincher GB. Molecular and cellular biology associated with endosperm mobilization in germinating cereal grains. Ann Rev Plant Physiol. 1989; 40:305–346.
2. Hooley R. Gibberellic acid controls specific acid-phosphatase isozymes in aleurone cells and protoplasts of *Avena fatua* L. Planta. 1984; 161:335–360.
3. Jacobsen JV, Beach L. Control of transcription of α-amylase and rRNA genes in barley aleurone protoplasts by gibberellin and abscisic acid. Nature. 1985; 316:275–277.

4. Chrispeels MJ, Varner JE. Gibberellic acid-enhanced synthesis and release of α-amylase and ribonuclease by isolated barley aleurone layers. Plant Physiol. 1967; 42:398–406.
5. Stuart IM, Loi L, Fincher GB. Development of $(1\rightarrow 3, 1\rightarrow 4)$-$\beta$-$D$-glucan endohydrolase isoenzymes in isolated scutella and aleurone layers of barley (*Hordeum vulgare*). Plant Physiol. 1986; 80:310–314.
6. Jones RL, Carbonell J. Regulation of the synthesis of barley aleurone α-amylase by gibberellic acid and calcium ions. Plant Physiol. 1984; 76:213–218.
7. Akazawa T, Hara-Nishimura I. Topographic aspects of biosynthesis, extracellular secretion, and intracellular storage of proteins in plant cells. Ann Rev Plant Phys. 1985; 36:441–472.
8. Akiyama T, Uchimiya H, Suzuki H. Gibberellic acid-induced increase in activity of a particular isozyme of acid phosphatase in wheat half-seeds. Plant Cell Physiol. 1981; 22:1023–1028.
9. Deikman J, Jones RL. Control of α-amylase mRNA accumulation by gibberellic acid and calcium in barley aleurone layers. Plant Physiol. 1985; 78:192–198.
10. Karn RC, Malacinski GM. The comparative biochemistry, physiology, and genetics of animal α-amylases. Adv Comp Physiol Biochem. 1978; 7:1–103.
11. Bush DS, Sticher L, Van Huystee RB, et al. The calcium requirement for stability and enzymatic activity of two isoforms of barley aleurone α-amylase. J Biol. Chem. 1989; 264:19392–19398.
12. Hepler PK, Wayne RO. Calcium and plant development. Ann Rev Plant Physiol. 1985; 36:397–439.
13. Poovaiah BW, Reddy ASN. Calcium messenger system in plants. CRC Crit Rev Plant Sci. 1987; 6:47–103.
14. Felle H. The fabrication of H^+ selective liquid-membrane micro-electrodes for use in plant cells. J Exp Bot. 1982; 37:1416–1428.
15. Bush DS, Jones RL. Measurement of cytoplasmic Ca^{2+} in barley aleurone protoplasts using indo-1 and fura-2. Cell Calcium. 1987; 8:455–472.
16. Jones RL, Bush DS, Biswas AK. Gibberellic acid and calcium participate in the synthesis of active α-amylase molecules. In: Pinfield NJ, Black M, eds. Growth regulators and seeds. British Plant Growth Regulator Group Monograph 15. 1987; p. 31–42.
17. Bush DS, Jones RL. Cytoplasmic calcium and α-amylase secretion from barley aleurone protoplasts. Eur J Cell Biol. 1988; 46:466–469.
18. Gilroy S, Hughes WA, Trewavas AJ. The measurement of intracellular calcium levels in protoplasts from higher plant cells. FEBS Lett. 1986; 199:217–221.
19. Lynch J, Polito V, Lauchli A. Salinity stress increases cytoplasmic calcium activity in maize root protoplasts. Plant Physiol. 1989; 90:1271–1274.
20. Jones RL, Price JM. Gibberellic acid and the fine structure of barley aleurone layers. III. Vacuolization of the aleurone cell during the phase of ribonuclease release. Planta. 1970; 94:191–202.
21. Bush DS, Biswas AK, Jones RL. Gibberellic acid-stimulated Ca^{2+} accumulation in barley endoplasmic reticulum. Planta. 1989; 178:411–420.

CHAPTER 12

Rice α-Amylase and Gibberellin Action— A Personal View

T. Akazawa, J. Yamaguchi, and M. Hayashi

1 Introduction

An American colleague many years ago remarked to me that "you are better off not working on plant hormone and photosynthesis; if you do, you will get yourself lost in the labyrinth!". We have not published papers dealing with the gibberellin (GA) effect on enzymatic breakdown of starch except for reporting a few negative results, but this is not necessarily because of this friend's advice. Rather, I had some fond memories of GA research during my youthful days in America. I was a member of the class of 1950 at the University of Tokyo. Some of my classmates were engaged in the isolation and structural characterization of GAs and had a really difficult time because of the awful laboratory conditions during the postwar period. Therefore, I tremendously appreciated the advancements on GA research made in various laboratories in the United States. When I was a graduate student at the University of California—Berkeley (1956–57), I had the opportunity to attend double-header seminars given by Bernie Phinney and Anton Lang. I was truly fascinated by their remarkable discoveries concerning the effect of GAs on plant growth. Later, H. Tamiya told me that a lecture given by Y. Sumiki at the International Congress of Microbiology in Brazil was a memorable one, arousing the participants' interests in this mysterious plant hormone hitherto unknown to scientists other than Japanese.

On my way back to Japan after two years of study in the United States, I stopped over at UCLA. Bernie Phinney took me to a cold room in the Chemistry Department and showed me a small test tube containing a tiny crystal of GA, isolated by C. West. Whenever I saw Bernie in later years measuring the length of maize mutant seedlings in the laboratory at UCLA, I was always reminded of the old times.

During my stay at the International Rice Research Institute (IRRI) in the Philippines (1962–64), I initiated a study of starch biosynthesis and enzymatic breakdown in rice plants. In the former project, we took full advantage of using rice seeds at the ripening stage as an experimental material and we competed keenly with the group of Leloir at Buenos Aires. With regard to the second project, it was undertaken just at the time when GA-induced biosynthesis of α-amylase in barley half-seeds was initiated independently by three groups. I started to follow carefully a series of investigations carried out by Varner and his associates.[1,2] However, I did not undertake similar kinds of experiments using rice, because it is obviously not practical at all to use tiny rice seeds in which the portion of aleurone layers is much smaller than in barley. However, we were able to demonstrate quite clearly that GA can elicit α-amylase formation in the embryoless rice half-seeds, just as in the case of barley.[3,4] Realizing that the mode of action of GA on α-amylase formation was analogous in rice and barley, I thought that we should tackle the problem from a different angle. As a Japanese plant biochemist, I regard rice as a precious and sacred crop plant in the Orient!

2 Aleurone Concept Versus Scutellum Concept

From the beginning of our study, I have held the skeptical view that the α-amylase formation in barley half-seeds soaked in a GA-containing bathing solution may not reflect enzyme synthesis in situ by germinating cereal seeds although it has been shown to be an extremely powerful system for the study of the GA action in vitro. Consequently, years later, an imaginative student of mine, K. Okamoto, was able to demonstrate the apparent synthesis and secretion of α-amylase in the scutellar epithelium cells of rice as well as other starchy cereals during their early stage of germination[5,6] (Fig. 1). We employed a simple histochemical starch-film method, and, based on such observations, we put forward the hypothesis that the scutellum is the initial site of α-amylase formation. This concept was in sharp contrast to the long-held popular "aleurone concept" so frequently described in many plant physiology textbooks.[7,8] Since this opposing view, together with the relevant historical background, has been treated in detail in several review articles,[9-11] it will not be repeated here.

It should be mentioned that there are analogous anatomical structures in epithelial surface cell layers of rice seed scutellum and the villi of the mammalian digestive organs.[9,12] We thought that the lamella-type structure of epithelium, as viewed under the scanning electron microscope, should be advantageous for the secretion of digestive enzymes such as α-amylase and proteinase into the endosperm and, at the same time, for the absorption by the scutellum (embryo) of low-molecular-weight metabolites from the endosperm. Another structural characteristic of the epithelial cells, typical of their secretory function, is the well-developed endoplasmic reticulum (ER) and Golgi bodies.[12] In contrast, such structural organization is not clearly recognizable in the aleurone layer proper, although the ultrastructural change of barley aleurone cells evoked in response to GA

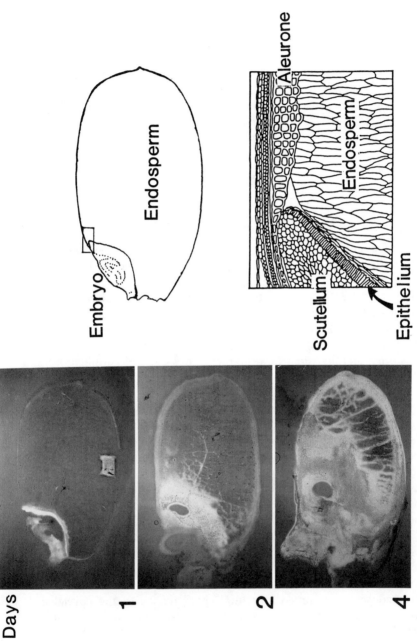

Fig. 1 Anatomical structure of rice grain and secretion site of α-amylase. Upper right panel: A longitudinal section containing embryo scutellum and endosperm. Bottom Lower right panel: A magnified part of upper right panel showing epithelium (interface of scutellum and endosperm) and aleurone. Left panel: Dissolution of endosperm starch resumes initially (1 day of germination) by the hydrolytic action of α-amylase secreted from the scutellum–endosperm interface (epithelium) of rice

treatment was extensively studied by Jones.[13] In later years, not only our own observations[12] but also an elaborate electron microscope study by Gram[14] in Carlsberg using barley showed that the scutellar epithelium is better equipped with the subcellular machineries for enzyme synthesis and secretion. Of course, we must bear in mind that endogenous GA in the scutellum may be crucially important for the enzyme formation.

3 Rice α-Amylase Is a Typical Secretory Glycoprotein

From our studies of the biosynthesis, intracellular transport, and secretion of α-amylase in rice seeds, it became evident that the enzyme is a typical secretory glycoprotein in plant cells which has been poorly characterized at this time. The basic experimental systems we used were: (a) intact rice seedlings; (b) freshly excised segments containing epithelial cells on the surface layer of scutellum; and (c) in vitro systems, e.g., mRNA, ribosomes, and microsomes.[15-19] In these studies, information available from previous investigations on the secretory pathway of the mammalian system was particularly useful to us.[20-22] Consequently, rice α-amylase was proved to be a glycoprotein bearing high-mannose type and complex type oligosaccharide moieties conjugated to the Asn residues of polypeptide chains. It should be mentioned that during our studies the use of specific inhibitors, such as tunicamycin and monensin, was a powerful tool to clarify the mechanism of the post-translational modification of the enzyme molecule with the oligosaccharide chain.

From the combined results, we have put forward a hypothetical scheme in which the Golgi apparatus plays a central role in carrying out the terminal carbohydrate conjugation reaction.[9-11] Furthermore, we found that in rice seeds there are two α-amylase isoforms, the R- and the S-type, which are distinguishable on the basis of their endo-β-H digestibilities of the oligosaccharide chains[18,19] and that their synthesis is affected by environmental factors, such as Ca^{2+} and temperature.* Consequently, we have postulated two secretion pathways, namely, the "one-route" and the "two-route" path of α-amylase biosynthesis.[10,11]

Our recent investigation of the biosynthesis of α-amylase employing the liquid-cultured cells of rice seeds demonstrated that the basic properties of the enzyme molecules secreted into the culture medium were identical to those characterized in the intact rice seedlings.[24] However, we found that, under the incubation conditions used in the culture medium, the α-amylase species solely secreted is the R-type, the S-type enzyme molecules being barely detectable (Fig. 2). Interestingly, we have found that the presence of Tris buffer caused a selective inhibitory effect on the terminal sugar

---

*References 11, 18, 19, and 23.

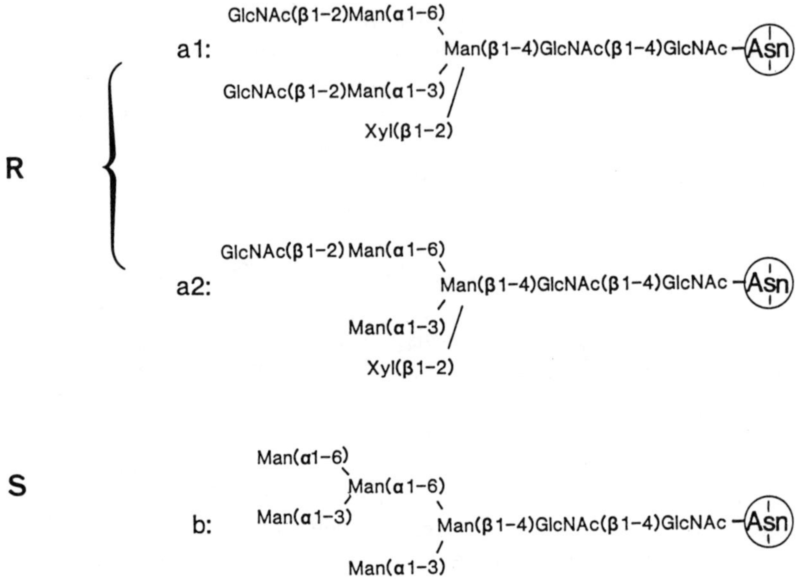

```
         GlcNAc(β1-2)Man(α1-6)
   a1:                        \
                               Man(β1-4)GlcNAc(β1-4)GlcNAc—(Asn)
         GlcNAc(β1-2)Man(α1-3)/
                      Xyl(β1-2)
R

         GlcNAc(β1-2)Man(α1-6)
   a2:                        \
                               Man(β1-4)GlcNAc(β1-4)GlcNAc—(Asn)
                  Man(α1-3)/
                      Xyl(β1-2)


         Man(α1-6)
                 \
                  Man(α1-6)
S                         \
   b:   Man(α1-3)/         Man(β1-4)GlcNAc(β1-4)GlcNAc—(Asn)
                          /
                 Man(α1-3)
```

Fig. 2. Oligosaccharide structure of α-amylase molecule secreted into the culture medium in the absence (a) or presence (b) of Tris: complex (R)-type and high-mannose (S)-type oligosaccharide chain of rice α-amylase[24]

conjugation reaction, resulting from an elevation of the pH of the environment on the medial face of the Golgi compartment (Fig. 3).

Another interesting observation, which was at variance with some previous reports, was that α-amylase molecules synthesized in both barley and wheat seeds are not glycosylated.[9,10] This is related to the above-described "one-route" secretory path of α-amylase involving the Golgi compartment, where the enzyme molecule contains no carbohydrate chain.[11,25]

4 Isoforms of Rice α-Amylase and Tissue-Specific Expression of Multigenes

It has long been known that there are various isoforms (isozymes) of α-amylase in cereal seeds, including rice, barley, and wheat, as demonstrable by gel electrophoresis.[3,26,27] The presence of multiple structural genes for α-amylase has been postulated, and the genetic implications have been presented. Indeed, with the advent of molecular biological techniques in recent years, work on the isolation and characterization of α-amylase gene families from cereal seeds (barley and wheat) has made remarkable progress.[11] At the same time, an effort has been made to explore the regulatory effect exerted by plant hormones, for example, GA and ABA, on

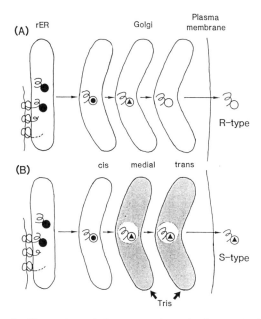

Fig. 3. Schematic illustration of the processing of oligosaccharide chains in α-amylase in rice scutellum. (A) Modification of oligosaccharide during the intracellular transport of α-amylase under normal conditions. High-mannose type oligosaccharide (●), presumably comprised of $Glc_3Man_9GlcNAc_2$ conjugated to the Asn residue of the putative α-amylase precursor formed in rER, is subsequently modified stepwise in the Golgi subcompartment, producing intermediary forms (◉) and (▲), and finally the complex type enzyme molecule (○) is discharged extracellularly. (B) Inhibition of oligosaccharide processing in the presence of Tris. Tris affects the acidity of the medial and trans faces of the Golgi, causing elevation of the pH of the environment (arrow). The inhibition of the oligosaccharide-processing enzymes under alkaline conditions presumably results in the secretion of the intermediary form of high-mannose type (S-type) α-amylase (▲) which can be found in the normal secretory pathway

the expression of these α-amylase genes, mostly using barley aleurone cells as an assay system.

However, as we have discussed already, there exist temporal and topographic sequences of α-amylase biosynthesis and secretion in the germinating cereal seeds at both the molecular and the cellular levels. With respect to the cell specificity, the secretion of α-amylase occurs predominantly from the scutellar epithelium in the early stage but progressively spreads into the aleurone parts. Therefore, a number of questions arise. (a) Does such a sequential change of the enzyme molecules and their secretory discharge represent regulatory mechanism(s) operating in the expression of the α-amylase gene? (b) Are there tissue- and/or cell-specific α-amylase genes and are they expressed in situ? (c) Are there tissue-specific GA species with differential effects, or is there a single GA species involved and acting universally? So far, none of these important and interesting questions has been explored experimentally and properly answered. Certainly, it should be extremely interesting to examine the tissue-specific expression of the α-amylase gene employing the in situ hybridization technique. In this context, recent achievements by Fincher and associates[28,29] concerning the development formation of β-glycanase and the sequential expression of the gene in barley seeds are most informative.

The possible presence of tissue-specific GA species and their differential effect on the α-amylase gene expression will be an interesting subject for future studies. It must be remembered that there is no assurance that GA_3, which has been almost invariably used in the previous investigations of α-amylase biosynthesis, is truly effective in inducing the α-amylase gene expression in situ in the scutellar epithelium or aleurone of any cereal seeds. It is likely that microanalytical assay of GA and ABA using specific antibodies can provide us with a powerful strategy toward this end.[30]

5 Transcriptional Regulation of α-Amylase Gene and Gibberellin Action

We are most interested in clarifying the mode of action of GA on α-amylase biosynthesis at the molecular level. While several mechanisms have been proposed to account for the stimulating effect of GA on α-amylase formation, many experimental data have been accumulated showing that it results from the enhancement of transcriptional activities.[11] Jacobsen and Beach[31] isolated nuclei from barley aleurone protoplasts that had been pretreated with GA_3 and determined their transcriptional activities by measuring the radioactive RNA transcripts that are hybridizable with the α-amylase gene probe. A similar experiment was also carried out by Zwar and Hooley[32] using protoplasts isolated from GA_4-treated oat aleurone cells.

From our current knowledge concerning the mechanism of action of steroid hormones in animal cells, we can hypothesize a mechanistic scheme for the mode of action of GA in eliciting α-amylase gene expression as illustrated in Fig. 4. How can we then test this working model experimentally? Since the initial transcription step involves the binding of RNA polymerase II to the gene, the involvement of a specific regulatory protein associated with GA in this reaction step can be inferred. We can assume that GA transported from the embryonic cells to the scutellar epithelium will bind with GA receptor (R) localized in either the plasma membrane or cytosol, resulting in the formation of the GA–R complex. The cytosolic GA–R complex will then move into the nucleus and bind at a specific site on the α-amylase gene, most likely at the promoter region. An

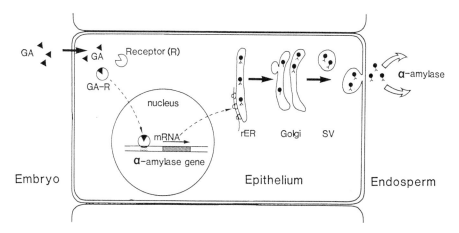

Fig. 4. Mechanistic scheme for the mode of action of GA in eliciting α-amylase gene expression. GA transported from embryonic cells to the scutellar epithelium will bind with the GA receptor localized in the cytosol, resulting in the formation of the GA–R complex. The cytosolic GA–R complex will then move into the nucleus and bind at a specific site on the α-amylase gene, most likely at the promoter region. An enhancement of the transcriptional activity will lead to the formation of α-amylase protein molecules and their final secretory discharge after post-translational modification of sugar moieties

enhancement of the transcriptional activity will lead to the formation of α-amylase protein molecules and their final secretory discharge after post-translational modification of sugar moieties as discussed above. The structural and functional characterization of R is necessary to substantiate the hypothetical scheme. Instead of examining the GA effect on α-amylase gene expression, we have attempted to determine the regulatory sequence (*cis*-acting elements) involved in the transcriptional regulation of the rice α-amylase gene possibly related to GA action, by developing a gene transfer system using the rice seed protoplasts. Employing this system, we first studied the transcriptional regulation of a chimeric gene comprising the 5′-flanking region of the α-amylase gene fused to a β-glucuronidase gene from *E. coli*. Functional analysis of the 5′-deletion experiments showed that the promoter activity was located between −239 and −93 in the upstream region of the rice α-amylase gene. Second, for the purpose of identifying the protein factor(s) (*trans*-acting factors) that will bind to the regulatory sequence of the 5′-flanking region of the gene, we tried to isolate the intact nuclei from rice germ. The nuclear extracts were then subjected to gel retardation assays (DNA mobility shift) to characterize protein factor(s) which can specifically interact with the *cis*-acting elements located in the upstream region. By employing the transient expression system, it was found that the protein can bind to the sequence between −239 and +27, causing the transcriptional enhancement.

In this connection, it should be noted that Wu and his associates[33] at Cornell have reported that a GA-induced factor interacts with the upstream region of the α-amylase gene in rice aleurone tissues. The basic experimental procedure employed by these investigators was different from ours, in that they prepared the extract of rice aleurone (plus scutellum) cells after imbibition for 2–3 days of 5 μM GA_3 solution. A portion of the extract was then incubated with a labeled α-amylase gene for the DNA mobility-shift assays as well as for the protection analysis using the exonuclease III digestion. They have thus claimed that the protein factor interacting with a specific sequence of the α-amylase gene is formed by the GA treatment of rice aleurone cells.

6 Conclusion

Our research interest concerning GA action, particularly in relation to its transcriptional regulation of the α-amylase gene, centers on the identification of the GA receptor and the clarification of the molecular mechanisms operating therein. Although our experimental evidence appears to support the role of a diffusible factor (receptor) in the germinating rice seeds, many intriguing questions remain to be answered before we can understand the true mechanism.

References

1. Varner JE. Gibberellic acid-controled synthesis of α-amylase in barley endosperm. Plant Physiol. 1964; 39:413–415.
2. Varner JE, Ho DT-H. In: Hormones. Bonner J, Varner JE, eds. New York, 1976: p. 713–770.
3. Tanaka Y, Ito T, Akazawa T. Enzymic mechanism of starch breakdown in germinating rice seeds. III α-amylase isozymes. Plant Physiol. 1970; 46:650–654.
4. Akazawa T. Gibberellic acid and α-amylase induction in germinating cereal seeds. In: Piras R, Pontis H, eds. Biochemistry of the glycosidic linkage. New York: Academic Press, 1972: p. 449–458.
5. Okamoto K, Akazawa T. Enzymic mechanism of starch breakdown in germinting cereal seeds 7. Amylase formation in the epithelium. Plant Physiol. 1970; 63:336–340.
6. Okamoto K, Kitano H, Akazawa T. Biosynthesis and secretion of hydrolases in germinating cereal seeds. Plant Cell Physiol. 1980; 21:201–204.
7. Curtis H. Plant hormones and plant responses. In: Biology. New York: Worth Publ., 1979: p. 545–566.
8. Laetsch WM. Regulation of development. In: Plants, Basic concepts in botany. Boston: Little, Brown, & Co., 1979: p. 197–222.

9. Akazawa T, Miyata S. Biosynthesis and secretion of α-amylase and other hydrolases in germinating cereal seeds. Essays in Biochem. 1982; 18:40–78.
10. Akazawa T, Hara-Nishimura I. Topographic aspects of biosynthesis, extracelluar secretion and intracellular storage of proteins in plant cells. Ann Rev Plant Physiol. 1985; 36:441–472.
11. Akazawa T, Mitsui T, Hayashi M. Recent progress in α-amylase biosynthesis. In: The biochemistry of plants, Vol. 14. Academic Press 1988: p. 465–492.
12. Okamoto K, Murai T, Eguchi G, et al. Enzymic mechanism of starch breakdown in germinating rice seeds 11. Ultrastractural changes in scultellar epithelium. Plant Physiol. 1982; 70:905–911.
13. Jones RL. Quantitative and qulitative changes in the endoplasmic reticulum of barley aleurone layers. Planta. 1980; 150:70–81.
14. Gram NH. The ultrastructure of germinating barley seeds I. Changes in the scutellum and the aleurone layer in Nordal barley. Carlsberg Res Commun. 1982; 47:143–162.
15. Miyata S, Akazawa T. Enzymic mechanism of starch breakdown in germinating rice seeds. 12. Biosynthesis of α-amylase in relation to protein glycosylation. Plant Physiol. 1982; 70:147–153.
16. Miyata S, Akazawa T. α-Amylase biosynthesis: Evidence for temporal sequence of NH_2-terminal peptide cleavage and protein glycosylation. Proc Natl Acad Sci USA. 1982; 79;6566–6568.
17. Miyata S, Akazawa T. Biosynthesis of rice seed α-amylase: Proteolytic processing glycolylation of precursor polypeptides by microsomes. J Cell Biol. 1983; 96:802–806.
18. Mitsui T, Christellar JT, Akazawa T, et al. Biosynthesis of rice seed α-amylase secretion by the scutellum. Arch Biochem Biophys. 1985; 241:315–328.
19. Mitsui T, Akazawa T. Secondary modification of carbohydrate chains in α-amylase molecules synthesized in rice scutellum. Physiol Veg. 1986; 24:629–638.
20. Palade GE. Intracellular aspects of the process of protein synthesis. Science. 1975; 189:347–358.
21. Farquhar MG, Palade GE. The Golgi apparatus (complex)—(1954–1981)—from artifact to center stage. J Cell Biol. 1981; 91:77S–103S.
22. Tartakoff AM. The secretory and endocytic paths. New York: John Wiley & Sons, 1987.
23. Mitsui T, Akazawa T. Preferential secretion of R-type α-amylase in rice seed scutellum at high temperatures. Plant Physiol. 1986; 82:880–884.
24. Hayashi M, Tsuru A, Takahashi N, et al. Structure and biosynthesis of the xylose-containing N-linked carbohydrate moiety of α-amylase secreted from suspension-cultured cells of rice. Eur J Biochem. 1990; in press.
25. Robinson DG. Synthesis and secretion of extracellular macromolecules. In: Plant membranes: Endo- and plasma membranes of plant cells. New York: John Wiley & Sons, 1985.
26. Daussant J, McGregor AW. Combined immunoadsorption and isoelectric focusing of barley and wheat amylases in polyacrylamide gel. Anal Biochem. 1979; 93: 261–266.
27. Daussant J, Miyata S, Mitsui T, et al. Enzyme mechanism of starch breakdown in germinating rice seeds 15. Immunochemical study on multiple forms of amylase. Plant Physiol. 1983; 71:88–95.

28. MacFadden FI, Ahluwalia B, Clarke AE, et al. Expression sites and developmental regulation of genes encoding $(1\rightarrow 3, 1\rightarrow 4)$ β-glucanases in germination barley. Planta. 1988; 173:500–508.
29. Fincher GB. Molecular and cellular biology associated with endosperm mobilization in germinating cereal grains. Ann Rev Plant Physiol. 1989; 40:305–346.
30. Weiler EW. Immunoassay of plant constituents. Biochem Soc Trans. 1983; 11:485–495.
31. Jacobsen JV, Beach LR. Control of transcription of α-amylase and rRNA genes in barley aleurone protoplasts by gibberellin and abscisic acid. Nature. 1985; 316:275–277.
32. Zwar JA, Hooley R. Hormonal regulation of α-amylase gene transcription in wild oat (*Avena fatua* L.) Plant Physiol. 1986; 80:459–463.
33. Ou-Lee TM, Turgeon R, Wu R. Interaction of gibberellin-induced factor with the upstream region of an α-amylase gene in rice aleurone tissue. Proc Natl Acad Sci USA. 1988; 85:6366–6369.

CHAPTER 13

Gibberellin Production and Action during Germination of Wheat

J.R. Lenton and N.E.J. Appleford

1 Introduction

The cereal aleurone layer has proved to be an ideal system in which to study the mechanism of gibberellin (GA) action at both the cellular and molecular levels. However, evidence that endogenous GAs are regulating the synthesis of hydrolytic enzymes in the aleurone during germination of intact grains is fragmentary and often contradictory (for reviews, see refs. 1–5). From work initiated in the 1960s, a relatively simple hormonal model for germinating cereal grains was proposed whereby GA_1 (or GA_3) produced in the embryo (either in the scutellum or embryonic axis) diffused to the aleurone layer where it initiated *de novo* synthesis of several hydrolytic enzymes, including α-amylase. These enzymes were secreted into the endosperm, and the products of reserve degradation were utilized by the growing embryo.

More recently, the validity of this model has been questioned both in terms of the nature and site of production of the hormonal stimulus and the site(s) of production of α-amylase and other hydrolytic enzymes. Concerning signal production and action, Trewavas[3] has argued that "GA is an essential but non-limiting and therefore non-regulatory factor in amylase production." More controversially, he suggested that "hydrolysis of GA-conjugates derived from the endosperm takes place in the embryo and represents the major source of free GA in early germination." However, evidence supporting such a notion is rather limited. In maize,[6] which does not require an embryo stimulus to initiate hydrolytic activity, conjugation of GA_{20} occurred during seed maturation, and the product was stored in the pericarp/aleurone layers. Upon imbibition, GA_{20}-conjugate moved to the endosperm and was released as the free acid during germination. In wheat,[7] GA-conjugates were detected by bioassay in the embryo of dry seed and were present as free acids 12 h after the start of imbibition.

In the dry seed of barley, GA_1 and traces of GA_3 and GA_4 were detected by immunoassay as the free acids but not as conjugates.[8] Evidence was presented showing that GA_4 was produced directly in aleurone tissue during germination and was associated with amylase synthesis. However, subsequent analysis of germinating barley by combined gas chromatography–mass spectrometry (GC–MS) has failed to detect GA_4, and inhibitor studies claiming a requirement for GA_4 production in aleurone have been questioned.[9]

The relative contribution of the scutellar epithelium and aleurone to the pattern of production of hydrolytic enzymes has been a matter of considerable debate.* Confusion appears to have arisen over (1) the anatomical relationship between aleurone and scutellum during grain development and in the mature seed, (2) the difficulty of separating the tissues in germinating wheat and barley, (3) methods used to localize enzymes, and (4) inherent differences between cereal species. The current consensus[5] is that hydrolytic enzymes are secreted initially from the scutellar epithelium and later from the aleurone layer, which, in wheat and barley, becomes the major source.

One objective of the work reported herein was to examine the relationship between endogenous hormonal status and tissue responsiveness. Germinating wheat grain was chosen because of the availability of (1) background information on endogenous GAs and effects of GA-biosynthesis inhibitors,[11] (2) near-isogenic lines of different GA-response mutants,[12] and (3) specific cDNA clones to α-Amy1 and α-Amy2 genes (kindly made available by Dr. D.C. Baulcombe, Sainsbury Laboratory, John Innes Institute, Norwich). The advantages of using cDNA probes are that they show, indirectly, effects of GA and abscisic acid (ABA) on gene transcription[13] and measure a response nearer the site of hormone perception rather than the secreted terminal protein.

2 GA Production During Germination

An initial objective was to identify the GAs present in dry seed and germinating grain by full-scan GC–MS. Several GAs of the early-13-hydroxylation pathway ($GA_{53} \rightarrow GA_{44} \rightarrow GA_{19} \rightarrow GA_{20} \rightarrow GA_1 \rightarrow GA_8$) were identified in an acidic fraction from mature wheat grain (Table 1). The main biologically active compound, GA_1, was also present in trace amounts but it gave only a weak mixed spectrum. There was no evidence for the occurrence of 1β-hydroxy-GAs[14,15] (GA_{54}, GA_{55}, GA_{60}, GA_{61}, and GA_{62}) in either the acidic or the conjugate fractions of mature grain, despite their abundance in developing wheat grain at the time of maximum fresh weight. However, a compound tentatively identified as 1β, 2β, 3β-trihydroxy-GA_9 which accumulated in large amounts in the later stages of grain growth was subsequently identified in the mature grain in the conju-

---

*References 1, 2, 5, and 10.

Table 1. Gibberellins identified by GC–MS in mature wheat grain (var. Maris Huntsman, 1985 harvest) (unpublished)

Acids[a]	GA_{19}	GA_{17}	GA_{20}	GA_{29}	GA_8
Conjugates[b]	GA_{20}	GA_8	GA_{79} ($1\beta, 2\beta, 3\beta$-trihydroxy-GA_9)		

[a] No evidence of 1β-hydroxy-GAs (GA_{54}, GA_{55}, GA_{61})
[b] Released following treatment of an acidic butanol fraction with cellulase. Mixed spectra of GA_{19}, GA_{17}, and GA_{29} were obtained

Table 2. Gibberellins identified by GC–MS in 4-day-old wheat seedlings (var. Maris Huntsman, 1985 harvest) (unpublished)

Gibberellin	Embryo		Endosperm	
	Acids	Conjugates	Acids	Conjugates
GA_{53}	– – –	+	– – –	+
GA_{44}	+++	– – –	+	– – –
GA_{19}	+++	+	++	++
GA_{17}	+	+	++	– – –
GA_{20}	+++	+++	+++	+++
GA_1	+++	+	+++	+++
GA_3	++	– – –	++	+
GA_4	+	– – –	– – –	– – –
GA_{34}	+	++	– – –	– – –
GA_{29}	++	+++	++	+++
GA_8	+++	+++	+++	+++

Symbols: +++, intense full spectrum; ++, contaminated spectrum; +, weak mixed spectrum; – – –, not detected

gate fraction (Table 1). The structure of this compound has now been confirmed by partial synthesis, and the compound assigned the number GA_{79} (C.L. Willis, personal communication). The presence of 2β-hydroxylated compounds (GA_8, GA_{79}) in the conjugate fraction of mature grain is unlikely to be of any biological significance even if they were hydrolyzed during seed germination. However, GA_{20}-conjugate, if hydrolyzed during germination, is a potential precursor of either GA_1 or GA_3 (via GA_5).

Additional members of the early-13-hydroxylation pathway were identified by full-scan GC–MS in both fractions of embryo and endosperm of 4-day-old germinating grain (Table 2). Mass spectral identifications were obtained for both GA_1 and GA_3 and the early C_{20} precursors, GA_{53} and GA_{44}. Two members of the early-non-hydroxylation pathway, GA_4 and its 2β-hydroxylated product, GA_{34}, were identified in the embryo but not the endosperm of these wheat seedlings. There was an increase in the number of GA-conjugates found in both the embryo and endosperm, compared with the dry seed, and GA_{20}-conjugate was also still present in both tissues

Table 3. Content (pg/part) of GAs and ABA in germinating wheat grains (var. Maris Huntsman, 1983 harvest) (unpublished)

Time (h)	Embryo						Endosperm					
	GA_{19}	GA_{20}	GA_1	GA_3	GA_8	ABA	GA_{19}	GA_{20}	GA_1	GA_3	GA_8	ABA
0	59	3	1	—	6	176	133	8	2	—	27	354
24	141	2	18	—	20	32	80	3	3	—	10	126
36	203	5	39	—	35	43	68	3	14	—	15	77
48	732	20	74	38	90	92	64	3	34	8	21	49
72	1311	73	302	91	325	140	148	8	182	34	20	37

(Table 2). Quantification of GA-conjugates, coupled with metabolic studies, is required before any reliable statements can be made about GA-conjugate production and turnover in germinating cereal grain. However, it seems reasonable to conclude that the additional GA-conjugates found in 4-day-old seedlings were produced as a consequence of GA synthesis and transport during germination.

Attention was focused on late members of the early-13-hydroxylation pathway, together with GA_4 and ABA, in grains germinating at 25°C in darkness. The compounds were quantified by GC–MS using heavy-isotope-labeled internal standards and selected ion monitoring (SIM). The increased selectivity and sensitivity of SIM permitted accurate quantification of the very small amounts of GA_1 present in both the embryo and endosperm of surface-sterilized mature seed (Table 3). The amount of GA_{19} in both the embryo and endosperm of mature seed was greater than that of the other GAs, and it accumulated rapidly in the embryo, particularly after 36 h, suggesting very active GA synthesis in this tissue. In contrast, the GA_{19} content of the endosperm remained relatively constant over 3 days. In the embryo, there was a 300-fold increase in GA_1 and a similar pattern of changes in amounts of its immediate precursor, GA_{20}, and product, GA_8, over the same time period. These observations suggested that the embryo was the site of GA production during the early stages of germination in wheat, although metabolic studies are required to confirm this proposition.

In addition to GA_{19}, the contents of GA_{20} and GA_8 remained relatively constant in the endosperm over 3 days (Table 3). However, the content of GA_1 rose rapidly after 1 day, suggesting diffusion from the embryo. Changes in the amounts of GA_3, which showed similar biological activity to GA_1 in a half-seed test, indicated later synthesis in the embryo and transport to the endosperm. Gibberellin A_4 was not detected in the endosperm up to 3 days but was present at 0.17 ng/g fr wt in the embryonic axis (minus scutellum) 3 days after the start of germination.

The content of ABA was greater in the endosperm than the embryo of mature seed and declined in both tissues, although more rapidly in the

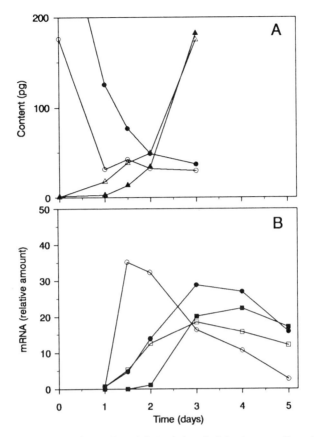

Fig. 1. Content of ABA (circles) and GA_1 (triangles) in the scutellum (open symbols) and endosperm (closed symbols) (A) and the relative amount of α-$Amy1$ mRNA (circles) and α-$Amy2$ mRNA (squares) in the scutellum (open symbols) and aleurone (closed symbols) (B) of germinating wheat grain (var. Maris Huntsman, 1983 harvest)

endosperm, up to 36 h (Table 3). After this time, the ABA content continued to decline in the endosperm but increased in the embryo. Dissection of the embryo after 24 h showed that the ABA content of the "scutellum" remained relatively constant between days 1 and 3 (Fig. 1A), indicating that the increase in ABA in the embryo as a whole must have been confined to the embryonic axis. After removal of the axis, grains were subdivided further into "scutellum" and "endosperm," consisting of approximately 20% and 80%, respectively, of the grain fresh weight. Thus, the "scutellum" would have contained some proximal aleurone and endosperm. Measurement of the changes in GA_1 content of these tissues (Fig. 1A) indicated diffusion of the embryo-produced signal to the more distal endosperm.

3 GA-Response System—α-Amylase Gene Transcription

There is good evidence that GA stimulates and ABA inhibits transcription of α-amylase genes in cereal aleurones (for reviews, see refs. 4, 5, 13, 16, and 17). Such changes in gene transcription can be measured indirectly as differences in steady-state mRNA levels by Northern hybridization using ^{32}P-labeled cDNA clones. Comparison of the signals obtained with those from a 25S rRNA clone can be used to quantify α-amylase mRNA.[18] In wheat, as in barley, there are two major α-amylase gene families, present on different chromosomes, which code for isozymes that can be resolved by isoelectrofocusing into high-pI (α-amy1) and low-pI (α-amy2) groups. The cDNA clones, which were specific for α-Amy1 (probe 2119) or α-Amy2 (probe 3'4868) genes,[19] were hybridized to RNA isolated from "scutellum" and the aleurone layer of "endosperm" from germinating grain.

In the "scutellum," the amount of mRNA for the high-pI isozymes (α-amy1) increased 35-fold between 24 and 36 h, then declined relatively rapidly over 4 days (Fig. 1B). In the aleurone, mRNA levels increased more gradually, reaching a peak at 3–4 days, then also declined rapidly. Messenger RNA for the low-pI isozymes (α-amy2) was detected first in the "scutellum" at 36 h, reached a peak at 3 days, and declined relatively slowly. By contrast, there was a further lag period of 12 h before mRNA for the low-pI isozymes was detected in aleurones (at 2 days); thereafter, amounts rose to a maximum at day 4 and declined slowly. It appeared that the synthesis of two α-amylase gene families was regulated coordinately in the "scutellum" but not in the more distal aleurone. In general, these tissue and temporal differences in mRNAs support the limited amount of information available, at the protein level, from germinating wheat grain.[12,20-22] It has been shown previously that α-Amy2 mRNA continues to accumulate for a longer period than α-Amy1 mRNA in distal half-grains treated with GA_3.[19]

4 Coupling GA Production and Response

Assuming that GA was promoting and ABA inhibiting α-amylase gene expression in germinating wheat, it appeared that ABA would have to be reduced to a low threshold concentration before transcription of α-Amy1 genes commenced, and expression of these genes would also have to be highly responsive to small amounts of GA_1 (Fig. 1). On the other hand, transcription of the α-Amy2 genes, coding for the low-pI isozyme group, appeared to be either more dependent on increasing GA_1 concentration or more responsive to small amounts of ABA. Such differences in responsiveness of the two gene families to the endogenous hormonal status might also account for their apparent temporal regulation in distal aleurone (Fig. 1B)

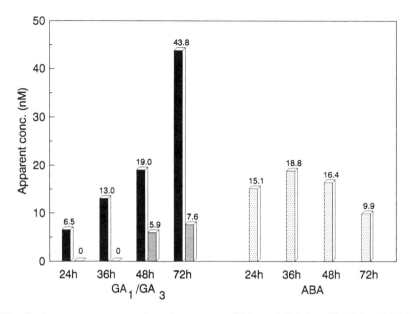

Fig. 2. Apparent concentration of promoters (GA_1 and GA_3) and inhibitor (ABA) in the scutellum of germinating wheat grains. The samples of scutellum contained some proximal endosperm

since this tissue was capable of responding to GA after 24 h (data not shown).

Were the changes in endogenous hormones sufficient to account for the observed responses? At present, it is difficult to give a direct answer to this question. However, it was possible, knowing the water content and assuming a uniform distribution of solute and solvent, to calculate an apparent concentration of hormones in the "scutellum" (Fig. 2). Similar estimates of the concentration of promoters and inhibitor in the endosperm were considered invalid since the peripheral layer of aleurone cells, the responsive tissue, formed such a small proportion of the total endosperm. On the basis of the estimates for the "scutellum," which contained some proximal endosperm, there was a four-fold increase in the concentration of promoting hormones (GA_1/GA_3) between 24 and 48 h, and this increase was against a relatively constant background of approximately 17 nM ABA. By 3 days, the concentration of promoters had increased to ~50 nM and ABA had declined. Obviously, since this tissue included the possible site of GA production, it would have been difficult to determine its responsiveness to applied GA unless it was made GA deficient either chemically or genetically. However, it was possible to test the responsiveness of the aleurone in distal half seeds challenged with GA_1 for 48 h.

An increase in transcription of α-*Amy1* genes was observed in aleurones

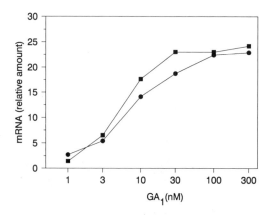

Fig. 3. Relative amounts of α-*Amy1* mRNA (circles) and α-*Amy2* mRNA (squares) produced in aleurones of distal half-grains of wheat incubated with GA_1 for 48 h

isolated from half-seeds treated with 1 nM GA_1, and the response saturated at 100 nM GA_1 (Fig. 3). Transcription of the α-*Amy2* genes appeared slightly more responsive to increasing GA_1 concentration and saturated at 30 nM GA_1. Thus, the apparent endogenous GA_1 concentrations calculated for the "scutellum," which would have included both the epithelial layer and some proximal aleurone, were probably within the GA-responsive range of the tissue (Figs. 2 and 3). One complicating factor was that the GA responsiveness of distal half-seeds was determined against an unknown endogenous ABA content of the aleurone layer, although the endosperm as a whole was known to contain a large amount of ABA initially (Table 3). The responsiveness of the aleurone to applied ABA was determined by incubating distal half-seeds in the presence of 30 nM GA_1 for 48 h. Preliminary evidence showed that transcription of the α-*Amy2* genes, coding for the low-pI isozymes, was inhibited by concentrations of racemic ABA greater than 30 nM, whereas transcription of α-*Amy1* genes was not responsive to ABA concentrations below 300 nM.

The responsiveness of the distal half-seed of wheat to GA and ABA was different from that observed with isolated aleurones of barley* but was consistent with predictions deduced from measurement of the hormonal and transcriptional changes in intact germinating wheat (Fig. 1). A causal relationship between hormone concentration and the response system has yet to be firmly established in wheat, and a crucial test will be to manipulate endogenous GA and/or ABA and determine the consequence for α-amylase gene transcription. Preliminary attempts to perturb GA produc-

---

*References 5, 17, 18, and 23.

tion with the *ent*-kaurene oxidase inhibitor 2*S*, 3*S*-paclobutrazol[24] have met with only partial success, and relatively little effect has been observed on α-amylase mRNA production in the scutellum although there was some reduction in the aleurone layer. Further metabolism of the pool of later GA intermediates, such as GA_{19}, present in mature seed could have accounted for such observations, although the potential hydrolysis of GA_{20}-conjugate should not be disregarded.

The apparent delay in the initiation of α-*Amy2* gene transcription in the aleurone of intact germinating grain (Fig. 1B) compared with the GA responsiveness of half-seeds (Fig. 3) may have been accounted for by the release of ABA from the cut surface of half-seeds. Such a proposal, if proved to be correct, would have important implications for the interpretation of results obtained previously with dissected tissues, such as isolated embryos and aleurones, from both wheat[22] and barley.[18,23] It would also highlight the biological significance of results obtained with intact germinating grain, as compared to deductions based on the responsiveness of isolated tissues.

5 Conclusions

Evidence has been presented showing that the original model of GA_1 (and later GA_3) production in the embryo and diffusion to the endosperm was valid for germinating wheat grain. However, the precise site and pathway of synthesis has yet to be determined. The pattern of increase of GAs during germination has been found to be associated with the initiation of α-amylase gene transcription in both "scutellum" and "endosperm," indicating that they are important components of the embryo-produced stimulus required for endosperm reserve mobilization. Although the quantitative relationships between GA production and response suggest that GAs are regulating transcription of α-amylase genes during germination, the possibility of GA-independent α-amylase production in the scutellar epithelium cannot be excluded. However, transcription of α-amylase genes in the scutellum of a GA-deficient mutant of barley has been shown to be GA dependent (P.M. Chandler, personal communication). The ABA status of the scutellum and endosperm of germinating wheat grain also appears to be important in relation to the initiation of α-amylase gene transcription, particularly for those genes coding for the low-pI isozymes. Overall, the results suggest that hormone concentration, together with the acquisition of tissue responsiveness, regulates the initiation of α-amylase gene expression in germinating wheat grain.

Acknowledgments. This paper is dedicated to the memories of the late Dr. Margaret Radley, who made significant contributions to the original hormonal model involving GAs in endosperm mobilization, and to Professor

Ken Treharne, whose sudden death robbed us of a champion of plant hormone research at Long Ashton. We thank the following for their generous gifts: Dr. D.C. Baulcombe (α-amylase cDNA clones) and Professors L.N. Mander and B.O. Phinney and Dr. R. Horgan (heavy-isotope-labeled internal standards). We are also extremely grateful to our colleagues Dr. R. Hooley, for the gift of a 25S rRNA probe and his continuous advice on molecular biology techniques, and Dr. P. Hedden and Mr. M.J. Lewis, for their help with GA identification and analysis.

References

1. Akazawa T, Miyata Y. Biosynthesis and secretion of α-amylase and other hydrolases in germinating cereal seeds. Essays in Biochem. 1982; 18:41–78.
2. Akazawa T, Hara-Nishimura I. Topographical aspects of biosynthesis, extracellular secretion, and intracellular storage of proteins in plant cells. Ann Rev Plant Physiol. 1985; 36:441–472.
3. Trewavas AJ. Growth substance sensitivity: The limiting factor in plant development. Physiol Plant. 1982; 55:60–72.
4. Baulcombe D, Lazarus C, Martienssen R. Gibberellins and gene control in cereal aleurone cells. J Embryol Exp Morph. 1984; 83:Supplement, 119–135.
5. Fincher GB. Molecular and cellular biology associated with endosperm mobilisation in germinating cereal grains. Ann Rev Plant Physiol Plant Mol Biol. 1989; 40:305–346.
6. Rood SB, Pharis RP, Koshioka M. Reversible conjugation of gibberellins *in situ* in maize. Plant Physiol. 1983; 73:340–346.
7. Thomas TH, Khan AA, O'Toole DF. The location of cytokinins and gibberellins in wheat seeds. Physiol Plant. 1978; 42:61–66.
8. Atzorn R, Weiler EW. The role of endogenous gibberellins in the formation of α-amylase by aleurone layers of germinating barley caryopses. Planta. 1983; 159:289–299.
9. Gilmour SJ, MacMillan J. Effect of inhibitors of gibberellin biosynthesis on the induction of α-amylase in embryoless caryopses of *Hordeum vulgare* cv. Himalaya. Planta. 1984; 162:89–90.
10. Hill RD, MacGregor AW. Cereal α-amylases in grain research and technology. In: Pomeranz Y, ed. Advances in cereal science and technology, Vol. 9. St. Paul, Minnesota: American Association of Cereal Chemists, 1988: p. 217–261.
11. Lenton JR, Hedden P, Gale MD. Gibberellin insensitivity and depletion in wheat—consequences for development. In: Hoad GV, Lenton JR, Jackson MB, Atkin RK, eds. Hormone action in plant development—A critical appraisal. London: Butterworths, 1987: p. 145–160.
12. Flintham JE, Gale MD. The Tom Thumb dwarfing gene, *Rht3* in wheat, 1. Reduced pre-harvest damage to breadmaking quality. Theor Appl Genet. 1982; 62:121–126.
13. Huttly AK, Baulcombe DC. Hormonal control of wheat α-amylase genes. In: Grierson D, Lycett G, eds. Genetic engineering of crop plants. London: Butterworths, 1989: p. 171–189.
14. Gaskin P, Kirkwood PS, Lenton JR, et al. Identification of gibberellins in developing wheat grain. Agric Biol Chem. 1980; 44:1589–1593.

15. Kirkwood P, MacMillan J. Gibberellins A_{60}, A_{61} and A_{62}: Partial syntheses and natural occurrence. J Chem Soc Perkin Trans 1. 1982; 689–697.
16. MacGregor EA, MacGregor AW. Studies of cereal α-amylase using cloned DNA. CRC Crit Rev Biotechnol. 1987; 5:129–142.
17. Jacobsen JV, Chandler PM. Gibberellin and abscisic acid in germinating cereals. In: Davies PJ, ed. Plant hormones and their role in plant growth and development. Dordrecht: Martinus Nijhoff Publishers, 1987: p. 164–193.
18. Chandler PM, Zwar JA, Jacobsen JV, et al. The effects of gibberellic acid and abscisic acid on α-amylase mRNA levels in barley aleurone layers; studies using an α-amylase cDNA clone. Plant Mol Biol. 1984; 3:407–418.
19. Lazarus CM, Baulcombe DC, Martienssen RA. α-Amylase genes of wheat are two multigene families which are differentially expressed. Plant Mol Biol. 1985; 5:13–24.
20. Sargeant JG, Walker TS. Adsorption of wheat alpha-amylase isozymes to wheat starch. Stärke. 1978; 30:160–163.
21. Daussant J, Renard HA, Skakoun A. Application of direct tissue rocket-line immunoelectrophoresis to the study of α-amylase production and its localisation in wheat seeds during the early stages of germination. Electrophoresis. 1982; 3:99–101.
22. Marchylo BA, Kruger JE, MacGregor AW. Production of multiple forms of alpha-amylase in germinated, incubated, whole, de-embryonated wheat kernels. Cereal Chem. 1984; 61:305–310.
23. Nolan RC, Ho T-H D. Hormonal regulation of gene expression in barley aleurone layers. Planta. 1988; 174:551–560.
24. Hedden P, Graebe JE. Inhibition of gibberellin biosynthesis by paclobutrazol in cell-free homogenates of *Cucurbita maxima* endosperm and *Malus pumila* embryos. Plant Growth Regul. 1985; 4:111–112.

CHAPTER 14

Probing Gibberellin Receptors in the *Avena fatua* Aleurone

R. Hooley, M.H. Beale, and S.J. Smith

1 Introduction

The cereal aleurone is a convenient model system for studying gibberellin (GA) action. Aleurone layers respond to application of exogenous GA by synthesizing and secreting a number of hydrolytic enzymes that are capable of hydrolyzing the stored reserves of the endosperm, and there are good reasons to believe that GA_1 produced by germinating wheat (Lenton and Appleford, Chapter 13 of this volume) barley,[1] and wild oat (P. Hedden, personal communication, 1988) embryos elicits a similar response in vivo.

The regulation of expression of genes encoding α-amylase, the most abundant of the hydrolases induced by GA in aleurone, has been studied in detail. It appears that one consequence of treating aleurone protoplasts with GA is a very substantial stimulation of the rates of transcription of α-amylase genes and that this leads to greatly elevated steady-state levels of α-amylase mRNA.[2,3]

DNA sequence elements likely to be involved in the GA regulation of expression of wheat α-amylase genes have been characterized by transient expression analysis of promoter constructs in oat aleurone protoplasts. It seems likely that elements determining GA regulation of expression of low-pI gene family members lie within 300 base pairs (bp) upstream of the start of transcription.[4] A putative *trans*-acting factor that may be involved in GA regulation of α-amylase gene expression in rice aleurone has been reported,[5] and similar DNA binding proteins are being sought in *Avena fatua* aleurone nuclei (P.J. Rushton, personal communication, 1989).

We anticipate that perception of GA by aleurone cells will involve an interaction between the hormone and a specific receptor. Binding studies using radiolabeled GAs have led to the discovery and partial purification of proteins from the cytosol of cucumber hypocotyls and pea epicotyls that have some of the characteristics expected of a GA receptor.[6-8] However, attempts made to identify GA-binding proteins in aleurones of wheat and barley by in vivo and in vitro binding studies with radiolabeled GAs[9,10] have not revealed a candidate GA receptor.

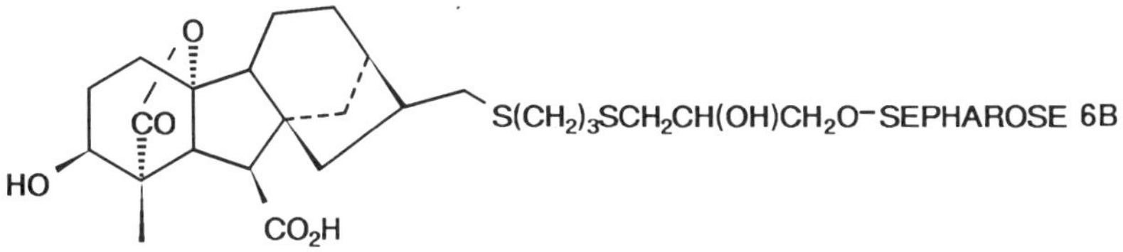

[Chemical structure with S(CH$_2$)$_3$SCH$_2$CH(OH)CH$_2$O–SEPHAROSE 6B substituent] (I)

HOOC(CH$_2$)$_2$SCH$_2$CH(OH)CH$_2$O – SEPHAROSE 6B (II)

Fig. 1. Structure of GA$_4$-17-Sepharose (I) and control Sepharose 6B (II)

We have been exploring novel ways of identifying GA receptors in GA-responsive isolated aleurone protoplasts of *A. fatua*[11] using novel GA derivatives as probes for receptors.[12,13] Functional assays for GA receptors have indicated that they may lie at the surface of aleurone protoplasts.[13] This paper is concerned mainly with the subcellular location of GA receptors in *A. fatua* aleurone protoplasts as determined by α-amylase induction using GA$_4$ immobilized on Sepharose beads. In addition, preliminary experiments that aim to fractionate aleurone membrane proteins by GA-affinity chromatography will be described.

2 Subcellular Location of GA Receptors

A number of GAs and GA derivatives have been bioassayed with *A. fatua* aleurone protoplasts. Data from these studies are consistent with the view that elements of the A/B ring area of the molecule determine biological activity of GAs.[12] Derivatives with additions at the C-17 position retain biological activity, and this has allowed us to explore the use of GAs with a range of additional groups at this position as affinity probes for GA receptors.[12,13]

A 17-thiol derivative of GA$_4$ has been linked to epoxy-activated Sepharose 6B[12] (Fig. 1, structure I). Thus, GA$_4$ has been immobilized on a support matrix in such a way that its "active site" is exposed. Gibberellin A$_4$-17-Sepharose has been employed in a functional assay for GA receptors in aleurone protoplasts[13] that is based on the premise that if GA receptors are located on the external face of the aleurone plasma membrane, then they may be capable of perceiving the immobilized GA$_4$ when aleurone protoplasts are presented with GA$_4$-17-Sepharose beads. As a control, the wall of the aleurone cell in the intact tissue would prevent immobilized GA$_4$ from coming into contact with the plasma membrane when GA$_4$-17-Sepharose is presented to aleurone layers.

Table 1. Induction of α-amylase by GA_4-17-Sepharose

Treatment	Response[a] (%)	
	Protoplasts	Cells
GA_4-17-Sepharose (100 μM)	85	10
GA_4-17-Sepharose (100 μM) plus GA_4 (1 μM)	100	91
Sepharose 6B	2	9
No additions	1	2

[a]Relative to the maximum α-amylase induced by 1 μM GA_4

We have described elsewhere evidence supporting this hypothesis.[13] Isolated aleurone protoplasts respond to GA_4-17-Sepharose by synthesizing and secreting α-amylase at levels comparable with the near maximal rates induced by 10^{-6} M free GA_4. Cells of the intact aleurone on the other hand do not respond significantly to GA_4-17-Sepharose (Table 1).

In order to induce a significant response, the amount of immobilized GA_4 present with the aleurone protoplasts has to be many times the concentration of free GA_4 that is needed to induce a maximum response. Dose–response data (Fig. 2) indicate that some 10,000 times more GA_4 immobilized on Sepharose is required to induce a given response in protoplasts compared with free GA_4. This may be due to the availability of the immobilized GA_4 to interact with the protoplast surface. We anticipate firstly that only a proportion of the GA_4 will be at the surface of the Sepharose beads while the majority of it will have been linked to epoxy groups within the bead matrix. Secondly, because GA_4-17-Sepharose beads are larger in diameter than aleurone protoplasts, it seems likely that, of the GA_4 at the surface, only a fraction will be involved in point-to-point contact with protoplasts. Thirdly, it is probable that during incubation not all protoplasts will come into contact with GA_4-17-Sepharose, and, for those that do, the time that the protoplast and immobilized GA_4 are in physical contact may or may not be sufficiently long to induce a response.

These factors may explain the substantially greater amounts of immobilized GA_4, compared with free GA_4, required to induce a response. Nevertheless, it is clear that if only 0.01% of the immobilized GA_4 were to be released from the Sepharose, for whatever reason, this would be capable of inducing a substantial response in aleurone protoplasts.

The fact that aleurone layers respond at only a very low level in the presence of GA_4-17-Sepharose argues against the compound being unstable, because the amount of free GA_4 required to induce the degree of response observed in protoplasts would produce a near maximum induction of α-amylase in aleurone layers as well. It is possible, however, that GA_4-17-Sepharose is stable in the presence of aleurone layers but becomes degraded, releasing free GA_4, when incubated with aleurone protoplasts.

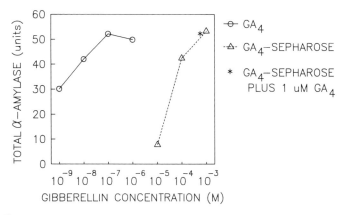

Fig. 2. Dose-response curves of aleurone protoplasts to GA_4 (——), GA_4-17-Sepharose (---), and GA_4-17-Sepharose plus 1 μM GA_4 (*)

Wild oat aleurone protoplasts are washed extensively during isolation,[11] and it seems unlikely that any cell wall hydrolyzing enzymes that might potentially degrade GA_4-17-Sepharose will be carried over into protoplast preparations. Alternatively, aleurone protoplasts may produce and secrete an enzyme capable of degrading GA_4-17-Sepharose and releasing free GA_4 and which either is an endogenous aleurone enzyme normally tightly bound in the aleurone cell wall or is produced by protoplasts as an artifact of protoplast isolation.

One way to resolve this question would be to coincubate aleurone layers and isolated aleurone protoplasts under identical conditions in the same flask with GA_4-17-Sepharose. If the observed response of aleurone protoplasts is brought about by GA_4 released from the Sepharose, by whatever means, then cells of the aleurone layers would perceive this GA_4 and respond. If, on the other hand, protoplasts are responding to immobilized GA_4 through contact with protoplast surface receptors, then only they would respond.

These experiments are not easy to perform and have required prudent attention to certain aspects of the tissue culture in order to establish conditions under which both isolated aleurone protoplasts and aleurone layers can be coincubated. After suitable culture conditions had been determined, aleurone layers and isolated aleurone protoplasts were coincubated with control Sepharose, GA_4-17-Sepharose, and GA_4-17-Sepharose plus 1 μM free GA_4. Aleurone layers and protoplasts were then separated from one another, and α-amylase mRNA levels in the respective tissues determined by Northern hybridization using the full length wild oat α-amylase cDNA clone Afa2 (R. Hooley and P.M. Chandler, unpublished).

Only very low levels of α-amylase mRNA are detected in coincubated aleurones and aleurone protoplasts treated with control Sepharose 6B

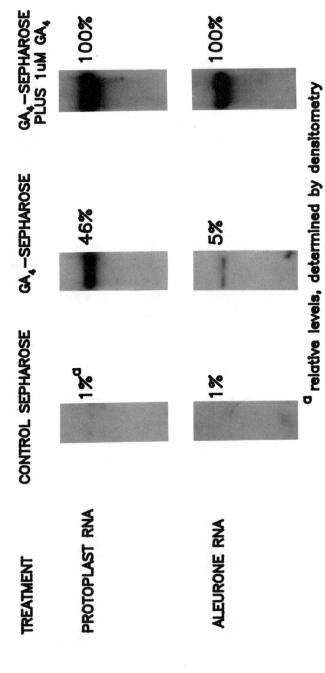

Fig. 3. α-Amylase mRNA levels in coincubated aleurones and protoplasts determined by Northern hybridization of protoplast and aleurone RNA to a full length *A. fatua* α-amylase cDNA insert, Afa2. Numbers alongside each track denote relative amounts of α-amylase mRNA determined from the autoradiograph by densitometry, as a percentage of the amount of α-amylase mRNA induced in each tissue in the presence of 1 μM GA_4

[a] relative levels, determined by densitometry

(Fig. 3). When GA_4-17-Sepharose and 1 μM free GA_4 are included in the medium, both cells and protoplasts produce substantial amounts of α-amylase mRNA. Aleurone protoplasts and aleurone layers coincubated in medium containing GA_4-17-Sepharose produce about 46% and 5%, respectively, of the α-amylase mRNA induced in the presence of GA_4-17-Sepharose and 1 μM free GA_4 (Fig. 3). Thus, under these conditions, the response of protoplasts to immobilized GA_4 is very much higher than that of aleurone layers incubated under identical conditions.

The slight response of aleurone layers to GA_4-17-Sepharose may reflect the release of very small amounts of GA_4 from the beads, perhaps as a result of physical or mechanical damage. The fact that the response of aleurone layers to GA_4-17-Sepharose, relative to the response that can be induced in the presence of 1 μM GA_4, is not significantly different whether the aleurone layers are incubated alone or in the presence of aleurone protoplasts (the latter producing substantial amounts of α-amylase mRNA under these conditions) (Fig. 4) is consistent with this view.

These data support the theory[13] that the response of aleurone protoplasts to GA_4-17-Sepharose is through an interaction between immobilized GA_4 and the protoplast surface. The implication, therefore, is that GA receptors are present at the external face of the aleurone plasma membrane.

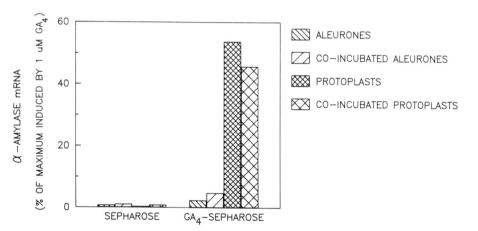

Fig. 4. α-Amylase mRNA levels in aleurones and protoplasts either coincubated or incubated as individual tissues. Relative amounts of α-amylase mRNA were determined by densitometry of autoradiographs from a Northern hybridization of protoplast and aleurone RNA probed with the α-amylase cDNA insert Afa2 and are presented as the percentage of the level of α-amylase mRNA induced in each tissue in the presence of 1 μM GA_4

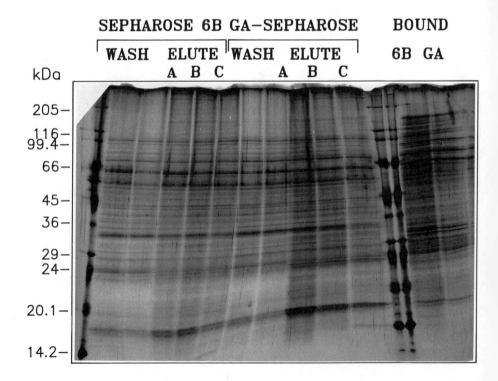

3 Affinity Chromatography of Solubilized Aleurone Membrane Proteins

Gibberellin A_4-17-Sepharose may be useful as an affinity medium for purifying GA receptors, although such an approach is not without its limitations. Solubilization of membrane proteins prior to affinity chromatography may eliminate biological activity. Also, we can anticipate that proteins which bind GA nonspecifically and quasispecifically may also bind to GA_4-17-Sepharose. In addition, some proteins may interact hydrophobically either with Sepharose or the spacer arm used to link the GA_4. By performing affinity chromatography experiments with both GA_4-17-Sepharose and a control Sepharose 6B with almost the entire length of the spacer arm (Fig. 1, structures I and II) and by eluting proteins with both the virtually inactive GA_3 methyl ester (GA_3-Me) and the active GA_3, it may be possible to minimize these problems.

Aleurone membrane proteins were solubilized from a crude aleurone protoplast membrane preparation using Triton X-100. Solubilized proteins were applied to columns of control Sepharose 6B and GA_4-17-Sepharose (GA_4 concentration 20 mM with respect to bed volume), and the columns were washed to remove unbound protein. Proteins were eluted sequentially with 50 mM GA_3-Me, 2 mM GA_3, and 50 mM GA_3. Aliquots of protein washed and eluted from the columns were analyzed by sodium dodecyl sulfate–polyacrylamide gel electrophoresis (SDS–PAGE) (Fig. 5a and b).

The patterns of polypeptides recovered from both the control and GA_4-17-Sepharose columns during washing were very similar. A number of polypeptides that were eluting during washes from both the control and GA_4-17-Sepharose columns were displaced by GA_3-Me, as were a number of other polypeptides that had not apparently eluted from the columns until this stage. There were no marked differences in the pattern of polypeptides eluting from the control and GA_4-17-Sepharose. The pattern of polypeptides eluting from the columns was not very different when GA_3 at 2 mM and 50 mM was applied. A 14.2-kDa polypeptide eluted from both control and GA_4-17-Sepharose, particularly at the higher concentrations of

◁─────────────

Fig. 5a and b. SDS–PAGE of Trition X-100-solubilized aleurone membrane polypeptides fractionated on control Sepharose 6B (Sepharose 6B) and GA_4-17-Sepharose (GA-Sepharose) affinity columns. (a) Columns were washed with buffer comprising 10mM MES (morpholineethane sulfonic acid), 50 μM EDTA, 50 μM DTT (dithiothreitol), 5 mM KCl, 50 mM NaCl, 0.5% Trition X-100, pH 6.0, and aliquots were analyzed on a 10–30% SDS polyacrylamide gradient gel that was silver stained. (b) After two further washes, columns were eluted with buffer containing 50 mM GA_3-Me (A), 2 mM GA_3 (B), and 50 mM GA_3 (C). The concentration of NaCl in these solutions was 0 mM, 48 mM, and 0 mM, respectively. Aliquots were analyzed by SDS–PAGE as above and silver stained in two cycles

GA_3. Any differences between the polypeptide patterns eluted from GA_4-17-Sepharose compared with control Sepharose 6B were subtle; a minor 29-kDa polypeptide eluted from GA_4-17-Sepharose but not control Sepharose 6B in the presence of 2 mM GA_3. A large number of proteins remained bound to both columns after elution with GA_3. The fact that the vast majority of these were common to both matrices indicated that they were bound to Sepharose or the spacer arm, perhaps by hydrophobic interactions.

4 Discussion

4.1 Subcellular Location of GA Receptors

Gibberellin A_4 immobilized to Sepharose beads in such a way that the A/B ring area is exposed has been employed in a functional assay for GA receptors using aleurone protoplasts. Evidence is presented here suggesting that immobilized GA_4 induces α-amylase gene expression in isolated aleurone protoplasts by interacting with the protoplast surface. These observations are consistent with the hypothesis[13] that GA receptors are located at the external face of the aleurone plasma membrane.

4.2 GA_4-Affinity Chromatography of Solubilized Aleurone Membrane Proteins

Numerous Triton X-100-solubilized aleurone membrane proteins bind loosely to the control Sepharose 6B and GA_4-17-Sepharose, and many remain bound after extensive washing in buffer containing 0.5% Triton X-100 and 50 mM NaCl followed by elution with GA_3-Me and GA_3. The majority of these proteins appear to bind to Sepharose and/or the spacer arm. It has not been possible to demonstrate either specific binding of particular polypeptides to the immobilized GA_4 or specific elution of polypeptides from GA_4-17-Sepharose by excess free biologically active GA_3.

One problem encountered when trying to purify a receptor that is an integral membrane protein is to retain its biological activity, or at least its ligand-binding characteristics, after solubilization. In the absence of a suitable GA-binding assay, we do not know if, in these preliminary experiments, Triton X-100 has solubilized GA-binding proteins from aleurone membrane preparations nor whether or not after solubilization they would have retained GA-binding activity.

In addition, because the relative abundance of GA receptors in aleurone is not known, the likelihood of seeing them as a band or bands on a silver-stained SDS gel, even if they did elute from GA_4-17-Sepharose with free ligand, cannot be assessed. What is clear, however, is that numerous aleurone polypeptides do have an affinity for the column matrix and this may limit the usefulness of GA-affinity chromatography for purifying receptors from crude protein preparations.

Currently, methods are being developed for the isolation of aleurone plasma membrane and the solubilization and purification by fast protein liquid chromatography and GA_4-affinity chromatography of aleurone plasma membrane proteins.

Acknowledgments. We thank Professor J. MacMillan for useful discussions. This article is dedicated to the late Professor K.J. Treharne.

References

1. Gaskin P, Gilmour SJ, Lenton JR, et al. Endogenous gibberellins and kaurenoids identified from developing and germinating barley grain. J Plant Growth Regul. 1984; 2:229–242.
2. Jacobsen JV, Beach LR. Control of transcription of α-amylase and rRNA genes in barley aleurone protoplasts by gibberellin and abscisic acid. Nature. 1985; 316:275–277.
3. Zwar JA, Hooley R. Hormonal regulation of α-amylase gene transcription in wild oat (*Avena fatua* L.) aleurone protoplasts. Plant Physiol. 1986; 80:459–463.
4. Huttly AK, Baulcombe DC. A wheat α-*Amy2* promoter is regulated by gibberellin in transformed oat aleurone protoplasts. EMBO J. 1989; 8:1907–1913.
5. Ou-Lee T-M, Turgeon R, Wu R. Interaction of a gibberellin-induced factor with the upstream region of an α-amylase gene in rice aleurone tissue. Proc Natl Acad Sci USA. 1988; 85:6366–6369.
6. Keith B, Foster NA, Bonettemaker M, Srivastava LM. *In vitro* gibberellin A_4 binding to extracts of cucumber hypocotyls. Plant Physiol. 1981; 68:344–348.
7. Keith B, Brown S, Srivastava LM. *In vitro* binding of gibberellin A_4 to extracts of cucumber measured using DEAE-cellulose filters. Proc Natl Acad Sci USA. 1982; 79:1515–1519.
8. Srivastava LM. The gibberellin receptor. In: Klambt D, ed. Plant hormone receptors. NATO ASI Series, Vol. H10. Heidelberg: Springer-Verlag, 1987: p. 199–227.
9. Keith B, Boal R, Srivastava LM. On the uptake, metabolism and retention of [^{3}H] gibberellin A_1 by barley aleurone layers at low temperatures. Plant Physiol. 1980; 66:956–961.
10. Jelsema CL, Ruddat M, Morre JD, Williamson FA. Specific binding of gibberellin A_1 to aleurone grain fractions from wheat endosperm. Plant Cell Physiol. 1977; 18:1009–1019.
11. Hooley R. Protoplasts isolated from aleurone layers of wild oat (*Avena fatua* L.) exhibit the classic response to gibberellic acid. Planta. 1982; 154:29–40.
12. Beale MH, Hooley R, MacMillan J. Gibberellins: structure–activity relationships and the design of molecular probes. In: Bopp M, ed. Plant growth substances 1985. Heidelberg: Springer-Verlag, 1986: p. 65–73.
13. Hooley R, Beale MH, Smith SJ, MacMillan J. Novel affinity probes for gibberellin receptors in aleurone protoplasts of *Avena fatua*. In: Pharis RP, Rood SB, eds. Plant growth substances 1988. Heidelberg: Springer-Verlag, 1989: In press.

CHAPTER 15

A Minireview on the Immunoassay for Gibberellins

I. Yamaguchi and E.W. Weiler

1 Introduction

An immunoassay for gibberellins (GAs) was first reported by Fuchs and Fuchs in 1969.[1] In this first report, modified bacteriophage T4 was used as a tracer, and several nanograms of GA_3 were required for detection. However, this immunoassay procedure did not attract the attention of GA scientists because it provided less information than combined gas chromatography–mass spectrometry (GC–MS) and insufficient information was available on the specificities of the antiserum. At that time, improvements in GC–MS were making it possible to detect submicrograms of GAs, and GC–MS was more conclusive as well as informative. The mass spectra of all known GAs were stored in MacMillan's (Bristol) and Takahashi's (Tokyo) laboratories for consultation.

In 1981, Weiler and Wieczorek[2] reported a radioimmunoassay (RIA) for GAs, which enabled detection of femtomole (picogram) amounts of GAs. Enzyme-linked immunosorbent assays (ELISAs) for GAs were also reported by Atzorn and Weiler.[3] The sensitivity of these RIAs and ELISAs was in the subnanogram range, even higher than that of GC–MS. In these reports,[2,3] the characteristics of antisera were studied in detail, and the specificities of the antibodies were high enough to make these immunoassays useful for plant analysis. Since then, several groups have prepared antibodies against GAs: Weiler and his co-workers reported the preparation of polyclonal antibodies against immunogens with haptens GA_1,[4] GA_3,[1,3,4] $GA_{4/7}$,[3] GA_9,[4] and GA_{24}[5]; Takahashi and his co-workers with GA_1,[6] GA_4,[7] GA_5,[8] GA_{13},[9] and GA_{20}[8]; and Oden et al.[10] with GA_1 and GA_9.

Monoclonal antibodies were also prepared by three groups headed by Weiler (GA_{13}),[11] MacMillan (GA_1, GA_4, GA_9),[12,13] and Takahashi (GA_4).[14] To make gibberellins immunogenic, several types of conjugation of GAs to carrier proteins were employed, and it became clear that there is some correlation between the position in the GA molecule used for the conjugation to carrier proteins and the recognition of epitopes on the GA

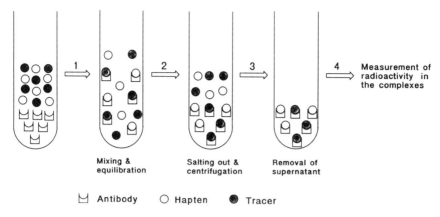

Fig. 1. Radioimmunoassay procedure

by the prepared antibodies. In this paper, the general immunoassay procedure is briefly described and the immunoassay of GAs is reviewed.

2 General Procedures of RIA and ELISA

The general procedure of RIA is shown in Fig. 1. The antibody, the tracer, and the antigen (authentic, sample, or blank) are mixed in a buffer and incubated for a certain period of time, usually at 4 to 25°C. After the antibody–antigen reaction has reached equilibrium, the antigen–antibody complex is separated from nonbound antigen. For this separation in the RIA of GAs, salting out of the complexes by 50% saturated ammonium sulfate is usually used, because immunoglobulins precipitate under these conditions but nonbound GAs do not. The radioactivity in the precipitate is measured to determine the inhibition of the tracer–antibody binding by added authentic GAs or samples.

Two types of calibration curves are used for quantification. One is the usual standard curve, shown in the left panel of Fig. 2, and the other is a logit–log-transformed plot, shown in the right panel of Fig. 2. In the calibration curve in the left panel of Fig. 2, the values are the percentage of radioactivity of the antigen–antibody complex in the presence of authentic nonlabeled GA divided by the radioactivity of the antigen–antibody complex in the absence of nonlabeled GA:

$$y = [(B - UB)/(B_o - UB)] \times 100(\%)$$

where B is the radioactivity in the individual precipitates, B_o is the radioactivity of the antigen–antibody complex in the absence of nonlabeled GA, and UB is the radioactivity unspecifically incorporated into the precipitate.

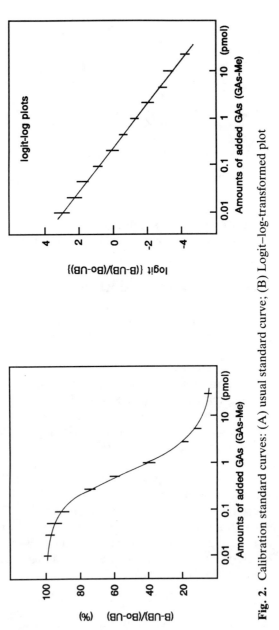

Fig. 2. Calibration standard curves: (A) usual standard curve; (B) Logit–log-transformed plot

In logit–log plots, the values of y are expressed by the following equation:

$$Y = \text{logit}[(B - UB)/(B_o - UB)]$$
$$= \ln([(B - UB)/(B_o - UB)]/\{1 - [(B - UB)/(B_o - UB)]\})$$

In ELISA, 96-well immunoplates are precoated first with antibodies that bind to the anti-GA antibodies to be used. As primary coating antibody, goat anti-rabbit Ig(subclass) is sometimes used for polyclonal rabbit anti-GA antibodies, and rabbit anti-mouse or anti-rat Ig(subclass) is usually used for monoclonal mouse or rat anti-GA antibodies. The precoated wells are then coated with anti-GA antibodies. When antibodies isolated from rabbit sera are used, it is common to bind them directly to the polystyrene due to their excellent, tight binding to the plastic. In the precoated wells, authentic GAs or samples are then incubated in the presence of the enzyme tracers for 3–4 h at 4°C, and the solutions in the wells are then discarded. After the wells are rinsed with water or buffer, the buffered enzyme substrate is added to each well and, after appropriate incubation, the amounts of the product formed are measured by a colorimeter. In this competitive ELISA system, the inhibition of the binding of antibodies and enzyme tracer by authentic GAs or samples is monitored. Therefore, the measured optical density (OD) data are processed in a way similar to that in which radioactivity data for the tracer–antibody complexes in RIA are processed.

3 Preparation of Immunogens

Gibberellins have to be coupled to a carrier protein to become immunogenic. The molar ratio of GAs coupled to a carrier protein may affect the immunogenicity. The smallest molar ratio of an immunogenic GA conjugate so far reported to have been successful in immunizing rabbits is 1:1.[5]

3.1 Immunogenic Conjugates of C_{19} GAs

The simplest way to conjugate GAs to the carrier protein is to use a carboxyl group to form an amide bond with the ϵ-amino group of lysine residues in the protein. As a carrier protein, bovine serum albumin (BSA), human serum albumin (HSA), or keyhole-limpet hemocyanin (KLH) is often used. To couple the carboxyl group in GA to ϵ-amino groups, the carboxyl group in GA is activated by an active ester method, a mixed anhydride method, or 1-ethyl-3-(3-dimethylaminopropyl) carbodiimide (EDC) (Fig. 3). In the mixed anhydride method,[2,4] isobutyl chlorocarbonate is added to the cooled dimethylformamide (DMF) solution of GA and tri-n-butylamine to form the mixed anhydride. For the preparation of active esters with N-hydroxysuccinimide or p-nitrophenol, GA and

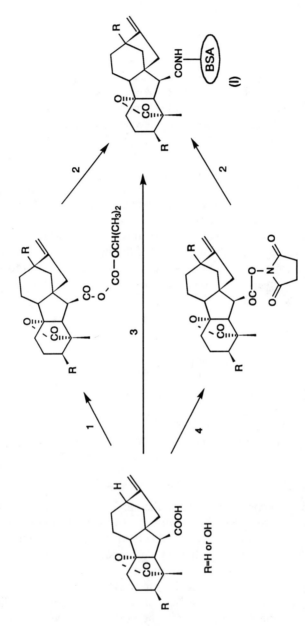

Fig. 3. Coupling of GAs to carrier protein via C-7 carboxyl group[2, 8]: 1. ClCOOCH(CH$_3$)$_2$, N(n-C$_4$H$_9$)$_3$; 2. BSA; 3. EDC (pH 4.5–6.5); 4. DCC, N-hydroxysuccinimide

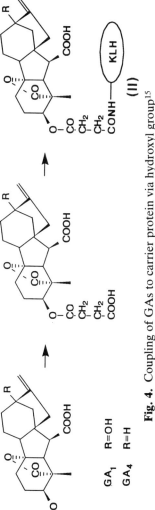

Fig. 4. Coupling of GAs to carrier protein via hydroxyl group[15]

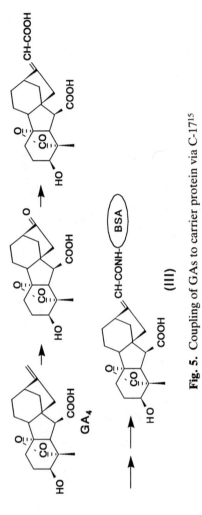

Fig. 5. Coupling of GAs to carrier protein via C-17[15]

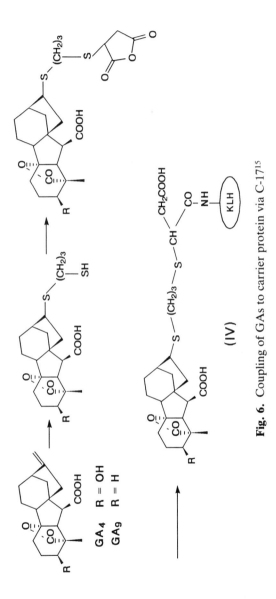

Fig. 6. Coupling of GAs to carrier protein via C-17[15]

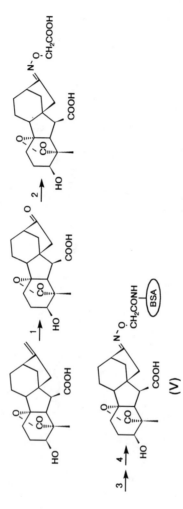

Fig. 7. Coupling of GA$_4$ to carrier protein via C-167: 1. OsO$_4$, NaIO$_4$; 2. H$_2$NOCH$_2$COOH; 3. DCC, N-hydroxysuccinimide; 4. BSA

N-hydroxysuccinimide (or p-nitrophenol) are mixed with dicyclohexylcarbodiimide (DCC) in a dry solvent such as acetonitrile, tetrahydrofuran, or DMF.[6-9] The formation of an active ester is accompanied by the precipitation of dicyclohexylurea. The mixed anhydride or the active ester thus formed is coupled to a carrier protein in a DMF–water (1:1) solution. In some cases, a spacer such as β-alanine is inserted between the GA molecule and the carrier protein.[6,8] The maximum molar ratio of GAs coupled to BSA by these methods is about 17:1.[8,15]

The hemisuccinate procedure takes advantage of a hydroxyl group on the GA molecule (Fig. 4).[14] A sterically nonhindered hydroxyl group on a GA molecule is easily converted to a hemisuccinate by treatment with succinic anhydride. The succinyl carboxyl group of the hemisuccinate can be selectively converted to an active ester or a mixed anhydride by treating the hemisuccinate with equimolar amounts of coupling reagents because the C-7 carboxyl group on GA is sterically hindered compared to the succinyl carboxyl group.

A third method is to introduce an additional functional group into GA as a hook for the coupling to a carrier protein. Beale et al.[15] prepared the 17-carboxyl and 17-thioalkylthiol derivatives (Figs. 5 and 6). Though details were not reported, the Peterson reaction was used in the preparation of the 17-carboxyl derivative of GA_4; in the preparation of the 17-thioalkylthiol derivatives of GA_4 and GA_9, the free-radical addition of $HS(CH_2)_nSH$, with azobisisobutyronitrile as an initiator, was used. The terminal SH group of the 17-thioalkylthiol derivatives was coupled to maleic anhydride, which was then coupled to the carrier protein.

Nakajima et al.[14] prepared 17-nor-GA_4-16-carboxymethoxime from GA_4 via GA_4-17-norketone (Fig. 7). Treatment of GA_4-17-norketone with carboxymethoxylamine in pyridine yielded the methoxime. The carboxyl group in the oxime moiety was selectively converted into the N-hydroxysuccinimidyl active ester by mixing with 0.9 molar equivalents of DCC and 0.9 molar equivalents of N-hydroxysuccinimide in DMF. The active ester was then conjugated to BSA.

3.2 Immunogenic Conjugates of C_{20} GAs

As C_{20} GAs contain two or three carboxyl groups, selective activation of a specific carboxyl group would be required if the group is to be used for the conjugation to a carrier protein. So far, there have been no reports on the preparation of immunogens using one of the carboxyl groups in C_{20} GAs for direct coupling to a carrier protein. In the preparation of immunogens of C_{20} GAs, suitable spacers have been used.

Kurogochi et al.[5] worked on the exocyclic methylene to introduce a hydroxyl group at C-17 by hydroboration of the dimethyl ester of GA_{24} by treatment with bis(3-methyl-2-butyl)borane, followed by addition of H_2O_2

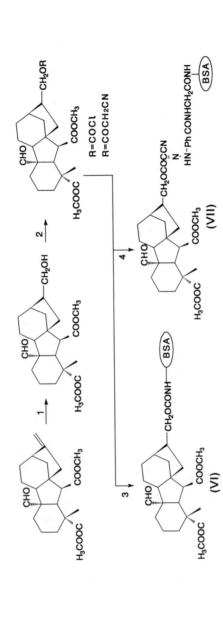

Fig. 8. Coupling of GA_{24}-Me to carrier protein via C-17[5]: 1. bis(3-methyl-2-butyl)borane, NaOH, H_2O_2; 2. Cl_2CO or $CH_2(CN)COCl$, $N(C_2H_5)_3$; 3. BSA; 4. $N_2C_6H_5CONHCH_2CONH$-BSA

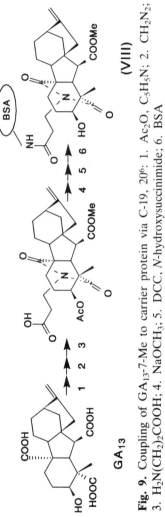

Fig. 9. Coupling of GA_{13}-7-Me to carrier protein via C-19, 20^6: 1. Ac_2O, C_5H_5N; 2. CH_2N_2; 3. $H_2N(CH_2)_2COOH$; 4. $NaOCH_3$; 5. DCC, *N*-hydroxysuccinimide; 6. BSA

(Fig. 8). The 16,17-dihydro-GA_{24}-Me-17-ol was converted to the chloroformate when treated with excess phosgene, and the chloroformate was coupled to BSA in the presence of a base. Alternatively, the 16,17-dihydro-GA_{24}-Me-17-ol was converted to the cyanoacetate by treatment with cyanoacetyl chloride and triethylamine. The cyanoacetate was coupled to the diazotized p-aminohippuric acid–BSA conjugate to form the immunogen.

Yamaguchi et al.[9] prepared an immunogen by taking advantage of two carboxyl groups (C-19 and C-20) of GA_{13} to form an imide ring. Treated with acetic anhydride, GA_{13} gave 3-acetyl-GA_{13}-19,20-anhydride. The anhydride was refluxed with β-alanine in acetic acid after monomethylation of the C-7 carboxyl group with diazomethane, and the imide was obtained (Fig. 9). The imide was conjugated to BSA by the active ester method using N-hydroxysuccinimide.

4 Preparation of Radiotracers and Enzyme Tracers

Radiotracers are usually prepared by introducing radioisotopes into GA molecules by chemical reactions such as catalytic tritiation using tritium gas and LiB^3H_4 reduction.

The methods used for the preparation of immunogens can be applied to the preparation of enzyme tracers for ELISA by coupling the active intermediates to enzymes, and they may also be applied to the preparation of radiotracers for RIA by coupling the active intermediates to radiolabeled amino acids with high specific activities. If a spacer is used for the preparation of an immunogen, sometimes a different spacer is used in the preparation of a tracer to avoid complications due to the presence of antibodies that recognize the spacer used in the immunogen.

For enzyme tracers, such stable enzymes as alkaline phosphatase, horseradish peroxidase, and β-galactosidase are most often used. Generally, the enzyme selected should be stable, be readily available at the highest sensitivity, withstand the coupling procedure without much loss in enzymatic activity, be compatible with the buffers used during the antigen–antibody reaction, and finally allow a simple and sensitive colorimetric assay of its activity. The enzyme tracer GA_1-alkaline phophatase prepared in our laboratory is stable for over five years when stored in Tris buffer saline (TBS)/glycerol (1:1) at $-20°C$.

Antibodies having high affinity for GA methyl esters permit the use of GA methyl esters carrying radioactive ester methyl groups as tracers. This is advantageous because the GA methyl ester tracer is easily prepared with high specific activity. Treated with [^{3}H]-labeled methyl iodide, sodium salts of GAs easily react to form GA methyl esters labeled at the ester methyl group in good yield.[5]

5 Titer Estimation for RIA

For RIA, antisera are used without further purification; for ELISA, the immunoglobulin fraction prepared by Hurn and Chantler's method[16] is often used.

In RIA, it is necessary to determine the titer of antiserum to be used. The titer is expressed as the final dilution of an antiserum in an assay mixture and is usually selected as the dilution that binds 30–50% of the amount (= radioactivity) of tracer added. As the titer is determined by the radioactivity, higher titer and, consequently, higher sensitivities are often obtained with tracers of higher specific activity.

The optimum condition of immobilized antibodies for ELISA is usually determined by coating the microtitration plates with various amounts of the immunoglobulin fraction and titrating against the dilution of the enzyme tracer. From the data, suitable concentrations of tracer dilutions and antibody dilutions can be obtained.

6 Relationship Between Specificities of Antibodies and Structures of Immunogens

Among the characteristics of antibodies, cross-reactivities are the most important factor for the immunoassay of GAs, because GAs are a big family, consisting of 79 members (as of 1989). Cross-reactivities are usually expressed as percent ratios of the molar amounts of GAs showing 50% inhibition of tracer–antibody binding, taking the molar amounts of the GA used for the tracer (standard GA) as 100%.

The cross-reactivities of polyclonal and monoclonal anti-GA antibodies reported so far are summarized in Tables 1a–c. When the antibodies are produced against immunogens in which GAs are coupled to carrier proteins via their carboxyl groups or in which their carboxyl groups are methylated, they show, in general, much less affinity for nonderivatized free carboxylic acid GAs than for GA methyl esters, and the antigen–antibody complex is not affected by pH in the range in which the antibodies are stable. On the other hand, antibodies prepared against immunogens in which the carboxyl groups in the GAs are not modified show high affinities for free acid GAs rather than for GA methyl esters, and the affinity is affected by pH. For example, in the case of one of the antibodies prepared in our laboratory, although the antibody is stable in the range of pH 3 to 8, the antigen–antibody complex is only stable in the range of pH 6 to 8; at pH values lower than 5.5, it is unstable and releases the tracer. This shows that the antibody recognizes the dissociated carboxylate but not the protonated associated form of the C-7 carboxyl group.

In general, the epitopes distant from the site of conjugation to the carrier

Table 1a. Cross-reactivities (%) of anti-GA antibodies prepared by Fuchs, Weiler, Atzorn, Eberle, and Nakajima

GA	Maker, Code,[a] and hapter[b]						
	Fuchs[1] HFP-1	Weiler[2] RWP-1	Atzorn[3,4]			Eberle[11] REM-3	Nakajima[7] TNP4-1
			RAP-1	RAP-2	RAP-3		
	GA_3 (C-7)	GA_3 (C-7)	$GA_{4/7}$ (C-7)	GA_1 (C-7)	GA_9 (C-7)	GA_{13}-7-Me (C-19, 20)	GA_4 (C-16)
GA_1	100						200.
GA_1-Me		12.9	13	100	<0.1	0.7	0.4
GA_3	100						16
GA_3-Me		100	<0.1	70	<0.1	76.4	
GA_4	0.4						100
GA_4-Me		6.3	100	40	3	100	
GA_5	100						
GA_5-Me		(<0.1)[c]	0.1	29	<0.1		
GA_7							58
GA_7	4						
GA_7-Me		(40)	80	70	1.1	764	
GA_8-Me		0	<0.1	11	<0.1	0	
GA_9	0.4						
GA_9-Me		0	<0.1	15	100	52.5	
GA_{12}-Me				<0.1	<0.1		
GA_{13}							14
GA_{13}-Me		0		<0.1	<0.1	14	
GA_{14}-Me						<0.3	
GA_{16}-Me		(<0.1)	<0.1	0.1	<0.1		
GA_{19}-Me		(0.1)	<0.1	<0.1	<0.1		
GA_{20}							4
GA_{20}-Me		(22)	1.5	55	<0.1	1.8	
GA_{24}-Me		(<0.1)	<0.1	<0.1	<0.1		
GA_{30}							14
GA_{32}-Me		(<0.1)	3.5	<0.1	<0.1		
GA_{34}-Me		(<0.1)	1.8	<0.1	<0.1		
GA_{35}							0.8
GA_{36}-Me						26.3	
GA_{37}							0.6
GA_{37}-Me						35.6	
GA_{42}-Me				<0.1	<0.1		
GA_{51}-Me						6.2	
GA_{53}-Me						4.2	
isoGA_3-Me				10	<0.1		
isoGA_7-Me				3	<0.1		

[a] Codes are temporarily assigned in this table
[b] Positions used for the conjugation to carrier proteins are given in parentheses
[c] The values in paretheses in this column are from ref. 3

Table 1b. Cross-reactivities (%) of anti-GA antibodies prepared by Kurogochi and Yamaguchi

GA	Maker, code,[a] and hapter[b]				
	Kurogochi[5]		Yamaguchi[6,8]		
	RKP-422 GA_{24}-Me (C-17)	RKP-423 GA_{24}-Me (C-17)	TYP5-1 GA_5 (C-7)	TYP20-1 GA_{20} (C-7)	TYP1-1 GA_1 (C-7)
GA_1					0.1
GA_1-Me	0	0	1.9	0.2	100
GA_2-Me	0.1	0			
GA_3-Me	0	0	0.89	0.05	22.5
GA_4-Me	0	0	0.25	0.09	48.1
GA_5-Me	0	0	100	17.5	0.1
GA_6-Me	0.1	0			
GA_7-Me	0	0.1	0.11	0.04	16.4
GA_8-Me	0.2	0.1	0.45	0.12	0.3
GA_9-ME	0.1	0.2	40.0	22.7	0.7
GA_{10}-Me	0	0.1			
GA_{12}-Me	24	26	<0.01		
GA_{13}-Me			<0.01	<0.04	<0.1
GA_{14}-Me			<0.01	<0.04	
GA_{15}-Me	0.2	0.2	0.16	<0.04	
GA_{16}-Me	0	0.2			0.2
GA_{17}-Me	0	0	<0.01	<0.04	<0.01
GA_{19}-Me	4.3	62	0.45		
GA_{20}-Me	0	0	47.8	100	1.0
GA_{21}-Me	0.1	0.1			
GA_{22}-Me	0	0.1			
GA_{23}-Me	0.1	0.1			
GA_{24}-Me	100	100	0.2	0.25	
GA_{27}-Me	0	0.2			
GA_{28}-Me	1.1	0.9			
GA_{30}-Me	0	0.1			0.1
GA_{31}-Me	0.1	0.2			
GA_{33}-Me	0	0			
GA_{34}-Me	0	0	0.08	<0.04	
GA_{35}-Me	0	0.2			0.5
GA_{36}-Me	2.5	0.4	0.03	<0.04	
GA_{37}-Me			<0.01	<0.04	0.6
GA_{38}-Me	0.3	0			
GA_{39}-Me	0.5	1.6			
GA_{40}-Me	0.1	0.1			
GA_{41}-Me	0	0.4			
GA_{44}-Me	0.1	0	<0.2	<0.2	
GA_{47}-Me	0	0			
GA_{53}-Me	0.1	54	<0.01	<0.04	
GA_1-GEs					0.11
GA_4-GEs					0.07

[a] Codes are temporarily assigned in this table
[b] Position used for the conjugation to carrier proteins are given in parentheses
[c] GA-GEs: GA-glucosyl esters

Table 1c. Cross-reactivities of anti-GA antibodies prepared by Knox[12,13]

GA	Code[a] and Hapten[b]							
	AFRC MAC-136 GA$_1$ (C-3)	-137 GA$_1$ (C-3)	-175 GA$_9$ (C-17)	-176 GA$_9$ (C-17)	-182 GA$_4$ (C-17)	-183 GA$_4$ (C-17)	-213 GA$_4$ (C-3)	-214 GA$_4$ (C-3)
GA$_1$	100	100	<0.02	6	48	50	0.1	0.3
GA$_1$-Me	0.1	0.1	<0.04		0.05			
GA$_2$	0.07	0.08	<0.04	30	50	24	1	0.02
GA$_3$	100	100	<0.02	1	6	4	0.1	0.1
GA$_4$	0.4	0.	0.1	100	100	100	100	100
GA$_4$-Me					0.16			
GA$_5$	48	67	3	2	0.8	2	0.3	0.5
GA$_6$	80	100	0.3	0.8	0.01	0.07	<0.06	8
GA$_7$	0.3	0.4	0.07	11	9	15	8	10
GA$_8$	53	80	<0.02	0.06	0.03	0.2	0.02	0.5
GA$_9$	0.2	0.5	100	100	0.9	4	25	24
GA$_9$-Me					<0.3			
GA$_{10}$	<0.09		26	20	0.9			
GA$_{12}$							84	0.2
GA$_{13}$							<0.5	<0.01
GA$_{14}$	0.7		<0.04	0.	0.01	<0.02	77	0.02
GA$_{15}$							7	0.43
GA$_{18}$	115	100	<0.04		0.05			
GA$_{20}$	100	100	16	9	0.5	3	0.04	0.05
GA$_{29}$	100	100	<0.5	<0.2	0.02		<0.2	1
GA$_{36}$							5	0.75
GA$_{37}$							26	0.5
GA$_{45}$	0.05	0.05	100	40	0.3		5	0.5
GA$_{51}$							13	29
GA$_{53}$	167	180	0.5		0.03			
GA$_{60}$	4	4	0.3		0.02			
GA$_{61}$	0.04	0.03	4	4	0.1	2	<0.2	
GA$_{63}$	0.05		0.07	27	11	89	67	
GA$_{67}$	0.05	0.04	6	2	0.05		7	3
GA$_{68}$	0.05		0.05	4	0.06			
GA$_{72}$	0.1	0.07	<0.04		3		0.05	<0.05
1β-MeGA$_4$	<0.05		0.05	130	245	136	33	4

[a] Codes are temporarily assigned in this table
[b] Position used for the conjugation to carrier proteins are given in parentheses

protein tend to be well recognized by antibodies, while the epitopes neighboring the coupling site tend to be less well recognized, but this generalization does not always hold true. For example, the antibody produced by the monoclonal cell line AFRC MAC 182 (Table 1c), which was produced against an immunogen (Fig. 6, **IV**: R = OH) containing 17-substituted GA$_4$, showed very low cross-reactivities with GA$_9$ and GA$_{20}$ because of the lack of a 3β-hydroxyl group in these two GAs, but exhibited higher cross-reactivity with 1β-methyl-GA_4 than with GA_4 in spite of the presence of an extra methyl group on C-1. Judging from the cross-reactivities, the antibodies seem to recognize parts of the GA molecule rather than the whole structure of the hapten. For example, antibodies with a high specificity for the A-ring structure of the GA molecule discriminate little between GAs with modifications in the C and D rings and especially at C-13. Antibodies raised against immunogens in which GAs are coupled via C-7 carboxyl groups tend to show a high specificity in recognition of the A ring. The existence and position of hydroxyl group(s) on the A ring and the difference between γ-lactone and σ-lactone are sharply discriminated; in the recognition of the 13-hydroxyl group, selectivity is poor.

When the cross-reactivities of polyclonal antibodies and monoclonal antibodies are compared, the polyclonal antibodies obtained from rabbits seem to show reasonably high specificities compared to the monoclonal antibodies obtained from mice or rats, though there is no example in which polyclonal and monoclonal antibodies were raised against the same immunogen.

The above generalizations are based on the specificities of available anti-GA antibodies (Tables 1a–c). The specificities are known to vary from animal to animal. In the case of monoclonal antibodies, it is possible to select a variety of cell lines producing antibodies with different specificities. Therefore, it is not reasonable to carry too far generalizations about the relationship between the specificities and structures of immunogens. It must be remembered that unexpected cross-reactions might occur in both polyclonal and monoclonal antibodies with compounds beyond the selected and/or available compounds for the specificity study.

7 Validation of Immunoassays for GAs

As the validation of immunoassays for plant hormones is described by Pengelly in *Plant Growth Substances 1985*,[17] it is not discussed in detail in this review.

As shown in Table 2, Yamaguchi et al.[8] checked the difference in quantification results from RIA of GA_5 and GA_{20} in immature seeds of *Pharbitis nil* at different purification stages and compared the results with those obtained by the reference method of GC–MS using deuterated internal standards. In the RIA of crude samples, overestimation occurred and could not be explained only by cross-reactivities among GAs. Thus, it is clear that suitable prepurification is required in order to obtain reliable results by immunoassay. The combination of high-performance liquid chromatography (HPLC) and immunoassay using highly specific antibodies is very effective but still not definitive for the assay of individual GAs within the family of 79 members (in 1989). Therefore, the immunoassay for GAs is expected to display its power of high sensitivity and reason-

Table 2. Quantification of GA_5 and GA_{20} in 16-day-old immature seeds of *P. nil* (ng/g fr wt)[8]

Purification step	RIA		GC–SIM	
	GA_5 equiv.	GA_{20} equiv.	GA_5	GA_{20}
AE fr.	527.9 ± 42.3	104.2 ± 14.2		
GPC $(A_5 + A_{20})$ fr.	92.7 ± 6.4	15.1 ± 2.8		
NMe_2				
A_5 fr.	21.4 ± 3.9		21.8 ± 2.6	
A_{20} fr.		21.9 ± 3.9		26.0 ± 1.9

AE, Acidic ethyl acetate; GPC, gel permeation chromatography (Shodex HF-2001); NMe_2, high-performance liquid chromatography (Nucleosil NMe_2)

able selectivity in studies of the fluctuation and localization of GAs in plants whose endogenous GAs have been identified previously by GC–MS.

8 Conclusions

Immunoassay for GAs has been proved to be a very effective technique with high sensitivity and reasonable selectivity if it is applied to the analysis of GAs with proper controls.

Another immunological technique using anti-GA antibodies is immunoaffinity purification. Immunoaffinity purifications of GAs (or GA methyl esters) were reported by Fuchs and German,[18] Durley et al.,[19] and Smith and MacMillan.[20] Immunoaffinity purification using antibodies with broad cross-reactivities or using a combination of several specific antibodies is expected to be a powerful and effective tool for the ultramicroanalysis of GAs in plant organs.[20]

Organ-specific production of GAs and the possibility that different GAs may be involved in the regulation of different physiological phenomena in plant growth have been suggested (see chapters 2–4 in this volume). It will therefore become important to study the localization of specific GAs in plant organs. If efficient methods for the fixation of GAs in plant tissues can be developed, immunohistochemistry and immunocytochemistry using specific antibodies will also become possible. It is expected that these techniques will give much information about the roles of GAs in plants.

References

1. Fuchs S, Fuchs Y. Immunological assay for plant hormones using specific antibodies to indoleacetic acid and gibberellic acid. Biochim Biophys Acta. 1969; 192:528–530.

2. Weiler EW, Wieczorek U. Determination of femtomol quantities of gibberellic acid by radioimmunoassay. Planta. 1981; 152:159–167.
3. Atzorn R, Weiler EW. The immunoassay of gibberellins. II. Quantification of GA_3, GA_4 and GA_7 by ultra-sensitive solid-phase enzyme immunoassays. Planta. 1983; 159:7–11.
4. Atzorn R, Weiler EW. The immunoassay of gibberellins. I. Radioimmunoassay for the gibberellins A_1, A_3, A_4, A_7, A_9 and A_{20}. Planta. 1983; 159:1–6.
5. Kurogochi S, Yamaguchi I, Murofushi N, et al. A radioimmunoassay for the dimethyl esters of GA_{24} and GA_{19}. Phytochemistry. 1987; 26:2895–2900.
6. Yamaguchi I, Nakazawa H, Nakagawa R, et al. submitted to Plant Cell Physiol 1990.
7. Nakajima M, Yamaguchi I, Kizawa S, et al. submitted to Plant Cell Physiol 1990.
8. Yamaguchi I, Nakagawa R, Kurogochi S, et al. Radioimmunoassay of gibberellins A_5 and A_{20}. Plant Cell Physiol. 1987; 28:815–824.
9. Yamaguchi I, Nakagawa R, Nakazawa H, et al. to be submitted to Agric. Biol. Chem. 1990.
10. Oden PC, Heide, OM. Detection and identification of gibberellins in extracts of *Begonia* leaves by bioassay, radioimmunoassay and gas chromatography–mass spectrometry. Physiol Plant. 1988; 73:445–450.
11. Eberle J, Yamaguchi I, Nakagawa R, et al. Monoclonal antibodies against GA_{13}-imide recognize the endogenous plant growth regulator, GA_4, and related gibberellins. FEBS Lett. 1986; 202:27–31.
12. Knox JP, Beale MH, Butcher GW, et al. Preparation and characterization of monoclonal antibodies which recognise different gibberellin epitopes. Planta. 1987; 170:86–91.
13. Knox JP, Beale MH, Butcher GW, et al. Monoclonal antibodies to 13-deoxygibberellins. Plant Physiol. 1988; 88:959–960.
14. Nakajima M, Yamaguchi I, Nagatani A, et al. submitted to Plant Cell Physiol. 1990.
15. Beale MH, Hooley R, MacMillan J. Gibberellins: Structure–activity relationship and the design of molecular probes. In: Bopp M, ed. Plant growth substances 1985. Heidelberg: Springer-Verlag, 1986: p. 65–73.
16. Hurn BAL, Chantler SM. Production of reagent antibodies. In: Vunakis HV, Langone JL, ed. Methods in enzymology, Vol. 70, Immunological techniques, Part A. London Academic Press, 1980: p. 105–142.
17. Pengelly WL. Validation of immunoassays. In: Bopp M, ed. Plant growth substances 1985. Heidelberg: Springer-Verlag, 1986: p. 35–43.
18. Fuchs Y, German E. Insoluble antibody column for isolation and quantitative determination of gibberellins. Plant Cell Physiol. 1974; 15:629–633.
19. Durley RC, Sharp CR, Maki SL, et al. Immunoaffinity techniques applied to the purification of gibberellins from plant extracts. Plant Physiol. 1989; 90:445–451.
20. Smith VA, MacMillan J. An immunological approach to gibberellin purification and quantification. Plant Physiol. 1989; 90:1148–1155.

CHAPTER 16

Physiology of Gibberellins in Relation to Floral Initiation and Early Floral Differentiation

R.P. Pharis

1 Introduction

The involvement of gibberellins (GAs) in flowering of a wide variety of higher plants has been reviewed by Zeevaart,[1] Pharis and King,[2] Bernier,[3] Looney and Pharis,[4] and Pharis et al.[5] In the six-volume *Handbook of Flowering* edited by Halevy,[6] most chapters detail effects, or lack thereof, of exogenously applied GAs on flowering. For conifers, reviews by Hashizume,[7] Dunberg and Oden,[8] Pharis and Kuo,[9] Pharis and King,[2] Nagao et al.,[10,11] Pharis and Ross,[12] and Pharis et al.[13] provide ample coverage of both applied and more basic approaches to the flowering physiology of this group of plants.

In the present paper, I will attempt to update the reader as to the possible causal role of less polar GAs in the flowering of conifers. For woody angiosperms, the negative roles (inhibition of flowering) attributed to GA_3 and GA_7 will be discussed, relative to the promotion of flowering observed with the use of GA_4. Finally, I will discuss relatively recent results which indicate that a GA known to be an "effector" in vegetative shoot elongation[14] (i.e., GA_1) has very weak florigenic activity, even at high doses in several higher plants. In contrast, GAs of a quite different structure (a C=C double bond in ring A, with hydroxyl groups at any or all of C-13, C-15β, and C-12) can be highly florigenic at very low doses, while having a minimal promotive effect on vegetative shoot elongation. The thesis that GAs with unique structural attributes are highly florigenic seems especially attractive for the long-day-requiring (LD) plant *Lolium temulentum*. However, it also appears relevant for at least some short-day-requiring (SD) plants. Evidence based on exogenous GA application and on analyses of endogenous GAs will be discussed in relation to this thesis.

2 Gibberellins and the Flowering of Conifers

The first evidence that GAs might be involved in strobilus (conebud) initiation/differentiation was provided by Kato et al.[15] and Hashizume.[16]

Fig. 1. Giant redwood [*Sequoia gigantea* (Lindl) Decne.], age 8 months, showing an ovulate strobilus which became visible at age 6 months after 16 weekly applications of GA_{13} at 125 μg per application (applied under long-day warm). Gibberellin A_3 was appreciably more effective than GA_{13}. This flowering response to application of bioactive (on vegetative growth assays) GAs is characteristic of virtually all members of the Cupressaceae and Taxodiaceae that have been tested, but does not generally occur for members of the Pinaceae (where only the less polar GAs are highly effective). The range of dose response can vary appreciably, however, between species—5 to 170 μg of GA_3 per strobilus for the redwoods, compared to as little as 0.002 μg per strobilus for Arizona cypress (*Cupressus arizonica*). A different photograph of the same plant was first published in Pharis and Morf[27]

Over the next fifteen years, Hashizume, Nagao, and Migita were notable among the many Japanese scientists who worked on both applied and basic aspects of GA-promoted flowering in conifers, especially in the Cupressaceae and Taxodiaceae (see Hashizume,[7] Nagao et al.,[10,11] and references cited therein). My interest in the physiology of flowering in conifers began in 1963 and continues to this day.

Graphic examples of the promotion of flowering in a Taxodiaceae species (giant redwood) and a Pinaceae species (white spruce) are shown in

Fig. 2. An example of the profuse female flowering that can occur when a $GA_{4/7}$ mixture (ca. 1:1) is used to treat grafted propagules of white spruce [*Picea glauca* (Moench) Voss] grown in containers in a heated plastic house. Female strobili are shown at the receptive stage the year following treatment with the hormone. Photograph courtesy of Dr. S. Ross, Research Division, British Columbia Forest Service, Victoria, B.C., Canada

Figs. 1 and 2. This success in manipulating flowering through application of GAs has allowed tree geneticists to speed up breeding programs and, more recently, is being used to promote seed production in seed orchards established with genetically superior propagules and families (cited in ref. 13).

It is generally correct that conifers in the Taxodiaceae and Cupressaceae will respond to most (if not all) bioactive GAs by initiating conebuds, whereas conifers in the Pinaceae generally respond only to application of the less polar GAs, most notably the $GA_{4/7}$ mixture and $GA_{4/7} + GA_9$ (a treatment which is synergistic) (Fig. 3). Although many conifer species respond to daylength, especially with regard to sex expression (cited in refs. 10, 11, and 17), application of a GA (or being subjected to sufficient "stress") is necessary to invoke the flowering response. Included in the

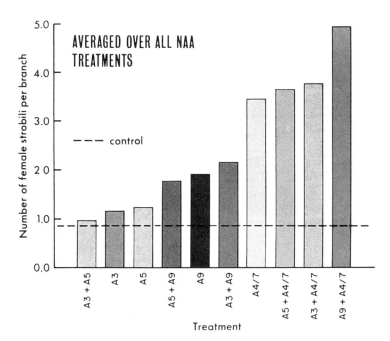

Fig. 3. Effects of gibberellins A_3, $A_{4/7}$ mixture (ca. 30:70), A_5, A_9, and their combinations on female flowering in 6-year-old seedlings of Douglas fir [*Pseudotsuga menziesii* (Mirb.) Franco]. The hormone treatments were applied 8 times between 12 April and 20 June, the period when strobilus differentiation is known to occur; thus, the hormones influenced the differentiation of preexisting primordia that were initiated prior to hormone treatment. Combination treatments comprised a total of 3.2 mg of GAs per branch, single GA treatments comprised 1.6 mg per branch. Results were averaged over all naphthaleneacetic acid (NAA) concentrations (0, 175, and 875 µg per branch). The $GA_{4/7}$ + GA_9 treatment differed significantly ($P = 0.05$) from all other treatments. The $GA_{4/7}$ treatment and combinations with GA_5 and GA_3 differed significantly ($P = 0.05$) from all other treatments, including ethanol controls (dashed horizontal line). Data adapted from Pharis et al[28]

term "stress" would be day-to-day living for an old, and generally large, tree (e.g., a plant which has reached the "ripeness to flower" stage[18]). In this latter respect, woody angiosperms and conifers share the responses—namely, stress will promote profuse flowering.

Thus, in researching the mechanism by which conifers flower in nature, the most useful manipulative tools (in addition to application of GAs) have been the judicious use of stress treatments, including rootpruning, girdling, drought, high temperature, and low N nutrition (see refs. 5, 10, 11, 13, and 16) and references cited therein). By using such stress treatments and examining the endogenous GAs and GA-like substances and the metabolism

Table 1. Concentrations of GAs in potential female primordia (excised on 7 July from the distal end of an elongated shoot) of 10-year-old Douglas fir (*Pseudotsuga menziesii*) seedlings (2 to 4 m in height)

Treatment	GA concentration (ng/g dry wt tissue)					% Reproductive conebuds 1982
	GA_1	GA_3	GA_4	GA_7	GA_9	
Control	114	93	188	16	210	0
Rootpruned	59	83	83	831	676	40
$GA_{4/7}$-treated	31	48	140	188	—[a]	0
Rootpruned + $GA_{4/7}$	190	249	13,472[b]	1264[c]	502	75

The trees had been given no treatment (Control), rootpruned (mid–late April, 1981), treated with $GA_{4/7}$ mixture (stem injection for 7 weeks, beginning 22 May with uptake of ~700 mg $GA_{4/7}$ per tree), or given rootpruning plus $GA_{4/7}$ (see refs. 25 and 26 for specific treatment details). Quantitation of GAs was accomplished by GC–MS–SIM using [$^2H_2$] GAs as internal standards. All GAs in the table are native to Douglas fir. Flowering results (1982) are noted for residual branches (not harvested for GA analyses) on the same trees and are expressed as percentage of lateral buds that were reproductive

Source: P. Doumas, R. Pharis, G. Thompson, J. N. Owens, S. Ross, J. Webber, and K. Takeno, unpublished results

[a] Could not be estimated because of overlap with contaminating ions
[b] Range on other harvest dates was 600 to 3000 ng/g dry wt
[c] Range on other harvest dates was 3600 to 4000 ng/g dry wt

of [^3H]GA_4, a number of workers have concluded that GAs and GA-like substances of a less polar nature build up in shoots of trees subjected to a floral-promotive stress [see Table 1 in Pharis et al.,[5] references cited therein, and references cited in Pharis et al.[13]).

Recent work using gas chromatography–mass spectrometry–selected ion monitoring (GC–MS–SIM), with quantitation effected by the use of ^2H-labeled GAs as internal standards, is confirmatory of the earlier bioassay and metabolic work (see Table 1). Differentiating primordia (next year's potentially female conebuds differentiate from previously initiated, but undetermined lateral primordia) of Douglas fir have appreciably enhanced GA_9 and GA_7 concentrations if the trees were stressed through rootpruning (Table 1), and this trend is extended to GA_4 when the $GA_{4/7}$ mixture is applied to rootpruned trees (Table 1). The trend shown in Table 1 was repeatable with other replicate samples (harvested in subsequent weeks) and in the shoot tissue adjacent to the differentiating primordia (P.R. Doumas, Pharis, G. Thompson, S. Ross, J. Webber, and J. Owens, unpublished research results).

Interestingly, in the Cupressaceae family, Arizona cypress (*Cupressus arizonica*), which is most responsive to the polar GA_3 in terms of flowering, has a buildup of less polar GA-like substances in the shoot tissues when caused to flower by low N nutrition (C.G. Kuo and R.P. Pharis, unpublished, cited in Pharis et al.[13]). Additionally, there is a massive increase in GA_7 in the apices and immediately surrounding tissue (analysis by GC–SIM) when Arizona cypress plants are caused to flower by being "pot-bound" and subjected to water stress (G. Smith, D. Pearce, and R. Pharis, unpublished results).

How the high concentrations of less polar GAs in the potential conebuds encourage sexual differentiation of the primordium remains unknown, but the mechanism may involve both increased assimilate diversion (see ref. 19 and Table III in ref. 5) and a separate morphogenic effect over and above that of assimilate diversion (e.g., only $GA_{4/7}$ promotes conebud differentiation in *Pinus radiata*, but both GA_3 and $GA_{4/7}$ promote mobilization of ^{14}C assimilate to potential female conebuds[19]).

To me, at least, the evidence in hand points toward a natural control of flowering in the woody perennial conifers by increased concentrations of one or more of the native less polar GAs, most likely, GA_7 and GA_9. This conclusion is substantiated by a number of physiological studies (mainly with Pinaceae species; cited above) and by a large number of examples[5,13] in the Pinaceae in which only less polar GAs (most notably the $GA_{4/7}$ mixture) promote flowering, with or without adjunct stress treatments. Whether the high efficacy of GA_3 in Cupressaceae and Taxodiaceae species is merely serendipitous or points to a different mechanism of control is presently unresolved.

3 Gibberellins and the Flowering of Woody Angiosperms

Although stresses similar to those that promote flowering in conifers also promote precocious and enhanced flowering in woody angiosperms, we do not have a long "history" in woody angiosperms of promotion of flowering by exogenously applied GAs as we do for the conifers. Quite the contrary, applied GA_3 and $GA_{4/7}$ (especially) have been shown to inhibit next year's flowering in fruit trees and other woody angiosperms (refs. 2, 4, and 20 and references cited therein). There is also no substantial body of evidence from analysis of endogenous GA-like substances showing a positive correlation between GAs and a high level of flowering in woody angiosperms.

There are, however, some reports in the literature that applied GAs (even GA_3) can promote flowering in woody angiosperms (cited in ref. 20), and one example of this, using GA_4, has been patented for use in apple[21] and is detailed in refs. 4, 20, and 5. Pharis and King[2] (reiterated in Pharis et al.[5]) speculate that the highly bioactive and/or persistent GAs (e.g., GA_3, GA_7) may reduce the number of nodes on the spur, thereby lengthening the plastochron and suppressing flower bud formation on the older organ primordia.

The promotive effect of GA_4 on next year's flowering in apple may thus be due to enhancement of early floral differentiation in previously induced apices of at least some spurs. Rapid metabolism of GA_4 may prevent it from acting in the inhibitory manner exhibited by GA_3 and GA_7 (both $\Delta^{1,2}$-GAs).[2,5]

Unfortunately, any role for GAs in flowering (initiation and/or early differentiation) of woody angiosperms must be assigned on the basis of response to application and bioassay results. To date, one would have to assign a negative role based on the published evidence. However, that negative role should perhaps be restricted to those GAs which are highly bioactive at low doses in vegetative growth promotion, and probably relatively persistent (e.g., GA_3, GA_7 type). The negative results with GA_3 and GA_7 (the latter is usually applied as a $GA_{4/7}$ mixture[2,4,20]), when contrasted with the promotive effects of GA_4 in apple,[4,21] imply that differing GA structures may yield very different responses (inhibition or promotion) on flowering in woody angiosperms.

4 Gibberellins and the Flowering of Short-Day Plants

Although there is some history of GA (GA_5, GA_{13}, and even GA_3 have been effective) promotion of flowering under noninductive SD (cited in refs. 1–3, 5, and 22), most of the highly responsive systems were quantitative SD plants. However, the ability of applied GAs to promote flowering in qualitative SD plants under marginally inductive long nights, especially in plants where endogenous GAs are reduced, has been more extensively researched (see ref. 22 and references therein). An example of the response in a near-isogenic dwarf line of *Pharbitis nil* (relative to that of a tall line) is shown in Fig. 4. As noted in King et al.,[22] a critical factor is when the GA is applied (for example, if it is applied after the inductive long night, flowering is inhibited). Dose is also critical, especially for GAs (such as GA_3) that are likely to be "long-lived" (again, too high a dose of such a GA applied too near the beginning of the marginally inductive long night will inhibit flowering[22]).

There are several intriguing aspects to the results shown in Fig. 4. First, there is no promotion of flowering in the tall line [in this trial; in some trials with cv. Violet, a promotion could be obtained [R. King, unpublished)]. The tall line presumably has near-optimal (for flowering purposes) endogenous GA concentrations. However, the applied GA_3 does promote stem elongation in the tall line at a dose as low as 0.02 µg per plant. Second, there is a significant promotion of flowering in the dwarf line of *Pharbitis* (which presumably has reduced endogenous GA concentrations since it is derived from a GA-responding dwarf, cv. Kidachi[23]), and this promotion occurs at a dose (0.02 µg per plant) that has no significant effect on stem elongation of the dwarf line. Higher concentrations of GA_3 (2 to 20+ µg per plant) give an inhibition of flowering, and those are the only doses which yield significantly increased vegetative elongation.

16. Roles of GAs in Floral Initiation   173

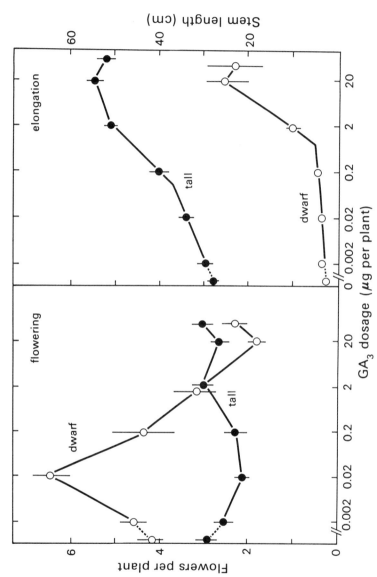

Fig. 4. Effect of the dosage of GA$_3$ on flowering and stem elongation (measured 10 days after GA application) of near-isogenic tall/dwarf lines of *Pharbitis*. The GA$_3$ was applied to the petiole in 95% ethanol 5 h prior to beginning the inductive dark period of 13.25 h, which began about 130 h after seed sowing. Printed with permission from King et al.[22]

Thus, the amount of GA_3 at a dose that significantly promotes flowering in the dwarf *Pharbitis* is approximately 100- to 1000-fold less than the amount at a dose that will significantly promote stem elongation. This differential florigenic:vegetative efficacy based on dose is very analogous to results discussed below for LD plants.

However, it must be kept in mind that *Pharbitis nil* and other qualitative SD plants have not, thus far, been induced to flower by applied GAs under totally noninductive photoperiods. Nonetheless, results by F. Lona, in 1972 (cited in refs. 1 and 24), indicated that both *Xanthium* and *Perilla* maintained under noninductive LD* could be induced to flower with an application of a crude acidic extract of developing peach seeds (which are known to be high in GA_{32}). In the context of the results noted below for LD plants, the possibility that a highly florigenic GA also exists for qualitative SD plants cannot be ruled out. However, the crude peach seed extract undoubtedly contained hundreds of other substances, some of which may have inhibited stem elongation and/or synergized any promotive effect caused by any GA_{32} that was present in the crude extract.

5 Gibberellins and the Flowering of Long-Day Plants

The earlier relevant work has been reviewed by Zeevaart,[1] Pharis and King,[2] and Bernier.[3] More recent experimental work, using *Lolium temulentum*, strain Ceres (which will flower in response to 1 LD of 16 or 24-h duration), is detailed in Pharis et al.,[5,24] and the work is ongoing (L. Evans, R. King, L. Mander, and R. Pharis, "*Planta*, in press 1990"). Using the *Lolium* system, a working hypothesis has been developed (detailed in ref. 5) which postulates "a specificity of GA structure in floral induction/initiation that differs from that in the promotion of shoot elongation," namely, that the presence of a double bond in ring A, and increasing hydroxylation, especially at C-13, C-15β, and C-12, yields increased florigenic activity, relative to any effect on stem elongation. An example of this is shown in Fig. 5 for GA_1, GA_3, and 15β-hydroxy-GA_3, all relative to a 24-h inductive LD.

Not only does addition of a double bond in ring A or increasing hydroxylation increase florigenicity, but it also lowers the threshold dose at which a flowering response can be obtained (Figs. 5 and 6). Thus, the highly florigenic 15β-hydroxy-GA_3 and GA_{32} are approximately 100- to 1000-fold more effective than GA_1, the known effector for shoot elongation in several higher plants, including maize.[14]

Evidence that these effects (Figs. 5 and 6) are not just pharmacological has been given in Pharis et al.[5,24] for endogenous GA-like substances in apices of LD-induced *Lolium* plants, relative to plants maintained under

---

*See note added in proof on page 178

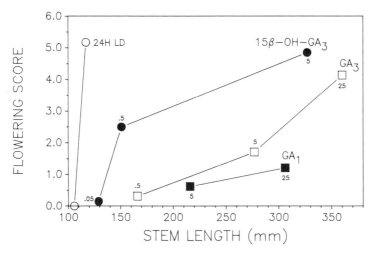

Fig. 5. Relative effect of two $\Delta^{1,2}$-GAs with two or three hydroxyl groups (GA_3 and 15β-hydroxy-GA_3) and of GA_1 (1,2-dihydro-GA_3) on flowering score and vegetative growth response (stem length) of *Lolium temulentum* plants. Flowering was assessed 3 weeks after inductive treatment when apices were dissected out (2 = double ridges present, 4 = glume primordia present, 5 = lemma present). Plants treated with GAs were held in noninductive SD at 25/20°C, as was the SD control (○; flowering score 0.0). The LD control was exposed to one long day of 24 h (○) duration. Gibberellin A_1 (■), GA_3 (□), and 15β-hydroxy-GA_3 (●) were applied to a single leaf at the doses (per plant) indicated. Adapted from Pharis et al.[5]

noninductive SD. Specifically, overall GA concentrations in the apices, over a broad range of polarity [based on reverse-phase C_{18} high-performance liquid chromatography (HPLC) retention time], are increased within 24–48 h of the inductive LD,[5,24] and this is especially notable for GA-like substances that elute at or near the retention time of GA_{32} (which has four hydroxyl groups). It was this finding (increased putative polyhydroxylated GA-like substances in the apices) that led us to test GA_{32} (ref. 24 and Fig. 6). Recent, unpublished work with induced leaves and phloem exudates of leaves is confirmatory of the changes alluded to above (R. Pharis, D. Pearce, S. Kim, L. Evans, R. King, and L. Mander, unpublished).

The work on *Lolium* continues, and extension to other LD plant systems (*Brassica*) is in progress, as are probes with several SD flowering systems. The very divergent effects of GA_1 and GA_{32}, for example, on flowering versus stem elongation in *Lolium* (Fig. 6), coupled with trends we have seen in endogenous GA changes after LD induction, imply, to me at least, that the structural requirements (at optimal GA concentrations) for floral induction/initiation may well be different from those for vegetative stem elongation.

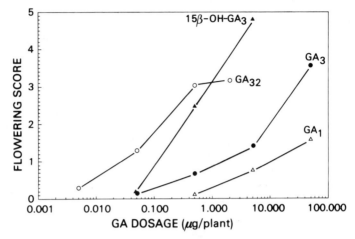

Fig. 6. Effect of dose of three $\Delta^{1,2}$-GAs with two, three, or four hydroxyl groups (GA_3, 15β-hydroxy-GA_3, GA_{32}) and of GA_1 (1,2-dihydro-GA_3) on flowering response of *Lolium temulentum* plants. Flowering was assessed 3 weeks after inductive treatment when apices were dissected out (2 = double ridges present, 4 = glume primordia present, 5 = lemma present). Plants were held in noninductive SD at 25/20°C. Adapted from Pharis et al.[5]

6 Conclusions

In a very diverse range of plant species, GAs have been implicated in flowering, not only in the stem elongation (bolting) that often follows floral initiation, but in the induction/initiation processes per se. Intriguingly, GA structures and doses that are highly florigenic often have nil or minimal effect on stem elongation. Conversely, for GAs known to be "effectors" of stem elongation, very high doses are often required to obtain an optimal flowering response.

The ability of chemists to synthesize GAs of divergent structures (L.N. Mander, personal communication) in amounts sufficient to test, together with the analytical capability of GC–MS and the availability of both radioactive- and stable-isotope-labeled GAs, may allow us to reopen this Pandora's box with regard to the possible causal role(s) of native plant GAs in floral evocation/early differentiation/later development.

References

1. Zeevaart JAD. Gibberellins and flowering. In: Crozier A, ed. The biochemistry and physiology of gibberellins. New York: Praeger, 1983: p. 333–374.
2. Pharis RP, King RW. Gibberellins and reproductive development in seed plants. Ann Rev Plant Physiol. 1985; 36:517–568.

3. Bernier G. The control of floral evocation and morphogenesis. Ann Rev Plant Physiol. 1988; 39:175–219.
4. Looney NE, Pharis RP. Gibberellins and reproductive development of tree fruits and grapes. Acta Hort. 1985; 179:59–72.
5. Pharis RP, Evans LT, King RW, et al. Gibberellins and flowering in higher plants—differing structures yield highly specific effects. In: Bernier G, Lord E, eds. Plant reproduction: From floral induction to pollination. American Society of Plant Physiologists Symposium Series, Vol. 1. 1989; p. 29–41.
6. Halevy AH, ed. Handbook of flowering, Vols. I–VI. Boca Raton, Florida: CRC Press, 1985–1989.
7. Hashizume H. The effct of gibberellin upon flower formation in *Cryptomeria japonica*. J Jpn For Soc. 1959; 41:375–381.
8. Dunberg A, Oden PC. Gibberellins and conifers. In: Crozier A, ed. The Biochemistry and physiology of gibberellins, Vol. 2. New York: Praeger, 1983: p. 211–296.
9. Pharis RP, Kuo CG. Physiology of gibberellins in conifers. Can J For Res. 1977; 7:299–325.
10. Nagao A, Sasaki S, Pharis RP. Flowering of *Cryptomeria* (*Cryptomeria japonica*). In: Halevy AH, ed. Handbook of flowering, Vol. 6. Boca Raton, Florida: CRC Press, 1989: p. 247–269.
11. Nagao A, Sasaki S, Pharis RP. Flowering of *Chamaecyparis*. In: Halevy AH, ed. Handbook of flowering, Boca Raton, Florida: CRC Press, 1989: p. 170–188.
12. Pharis RP, Ross SD. Hormonal promotion of flowering in *Pinaceae* family conifers. In: Halevy AH, ed. Handbook of flowering, Vol. 5. Boca Raton, Florida: CRC Press, 1986: p. 269–286.
13. Pharis RP, Webber JE, Ross SD. The promotion in forest trees by gibberellin $A_{4/7}$ and cultural treatments: A review of the possible mechanisms. For Ecol Manage. 1987; 19:65–84.
14. MacMillan J, Phinney BO. Biochemical genetics and the regulation of stem elongation by gibberellins. In: Cosgrove DJ, Knievel DP, eds. Physiology of cell expansion during plant growth. Bethesda, Maryland: The American Society of Plant Physiologists, 1987: p. 156–171.
15. Kato Y, Miyake I, Ishikawa H. Initiation of flower bud by gibberellin in *Cryptomeria japonica*. J Jpn For Soc. 1958; 40:35–36.
16. Hashizume H. Studies on flower bud formation, flower sex differentiation and their control in conifers. Bull Tottori Univ For. 1973; 7:1–139.
17. Ross SD, Pharis RP. Control of sex expression in conifers. Plant Growth Regul. 1987; 6:37–60.
18. Zimmerman RH, Hackett WP, Pharis RP. Hormonal aspects of phase change and precocious flowering. In: Pharis RP, Reid DM, eds. Encyclopedia of plant physiology, New series, Vol. 11. New York: Springer-Verlag, 1985: p. 79–115.
19. Ross SD, Bollman MP, Pharis RP, Sweet GB. Gibberellin $A_{4/7}$ and the promotion of flowering in *Pinus radiata*: Effects of partitioning of photoassimilate within the bud during primordia differentiation. Plant Physiol. 1984; 76:326–330.
20. Looney NE, Pharis RP, Noma M. Promotion of flowering in apple trees with gibberellin A_4 and C-3 epi-gibberellin A_4. Planta. 1985; 165:292–294.
21. Pharis RP, Looney NE, Mander LN. Promotion of flowering in woody angio-

sperms. UK Patent Appl. No. 8502424, filed January 31, 1985, updated January 1986.
22. King RW, Pharis RP, Mander LN. Gibberellins in relation to growth and flowering in *Pharbitis nil* Chois. Plant Physiol. 1987; 84:1126–1131.
23. Barendse GWM, Lang A. Comparison of endogenous gibberellins and of the fate of applied radioactive gibberellin A_1 in a normal and dwarf strain of Japanese morning glory. Plant Physiol. 1972; 49:836–841.
24. Pharis RP, Evans LT, King RW, et al. Gibberellins, endogenous and applied, in relation to flower induction in the long-day plant, *Lolium temulentum*. Plant Physiol. 1987; 84:1132–1138.
25. Ross SD, Webber JE, Pharis RP, et al. Interaction between gibberellin $A_{4/7}$ and rootpruning on the reproductive and vegetative process in Douglas fir. I. Effects on flowering. Can J For Res. 1985; 15:341–347.
26. Webber JE, Ross SD, Pharis RP, et al. Interaction between gibberellin $A_{4/7}$ and rootpruning on the reproductive and vegetative process in Douglas-fir. II. Effects on shoot growth. Can J For Res. 1985; 15:348–353.
27. Pharis RP, Morf W. Precocious flowering of coastal and giant redwood with gibberellins A_3, $A_{4/7}$, and A_{13}. BioScience. 1969; 19:719–720.
28. Pharis RP, Ross SD, McMullan E. Promotion of flowering in the Pinaceae by gibberellins. III. Seedlings of Douglas fir. Physiol Plant. 1980; 50:119–126.

*Note added in proof: A re-reading of the 1972 paper by Lona (pp. 423–429, In: Hormonal Regulation in Plant Growth and Development, Kaldewey H., and Vardar, Y., eds, Verlag Chemie, Weinheim, 1972; cited in ref. 1 and 24) now leads me to believe that Lona used a marginally inductive long night for the *Xanthium* and *Perilla* plants. This is reinforced by a photograph showing a *Xanthium* inflorescence produced by GA_3 application under the same conditions where the extract of peach seeds had also promoted flowering. However, the peach seed extract promoted flowering without the stem elongation that occurred with floral promotion by the GA_3 treatment.

CHAPTER 17

Role of Endogenous Gibberellins During Fruit and Seed Development

G.W.M. Barendse, C.M. Karssen, and M. Koornneef

1 Introduction

Fruit and seed development are closely related. In many species, fruit tissue weight increases with an increase in the number of well-developed seeds, and in some cases seeds can be removed or killed and replaced by hormones to restore normal fruit development (e.g., in *Pisum sativum*[1] and *Rosa arvensis*[2]).

In addition, plant hormones such as auxins and gibberellins (GAs) are able to induce parthenocarpic fruits in several plant species.[3]

Immature seeds, in particular, are rich sources of GAs and have been used in many studies involving quantification and characterization of endogenous GAs.[4-6] Much knowledge concerning GA biosynthesis and metabolism is derived from studies with immature seeds[7] or cell-free extracts from seeds.[8] The role of endogenous GAs during seed development[9] has been deduced from studies in which changes in GA-like activity are correlated with particular stages in seed development. However, care must be taken since not all GAs that are active in bioassays necessarily control seed development. Another approach to study the role of GAs has been to determine the *ent*-kaurene-synthesizing capacity in cell-free extracts that convert mevalonate into *ent*-kaurene, the first kaurenoid precursor for the GAs; this *ent*-kaurene-synthesizing capacity thus represents a direct estimate of the overall GA-synthesizing activity of a particular tissue.

We have used this latter approach with developing seeds of *Pharbitis nil*[10] in order to determine the possible role of GAs in seed development.

Studies on the role of GAs during fruit and seed development have also used GA-deficient mutants that have little or no endogenous GAs due to genetic blocks in the GA biosynthetic pathway.

In *Arabidopsis* and tomato, GA-deficient mutants have been isolated by the selection of seeds that do not germinate without application of GAs.[11-14] The seedlings developing from these GA-treated seeds show a dwarf growth which can be reverted to normal growth and flower development by additional GA application. These mutants have been used to study the role of GAs in fruit and seed development.[14-16]

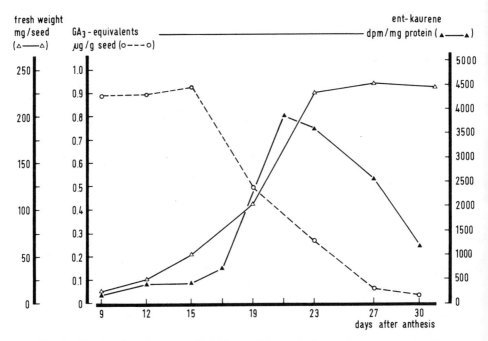

Fig. 1. The total endogenous GA-like activity and the *ent*-kaurene-synthesizing capacity in cell-free extracts in relation to the increase in fresh weight of developing *Pharbitis* seeds. Redrawn from Barendse et al.[10]

2 Results

2.1 Gibberellin Biosynthesis in *Pharbitis nil* During Seed Development

The endogenous levels of GA-like substances from the developing seeds of *Pharbitis nil* cv. Violet have been determined using high-performance liquid chromatography (HPLC) and the dwarf maize (d_5) bioassay.[10] Three GA fractions were obtained, their retention times being those for GA_3, GA_5 or GA_{20}, and GA_{19} and/or GA_{44}. The bioactivity of all three fractions increased during early seed development, followed by a decline during seed maturation. The total endogenous levels reached a maximum at 19 days after anthesis and pollination, which is just prior to the maximal fresh weight of seeds, reached about 23 days after anthesis (Fig. 1).

Ent-kaurene is a key intermediate in the GA biosynthetic pathway. Its formation has been studied in cell-free extracts from different organs of higher plants including seeds and fruits, e.g., endosperm of *Echinocystis macrocarpa* (wild cucumber)[17-20] and of *Cucurbita maxima* (pumpkin),[21] immature fruits and seeds of *Pisum sativum* (peas),[22-25] and developing seeds of *Pharbitis nil* cv. Violet (Japanese morning glory).[26] Coolbaugh and Moore[27] reported evidence for changes in the rate of *ent*-kaurene synthesis in cell-free extracts from pea seeds during seed development and related these changes to previously reported levels of endogenous GA-like substances in pea seeds. Barendse et al.[10] Investigated the relationship between in vitro *ent*-kaurene synthesis and levels of GAs extracted during the development of seeds of Japanese morning glory. The endogenous levels of GA-like substances were shown to be correlated with the *ent*-kaurene-synthesizing capacity during seed development.

Similarly, the *ent*-kaurene-synthesizing capacity showed a large increase during the log growth phase of seeds, followed by a decline during seed maturation (Fig. 1).

2.2 Gibberellins in Fruit and Seed Development of *Arabidopsis thaliana*

The three *Arabidopsis* dwarf mutants *ga*-1, *ga*-2, and *ga*-3 have similar phenotypes. They are simple recessives and nonallelic to each other. Each mutant has an absolute GA requirement for both seed germination and subsequent normal development, including flower formation.[11]

Such mutants enable a genetic approach to the study of the role of GAs, present in the different flower parts, in fruit and seed development. Mutants at the *ga*-1 locus were used to analyze the effects of the mutation in some detail and elucidate the effect of GA biosynthesis on siliqua growth in *Arabidopsis*.[15]

Siliqua of the wild type harvested at different stages after pollination showed highest *ent*-kaurene-synthesizing capacity in the younger siliqua, i.e., 4–6 days after pollination, while, this capacity was low in older siliqua. In siliqua of the *ga*-1 mutant, the *ent*-kaurene-synthesizing capacity was either very low or absent.

Similarly, young siliqua of the wild type contained significant amounts of endogenous GA-like substances, as measured by the dwarf maize bioassay, while the *ga*-1 mutant contained no detectable GA-like substances.

Reciprocal crosses between wild type (*GaGa*; short for *Ga*-1/*Ga*-1) and the mutant (*gaga*; short for *ga*-1/*ga*-1) revealed that the *ga*-1 mutation leads to reduced seed set and ultimate siliqua length, but at the same time more seeds per unit of siliqua length. As a result, the latter siliqua became "crowded" with seeds.

The effect on fruit length (Fig. 2) is determined by both the genotype of the maternal tissue (compare *GaGa* × *gaga* vs. *gaga* × *GaGa*) and the genotype of the embryo (compare *gaga* × *GaGa* vs. *gaga* × *gaga*). Exogenous GA application to the *ga*-1 mutant prior to pollination clearly affects siliqua growth. This treatment does not completely overcome the effect of the mutation on siliquas or embryos, which indicates that the *ga*-1 gene is also expressed after pollination.

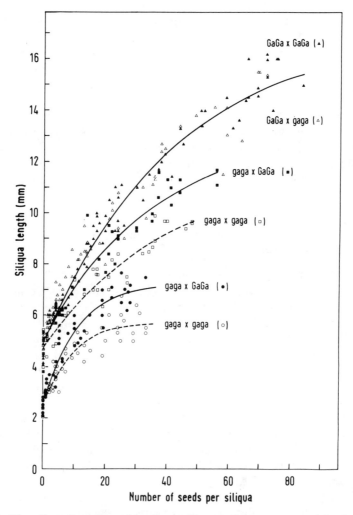

Fig. 2. The effect of number of seeds per siliqua and the genotype with respect to the *ga*-1 gene of both the mother plant and the embryo on siliqua lenth after reciprocal crosses and self-pollination with wild type (*GaGa*) and *ga*-1 mutant (*gaga*). The GA-sensitive mutant *ga*-1 was treated with GA_{4+7} once (●, ○) or three times (■, □). From Barendse et al.[15]

A time course study on siliqua development in the progeny from reciprocal crosses between the wild type and the *ga*-1 mutant revealed that in all genotypes siliqua growth occurs and is completed during the first week after pollination.[15] The differences in ultimate fruit lengths of the different genotypes are the result of their different growth rates during this first week of growth.

In addition to the effect due to genotype, the rate of siliqua growth is

Table 1. Germination of seeds derived from self-pollination of either homozygous wild-type (*GaGa*), mutant (*gaga*), or heterozygous (*Gaga*) plants or from the cross *gaga* (♀) × *Gaga* (♂)

Parent(s)	Progeny	Phenotype seedling	Germination in:		Not germinated	Total
			Water	GA_{4+7}		
GaGa Selfing	*GaGa*	Normal	145	0	5	150
gaga Selfing	*gaga*	Dwarf	0	150	0	150
gaga × *GaGa*	*Gaga*	Normal	144	0	6	150
Gaga Selfing	*GaGa*	Normal	185	0	0	185[a]
	Gaga	Normal				
	gaga	Dwarf	0	53	0	53[a]
	Unknown (died)		8	2	2	12

Seeds were imbibed in water, irradiated for 1 day under intermittent red light, and then incubated at 2°C for 5 days. After this dormancy-breaking treatment, seeds were incubated in water at 26°C in the dark for 7 days. Seeds that had not germinated in water were further treated with 10 μM GA_{4+7}. From Groot et al.[14]
[a] x^2 (normal:dwarf = 3:1) = 0.947, df. = 1, $P = 0.33$

also determined by the number of seeds per siliqua; i.e., more seeds per siliqua enhances siliqua length.

The average number of seeds present in a siliqua is approximately the same in the wild type and the *ga-1* mutant. However, seed set is highest in the self-pollinated wild type and lowest in the self-pollinated mutant, while progeny from the reciprocal crosses show intermediate seed set. Thus, seed set is affected by the genotype of the mother plant as well as by the genotype of pollen.

2.3 Gibberellins in Seed and Fruit Development of Tomato

The wild-type tomato *Lycopersicon esculentum* Mill. cv. Moneymaker and the homozygous GA-deficient mutant *ga-1* have been used to study the role of GAs in seed and fruit development.[14] As in the case of the *Arabidopsis* mutants, the germination of the *ga-1* mutant of tomato requires exogenous GA. These germination characteristics are determined exclusively by the genotype of the embryo, as was shown in seeds derived from self-pollination of either homozygous wild type (*GaGa*) or mutant (*gaga*) and seeds of heterozygous plants (*Gaga*) (Table 1).

Dwarf *ga-1* plants produce no normal flowers, but do initiate flower buds, which do not develop further. A single application of GA_{4+7} restores normal flower development and seed set resulting from pollination.

The *ga-1* mutation also affects the fresh weight of the mature fruit.[14] However, the fresh weight of fruits is mainly determined by the number of seeds per fruit. This increase in weight due to seed number is greater in wild-type seeds than in the mutant.

3 Discussion

Changes in the rate of *ent*-kaurene synthesis in cell-free extracts from developing pea seeds are related to levels of endogenous GA-like substances.[27] Similarly, it was shown for developing seeds of Japanese morning glory that both the *ent*-kaurene biosynthetic capacity in cell-free extracts and the levels of endogenous GA-like substances are related to seed development; that is, they increased rapidly during the phase of rapid seed development and decreased during seed maturation.[10] However, these endogenous GA-like substances comprise only the biologically active, free GA-like substances. It is known, however, that part of the GAs in immature seeds of Japanese morning glory occur as conjugates.[28-30] Barendse[31] has shown that the conjugated GA-like levels of seeds are low, much lower than the free GA-like levels, during the early stages of seed development. As seed development progresses, the GA-conjugate content increases, surpassing the free GA content when the maximum fresh weight of seeds is reached, indicating a progressive conjugation of GAs during seed development in *Pharbitis nil*.

The apparent relation between seed development and GA biosynthesis in *Pharbitis* would suggest a regulatory role for GAs with respect to seed development. However, Zeevaart[32] has shown that a reduction of the GA content of *Pharbitis* seeds by the growth retardant CCC did not affect seed development but only caused a delay in seed germination and subsequent development, suggesting a role for GAs in seed germination only. The *ga*-1 mutant of *Arabidopsis*, which has an absolute GA requirement for germination and subsequent development, also shows reduced siliqua growth compared to the wild type.[15] Siliquas of this mutant possess very reduced *ent*-kaurene-synthesizing capacity and no detectable endogenous GA-like substances. Together with the observation that exogenous GAs enhance siliqua growth in the mutant, this indicates a role for endogenous GAs in siliqua growth. Since seeds are known sites of active GA biosynthesis, this suggested that GAs synthesized in seeds would influence the growth of the surrounding maternal tissue in *Arabidopsis* siliqua. However, it appeared that siliqua growth is also affected by the mother plant genotype and thus not solely determined by embryonic GAs.

In all genotypes, the ultimate, siliqua lenths are reached 7–8 days after pollination. Müller[33] also showed that *Arabidopsis* embryos attain their maximal development in the same period, which also suggests a relation between embryo development and siliqua length.

Since the GA-deficient genotype (*ga*-1) shows a positive correlation between seed numbers and siliqua growth, apparently other factors produced by seeds are also responsible for siliqua growth.[15] The effect of the *ga*-1 mutation on seed set appears not to be due to a reduction in ovule numbers but to both reduced male and female fertility, which may be due to the fact that in nontreated *ga*-1 mutants the anthers also remain small.[11]

The seeds of GA-deficient mutants reach the same ultimate fresh weights as the seeds of the wild type, which suggests that the GAs produced by *Arabidopsis* seeds are not essential for their own development, which is in accordance with the conclusion drawn for *Pharbitis* seeds.[32]

The absolute GA requirement for germination in the *ga*-1 mutant also provides a strong argument for a general role of GA in the germination process in *Arabidopsis*.

In tomato, the influence of GA deficiency differs during flower development.[14] Whereas flower fertility and seed germination are completely blocked in the absence of endogenous GA, the growth of seeds and fruits still proceeds. Development of the seeds and fruits of the mutants occurs after a single treatment on flower buds with GA_{4+7}. The occasional development of small parthenocarpic fruits on *ga*-1 plants without a GA treatment indicates that GAs are not an absolute requirement for fruit development in tomatoes.

Tomato fruits and seeds may develop in the absence of endogenous GAs. However, GAs produced by the seeds increase the fresh weight of both fruits and seeds. This is not only true for the stimulation of parthenocarpic fruit growth by maternal GA, but also for fruit and seed development under the influence of seed-produced GA. Germination of *gaga* seeds completely depends on application of GAs, irrespective of whether the seeds develop in GA-deficient or GA-producing fruits (Table 1). Thus, maternal GAs apparently have no role in the germination capacity of seeds. Germination apparently requires seed-synthesized GAs.

4 Conclusions

Seeds, in particular developing seeds, constitute active sites of GA biosynthesis. The endogenous GAs may stimulate seed and fruit development, but their presence is not essential, except perhaps immediately after fertilization. In contrast, however, the development of fertile flowers and the germination of seeds are absolutely dependent on the presence of endogenous GA.

References

1. Eeuwens CJ, Schwabe WW. Seed and pod wall developments in *Pisum sativum* L. in relation to extracted and applied hormones. J Exp Bot. 1975; 26:1–14.
2. Jackson GAD. Hormonal control of fruit development, seed dormancy and germination with particular reference to *Rosa*. Soc Chem Ind (Lond) Monogr. 1968; 30:127–156.
3. Naylor AW. Functions of hormones at the organ level of organization. In: Scott TK, ed. Encyclopedia of plant physiology, New series, Vol. 10. Berlin: Springer-Verlag, 1984: p. 172–218.

4. Khan AA. Gibberellins and seed development. In: Khan AA, ed. The physiology and biochemistry of seed development, dormancy and germination. Amsterdam: Elsevier, 1982: p. 111-135.
5. MacMillan J. Analysis of plant hormones and metabolism of gibberellins. In: Crozier A, Hillman JR, eds. The biosynthesis and metabolism of plant hormones. Cambridge: Cambridge University Press, 1984: p. 1-16.
6. Pharis RP, King RW. Gibberellins and reproductive development in seed plants. Ann Rev Plant Physiol. 1985; 36:517-568.
7. Sponsel VM. Gibberellins in *Pisum sativum*—their nature, distribution and involvement in growth and development of the plant. Physiol Plant. 1985; 65:533-538.
8. Graebe JE, Hedden P, Rademacher W. Gibberellin biosynthesis. In: Lenton JR, ed. Gibberellins—chemistry, physiology and use. Wantage: British Plant Growth Regulator Group, 1980: p. 31-48.
9. Moore TC, Ecklund PR. Role of gibberellins in the development of fruits and seeds. In: Krishnamoorthy HN, ed. Gibberellins and plant growth. New York: Halsted Press. 1975: p. 145-182.
10. Barendse GWM, Dijkstra A, Moore TC. The biosynthesis of the gibberellin precursor *ent*-kaurene in cell-free extracts and the endogenous gibberellins of Japanese morning glory in relation to seed development. J Plant Growth Regul. 1983; 2:165-175.
11. Koornneef M, Van der Veen JH. Induction and analysis of gibberellin-sensitive mutants in *Arabidopsis thaliana* (L.) Heynh. Theor Appl Genet. 1980; 58:257-63.
12. Koornneef M, Van der Veen JH, Spruit CJP, et al. The isolation and use of mutants with an altered germination behaviour in *Arabidopsis thaliana* and tomato. In: Kitto PH, ed. Induced mutation—a tool in plant research. Vienna: 1981: IAEA-SM p. 227-232.
13. Karssen CM, Groot SPC, Koornneef M. Hormone mutants and seed dormancy in *Arabidopsis* and tomato. In: Thomas H, Grierson D, eds. Developmental mutants in higher plants. Cambridge: Cambridge University Press, 1987: p. 119-134.
14. Groot SPC, Bruinsma J, Karssen CM. The role of endogenous gibberellin in seed and fruit development of tomato: Studies with a gibberellin-deficient mutant. Physiol Plant. 1987; 71:184-190.
15. Barendse GWM, Kepczynski J, Karssen CM, et al. The role of endogenous gibberellins during fruit and seed development: Studies on gibberellin-deficient genotypes of *Arabidopsis thaliana*. Physiol Plant. 1986; 67:315-319.
16. Groot SPC, Karssen CM. Gibberellins regulate seed germination in tomato by endosperm weakening: A study with gibberellin-deficient mutants. Planta. 1987; 171:525-531.
17. Graebe JE, Dennis DT, Upper CD, et al. Biosyntheis of gibberellins. I. The biosynthesis of (−)-kaurene, (−)-kaurenol, and *trans*-geranylgeraniol in endosperm nucellus of *Echinocystis macrocarpa* Greene. J Biol Chem. 1965; 240:1847-1854.
18. Upper CD, West CA. Biosyntheis of gibberellins. II. Enzymic cyclization of geranylgeranyl pyrophosphate to kaurene. J Biol Chem. 1967; 242:3285-3292.
19. Oster MO, West CA. Biosynthesis of trans-geranylgeranyl pyro-phosphate in

endosperm of *Echinocystis macrocarpa* Greene. Arch Biochem Biophys. 1968; 127:112–123.
20. Coolbaugh RC, Hamilton R. Inhibition of *ent*-kaurene oxidation and growth by α-cyclorpropyl-α-(*p*-methoxyphenyl)-5-pyrimidine methyl alcohol. Plant Physiol. 1976; 57:245–248.
21. Graebe JE. The enzymic preparation of [^{14}C]-kaurene. Planta. 1969; 85:171–174.
22. Anderson JD, Moore TC. Biosynthesis of (−)-kaurene in cell-free extracts of immature pea seeds. Plant Physiol. 1967; 42:1527–1534.
23. Graebe JE. Biosynthesis of kaurene, squalene and phytoene from mevalonate-2-^{14}C in a cell-free system from pea fruits. Phytochemistry. 1968; 7:2003–2020.
24. Coolbaugh RC, Moore TC. Localization of enzymes catalysing kaurene biosynthesis in immature pea seeds. Phytochemistry. 1971; 10:2395–2400.
25. Coolbaugh RC, Moore TC. Metabolism of kaurene in cell-free extracts of immature seeds. Phytochemistry. 1971; 10:2401–2412.
26. Barendse GWM, Moore TC. Biosynthesis of the gibberellin precursor *ent*-kaurene in cell-free extracts of Japanese morning glory in relation to seed and seedling development. Proceedings of the 8th Annual Plant Growth Regulator Society of America Meeting. 1981: p. 124–131.
27. Coolbaugh RC, Moore TC. Apparent changes in rate of kaurene biosynthesis during development of pea seeds. Plant Physiol. 1969; 44:1364–1367.
28. Yokota T, Murofushi N, Takahashi N. Structure of new gibberellin glycosides in immature seeds of *Pharbitis nil*. Tetrahedron Lett. 1969; 2081–2084.
29. Yokota T, Takahashi N, Murofushi N, et al. Isolation of gibberellins A_{26} and A_{27} and their glycosides from immature seeds of *Pharbitis nil*. Planta. 1969; 87:180–184.
30. Yokota T, Murofushi N, Takahashi N. Structures of a new gibberellin glucoside in immature seeds of *Pharbitis nil* (3-*O*-β-glucosyl-gibberellin A_{29}; acid hydrolysis; enzymatic hydrolysis). Tetrahedron Lett. 1970; 1489–1492.
31. Barendse GWM. Formation of bound gibberellins in *Pharbitis nil*. Planta 1971; 99:290–301.
32. Zeevaart JAD. Reduction of the gibberellin content of *Pharbitis* seeds by CCC and after-effects in the progeny. Plant Physiol. 1966; 41:856–861.
33. Müller AJ. Embryonentest zum Nachweis recessiver Lethalfaktoren bei *Arabidopsis thaliana*. Biol Zentralbl. 1963; 82:133–163.

CHAPTER 18

Correlations Between Apparent Rates of *ent*-Kaurene Biosynthesis and Parameters of Growth and Development in *Pisum sativum*

T.C. Moore and R.C. Coolbaugh

1 Introduction

The extent to which growth and development of plants and their organs are correlated with (a) changes in concentrations of endogenous gibberellins (GAs) and (b) changes in the sensitivity of tissues and organs to GAs is incompletely understood.[1-3] Previous work in our laboratories with various cultivars of pea (*Pisum sativum* L.) has addressed this topic extensively. That research consisted partially of studies of the effects of exogenous GA_3 and inhibitors of GA biosynthesis on growth and development. Extensive investigations were performed as well on the biosynthesis of *ent*-kaurene, the first-formed tetracyclic, diterpenoid precursor in the GA biosynthetic pathway, in cell-free enzyme extracts in relation to certain parameters of growth and development. Reexamination of some of the data from those investigations has enabled the making of some generalizations regarding GA relationships in pea.

2 Periods of Sensitivity to Exogenous GA in Pea Shoots

Ecklund and Moore[4] and Moore[5] conducted experiments on the responses of Alaska (tall or normal cultivar) pea shoots to exogenous GA_3 and AMO-1618. They found that whereas this nondwarf cultivar generally has been regarded as unresponsive to exogenous GA_3, it actually is sensitive at two stages of ontogeny. That is, there was a positive shoot growth response to exogenous GA_3 when the hormone was applied during the first 12 days after germination (Fig. 1) and a second period of sensitivity during declining shoot growth and onset of senescence, when the plants again responded to exogenous GA_3 (Fig. 2). Sensitivity to AMO-1618, an inhibitor of GA biosynthesis, varied inversely with the rate of shoot growth up to the time when the maximum growth rate was attained (Fig. 3). Previous research in our laboratories yielded evidence that *ent*-kaurene biosynthesis is rate-

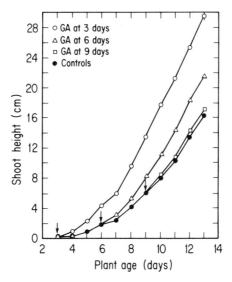

Fig. 1. Stem elongation responses in Alaska pea seedlings to which single growth-saturating doses of GA$_3$ (10 µg) were applied on the shoot tips on the days indicated by the arrows. Redrawn with modifications from ref. 5

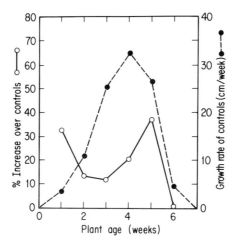

Fig. 2. Ontogenetic changes in the sensitivity of Alaska pea shoots to single shoot-tip applications of 1 µg of GA$_3$ per plant. Sensitivity to GA was measured as the percent increase in stem elongation over that of untreated plants 1 week after the treatment. Redrawn with modifications from ref. 4

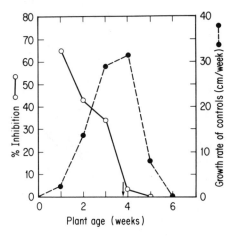

Fig. 3. Changes in the sensitivity of Alaska pea shoots to single shoot-tip applications of 50 μg of AMO-1618 per plant. Sensitivity to AMO-1618 was measured as the percent inhibition of stem elongation as compared with untreated plants 1 week after the treatment. The arrow denotes the time of anthesis. Redrawn with modifications from ref. 4

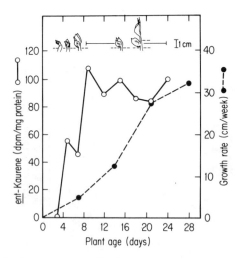

Fig. 4. Ontogenetic change in growth rate and in *ent*-kaurene-synthesizing capacity in cell-free enzyme extracts from Alaska pea shoot tips[12, 13]

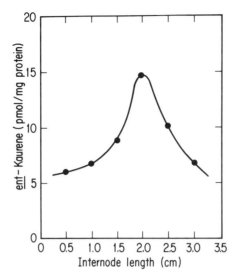

Fig. 5. Capacity for *ent*-kaurene biosynthesis from mevalonate in cell-free enzyme extracts from elongating 5th internodes of 9-to 11-day-old Alaska pea seedlings at different times during elongation. Redrawn with modifications from ref. 9 by permission of the American Society of Plant Physiologists

limiting for GA biosynthesis in pea. Hence, it was interesting to note that the capacity for synthesis of [^{14}C]*ent*-kaurene from [^{14}C]mevalonate in cell-free enzyme extracts prepared from shoot tips increased rapidly from an undetectable level in seedlings 1 to 3 days old to a maximum value in seedlings 9 days old and varied around that value through the 24th day of growth (Fig. 4).

Together these data are interpreted to mean that the rate of shoot elongation in Alaska peas is correlated directly with a changing rate of *ent*-kaurene (and presumably GA) biosynthesis throughout ontogeny. There may also be changes in sensitivity of the shoots to GA (endogenously or exogenously supplied). However, if a change in sensitivity is involved, that sensitivity change would have to be confined to the few youngest internodes that actually respond to the GA at whatever developmental age the hormone is applied. There are no data to reveal whether or not this specific change of internodes to GA occurs.

Clearly, a particular internode exhibits a normal growth-rate curve during its maturation. Also, the capacity for the biosynthesis of [^{14}C]*ent*-kaurene from [^{14}C]mevalonate (and presumably GA biosynthesis) in cell-free enzyme extracts follows a near-normal curve (Fig. 5). However, there is no reason to think that young internodes with equivalent growth potential at different stages of shoot development differ in their sensitivity to GA.

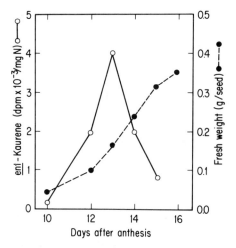

Fig. 6. Changes in rate of *ent*-kaurene biosynthesis in cell-free enzyme extracts of Alaska pea seeds at various times after anthesis. Redrawn with modifications from ref. 6 by permission of the American Society of Plant Physiologists

3 Change in Capacity for *ent*-Kaurene Biosynthesis During Pea Seed Development

The capacity for *ent*-kaurene (and presumably GA) biosynthesis, as measured by the formation of [^{14}C]*ent*-kaurene from [^{14}C]mevalonate in cell-free enzyme extracts, changes dramatically during the development of Alaska pea seeds (Fig. 6).[6] It exhibits a sharp increase in seeds harvested 10 days after anthesis to a maximum value in seeds harvested 13 days after anthesis and then a rapid decline in seeds harvested 15 days after anthesis. These data agree very well with many data on the changes in GA content of legume and other seeds during development. In fact, this particular close correlation is one of the major bases for believing that *ent*-kaurene biosynthesis often is rate-limiting for GA biosynthesis.

4 Biosynthesis of *ent*-Kaurene in Individual Organs of Pea Seedlings

Coolbaugh[7] investigated the biosynthesis of [^{14}C]*ent*-kaurene from [^{14}C]mevalonate in cell-free enzyme extracts of different vegetative organs of 10-day-old Alaska pea seedlings. The amounts of [^{14}C]*ent*-kaurene synthesized in 60-min incubations varied from a barely detectable 0.16 pmol/mg protein for root tips to 8.78 pmol/mg protein for young fifth internodes. The greatest amounts were obtained with extracts prepared from

Alaska
13-day-old seedling

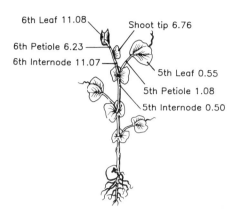

Fig. 7. Amounts (pmol/mg protein) of *ent*-kaurene synthesized by different organs of 13-day-old Alaska (tall or normal) seedlings during 60-min incubations of cell-free enzyme extracts. Redrawn with modifications from ref. 9 by permission of the American Society of Plant Physiologists

the fifth (youngest elongating) internode (8.78), fourth (youngest expanded) leaflets (3.32), and the shoot tip (2.02).

In general, the greatest capacities for *ent*-kaurene (and presumably GA) biosynthesis in shoot organs were found in those organs with the greatest potential for growth. The relatively very low activity observed for enzyme extracts of root tips is consistent with what little is known about GAs in roots and the requirement of roots for GAs. It is interesting to compare these data with the closely analogous data for [^{14}C]indoleacetic acid biosynthesis from [^{14}C]tryptophan in cell-free enzyme extracts of different organs of pea plants.[8]

The biosynthesis of *ent*-kaurene also was compared in cell-free extracts of shoot tips of dwarf and normal pea seedlings. Synthesis of [^{14}C]*ent*-kaurene from [^{14}C]mevalonate in cell-free enzyme extracts from different organs of dwarf (Progress No. 9) and tall or normal (Alaska) pea plants was investigated at ages 10, 13, and 16 days.[9] The data for 13-day-old plants are illustrative of what was learned (Figs. 7 and 8). Biosynthesis of *ent*-kaurene occurred in cell-free enzyme extracts prepared from shoot tips, leaves (and their parts), and internodes of both cultivars at all three stages of development, with the data for Alaska seedlings corroborating the results of earlier investigations. Little or no activity was detected in enzyme extracts of senescent cotyledons or root tips. Activity decreased in organs as they matured. The capacity for *ent*-kaurene biosynthesis generally was lower in ex-

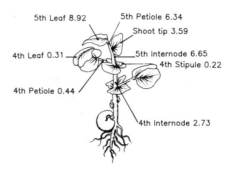

Fig. 8. Amounts (pmol/mg protein) of *ent*-kaurene synthesized by different organs of 13-day-old Progress No. 9 (dwarf) seedlings during 60-min incubations of cell-free enzyme extracts. Redrawn with modifications from ref. 9 by permission of the American Society of Plant Physiologists

tracts of dwarf shoot tips and internodes than in equivalent extracts from tall or normal plants, except that the activity in leaflets and stipules was somewhat higher in dwarf plants than in tall or normal plants. With the exception of root tips, there was a strong correlation between the capacity for *ent*-kaurene biosynthesis in extracts of an organ and the growth potential of that organ. The available data agree well with what is known about the comparative GA contents in herbaceous shoots[10] and the organs of GA biosynthesis.[10,11]

5 Biosynthesis of *ent*-Kaurene in Etiolated and Green Pea Seedlings

A comparison was made of the capacities for [^{14}C]*ent*-kaurene biosynthesis from [^{14}C]mevalonate in cell-free enzyme extracts of shoot tips of etiolated and light-grown tall or normal (Alaska) and dwarf (Progress No. 9) pea seedlings[12,13] (Table 1). Extracts of either light-grown cultivar displayed greater *ent*-kaurene synthesizing capacity than was observed with extracts from their dark-grown counterparts. Activity was higher in extracts of light-grown Alaska pea shoot tips than in extracts of light-grown Progress No. 9 shoot tips.

Investigating the difference between etiolated and green seedlings further, *ent*-kaurene-synthesizing capacity was measured in cell-free enzyme extracts of the shoot tips of etiolated Alaska seedlings as they underwent photomorphogenesis in high-intensity white light (Fig. 9). An approximately exponential increase in activity occurred between the 3rd and 12th

Table 1. Comparisons of *ent*-kaurene-synthesizing capacities in cell-free enzyme extracts from shoot tips of light- and dark-grown Alaska (tall or normal) and Progress No. 9 (dwarf) pea seedlings 10 to 14 days old[12,13]

Experiment	Enzyme source			
	Alaska, light-grown	Alaska, dark-grown	Progress No. 9, light-grown	Progress No. 9, dark-grown
	dpm in *ent*-kaurene/mg protein			
1	59		36	
2	86		17	
3	88	22	27	
4	144	23	82	35
5	45	7		
	dpm in *ent*-kaurene/g fr wt			
1	760		530	
2	1360		340	
3	860	160	240	
4	1240	280	1120	350
5	500	70		

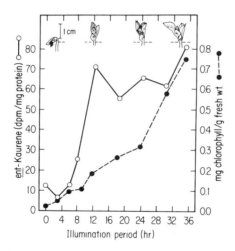

Fig. 9. Change in chlorophyll content and *ent*-kaurene-synthesizing activity in cell-free extracts of Alaska pea shoot tips during irradiation of 10-day-old etiolated seedlings with high-intensity white light[12,13]

hour of irradiation, with the activity reaching a level comparable to that in extracts of shoot tips of fully green seedlings of the same age. From the 12th hour through the 36th hour of greening, activity varied around the 12-hour, maximum level.[12,13]

6 Discussion

The concept of plant hormone control by changing concentrations was seriously challenged by Trewavas.[1] He observed that evidence frequently is lacking that hormones act via changes in amount or concentration and that, therefore, changes in responses to plant hormones must be attributable to changes in the sensitivity of the tissues to those hormones. Moreover, Trewavas noted, growth often is proportional to the logarithm of the concentration of exogenous hormone, and there may be an increasing response over three orders of magnitude in concentration. Thus, changes in the concentrations of endogenous hormones in tissues often are less than would be expected.

In the present paper, evidence is presented that changes in concentrations of endogenous GAs are sometimes as much as one order of magnitude or more. Sensitivity to AMO-1618, for example, varies from about 5% inhibition to 65% inhibition, for a 13-fold difference (Fig. 3). Capacity for *ent*-kaurene biosynthesis in enzyme extracts of greening pea seedling shoot tips varies nearly exponentially from approximately 5 to 65 dpm/mg protein between the 3rd and 12th hours of irradiation (Fig. 9). There is more than a 16-fold difference in *ent*-kaurene-synthesizing capacity during the development of pea seeds (Fig. 6). Finally, there are 10-fold and greater differences in *ent*-kaurene-synthesizing capacities of different organs of pea seedlings (Figs. 7 and 8). If *ent*-kaurene-biosynthesizing activity is rate-limiting for endogenous GA biosynthesis, then 10-fold and greater differences in concentration of endogenous GAs should be expected in these cases. Of course, there may be additional sites of regulation of the GA biosynthetic pathway in peas besides that part of the pathway leading from mevalonate to *ent*-kaurene, including perhaps reactions closer to the synthesis of growth-active GAs.

Thus, it is concluded that differences in concentration of endogenous GAs are correlated with different parameters of growth in peas. Evidence for specific correlations between endogenous amounts of particular GAs and stem elongation has been reported for peas and other species.[14-16] Metzger and Zeevaart[14] hypothesized that one major aspect of photoperiodic control of stem elongation in the long-day plant *Spinacia oleracea* is the availability of GA_{20} through regulation of the conversion of GA_{19} to GA_{20}. Fujioka et al.[15] showed in extensive qualitative and quantitative analyses of GAs in vegetative shoots of dwarf mutants and normal seedlings of *Zea mays* that it is not the total GA level that controls elongation growth in maize shoots but the level of one specific GA, namely, GA_1, much lesser amounts (1 to 2%) of which were found in vegetative shoots of four dwarfs than in normal plants. Ingram et al.[16] reported a quantitative relationship between GA_1 and internode length in peas.

There are also well-known differences in the sensitivity of pea shoot tissues and organs to GA. The case which has received the most attention is the difference between etiolated and light-grown seedlings, according to which the green tissues are less sensitive than etiolated tissues. However, interestingly, it appears that the plant has a compensatory mechanism for adjusting to the change in sensitivity to GA during de-etiolation by increasing its biosynthesis of GA (Fig. 9) and thereby changing the internal concentration of the hormone.

7 Conclusions

From the data presented in this paper, it is concluded that:

1. The level of *ent*-kaurene-synthesizing activity in pea seeds and shoots is directly correlated with, and probably rate-limiting for, GA biosynthesis.
2. There is a strong, direct correlation between the growth potential of organs and whole shoots of pea plants and the rates of *ent*-kaurene (and presumably GA) biosynthesis in the respective organs and shoots.
3. Growth parameters in peas appear to be directly correlated with changes in concentration of endogenous GAs. Changes in the sensitivity of green tissues and organs to GAs are not excluded; however, the only known changes in sensitivity occur during de-etiolation.
4. The decrease in sensitivity to GA that evidently occurs during the greening of etiolated pea seedlings appears to be correlated with an increase in endogenous GA biosynthesis.

Acknowledgments. This work was supported in part by the National Science Foundation under grants PCM 8016237 and PCM 8415924 to R.C. Coolbaugh and by the Oregon Agricultural Experiment Station (T.C. Moore), from which this is Technical Paper 8948.

References

1. Trewavas A. How do plant growth substances work? Plant Cell Environ. 1981; 4:203-228.
2. Trewavas A, Cleland RE. Is plant development regulated by changes in the concentration of growth substances or by changes in the sensitivity to growth substances? Trends Biochem Sci. 1983; 8:354-357.
3. Firn RD. Growth substance sensitivity: The need for clearer ideas, precise terms and purposeful experiments. Physiol Plantarum. 1986; 67:267-272.

4. Ecklund PR, Moore TC. Quantitative changes in gibberellin and RNA correlated with senescence of the shoot apex in the 'Alaska' pea. Amer J Bot. 1968; 55:494–503.
5. Moore TC. Gibberellin relationships in the 'Alaska' pea (*Pisum sativum*). Amer J Bot. 1967; 54:262–269.
6. Coolbaugh RC, Moore TC. Apparent changes in rate of kaurene biosynthesis during the development of pea seeds. Plant Physiol. 1969; 44:1364–1367.
7. Coolbaugh RC. Sites of gibberellin biosynthesis in pea seedlings. Plant Physiol. 1985; 78:655–657.
8. Moore TC. Comparative net biosynthesis of indoleacetic acid from tryptophan in cell-free extracts of different parts of *Pisum sativum* plants. Phytochemistry. 1969; 8:1109–1120.
9. Chung CH, Coolbaugh RC. *ent*-Kaurene biosynthesis in cell-free extracts of excised parts of tall and dwarf pea seedlings. Plant Physiol. 1986; 80:544–548.
10. Jones RL, Phillips IDJ. Organs of gibberellin synthesis in light-grown sunflower plants. Plant Physiol. 1966; 41:1381–1386.
11. Lockhart JA. Studies on the organ of production of the natural gibberellin factor in higher plants. Plant Physiol. 1957; 32:204–207.
12. Ecklund PR, Moore TC. Correlations of growth rate and de-etiolation with rate of *ent*-kaurene biosynthesis in pea (*Pisum sativum* L.). Plant Physiol. 1974; 53:5–10.
13. Moore TC, Ecklund PR. Biosynthesis of *ent*-kaurene in cell-free extracts of pea shoots. In: Plant growth substances 1973. Tokyo: Hirokawa Publishing Co., 1973: p. 252–259.
14. Metzger JD, Zeevaart JAD. Effect of photoperiod on the levels of endogenous gibberellins in spinach as measured by combined gas chromatography–selected ion current monitoring. Plant Physiol. 1980; 66:844–846.
15. Fujioka S, Yamane H, Spray CR, Gaskin P, MacMillan J, Phinney BO, Takahashi N. Qualitative and quantitative analyses of gibberellins in vegetative shoots of normal, dwarf-1, dwarf-2, dwarf-3, and dwarf-5 seedlings of *Zea mays* L. Plant Physiol. 1988; 88:1367–1372.
16. Ingram TJ, Reid JB, MacMillan J. The quantitative relationship between gibberellin A_1 and internode length in *Pisum sativum* L. Planta 1986; 168:414–420.

CHAPTER 19

Gibberellins and the Regulation of Shoot Elongation in Woody Plants

O. Junttila

1 Introduction

Perennial, temperate-zone woody plants display a great variety of patterns of shoot elongation. Anatomically, height growth in trees is due to an activity of apical meristem (eumeristem) and subapical meristem (rib meristem, primary elongating meristem). The apical meristem is responsible for most of the organogenic phenomena normally associated with shoot morphogenesis, while the subapical meristem is the major site of cell division and elongation contributing to stem extension.[1] Morphologically, shoot growth consists of formation of leaf and node initials and elongation of internodes. In the *free growth* pattern, characteristic of juvenile stages of many temperate-zone dicotyledonous as well as some coniferous species, these two processes can occur simultaneously. On the other hand, mature stages of most conifers and some deciduous trees have a *fixed* or *determined growth* pattern in which stem unit initiation and elongation are separated in time and shoot extension is a result of elongation of these preformed stem units.[2,3] The growth pattern is an inherent characteristic of the plant, and species showing various combinations of free and fixed growth can also be found. In addition, the growth pattern of a species normally changes with ontogeny, from free to more or less fixed growth with increasing maturity of the plant.

Because of their long life and large size, trees are exposed to a seasonal variation of climatic conditions, and mechanisms for tolerance of such variations must have been developed. Adaptation to survive low temperature is of primary importance for temperate-zone tree species. Consequently, shoot growth processes such as budbreak, shoot elongation, cessation of growth, and bud development must be closely synchronized with the seasonal climatic variations. This synchronization is brought about through an interaction of environmental signals and internal regulation mechanisms.

Given this variety of growth patterns and the range of interactions between environmental and internal factors, shoot growth in woody plants is a complex and challenging field of study and cannot be covered completely in a short review. The purpose of this paper is to focus on the relationship between gibberellins (GAs) and shoot growth in trees. This is just one small aspect of the whole picture, but the data accumulated so far clearly indicate that GAs may have an important regulatory function in control of shoot growth in woody species.

2 Growth Responses of Woody Plants to Exogenous Gibberellins

Most of the early studies with GAs on woody plants focused on the effects of exogenous GA_3 on growth and development. A great number of such experiments were carried out during the late 1950s and early 1960s, and results from those studies have been reviewed.[4-6] The general observation was that GA_3 greatly promoted stem elongation in many hardwoods but had a minor effect, if any positive effect at all, on conifers. The highest responses were usually obtained with plants showing a typical free growth pattern. Such species, for example, *Liriodendron*, *Platanus*, and *Populus*, exhibit a continuous growth under long-day (LD) and stop growing under short-day (SD) conditions (group A in the classification proposed by Nitsch[7]). However, species showing repeated flushing under LD (e.g., *Quercus*, *Syringa*) also responded to GA_3.[5,8] In spite of those sometimes dramatic effects, these early studies are of limited value for understanding the physiological role of GA in the regulation of shoot elongation. They demonstrated, however, that GA_3 in many cases enhanced shoot growth by increasing the internode length and that shoot elongation often was stimulated at the expense of root and leaf growth.[5,8]

Although results from the first experiments with GA_3 on conifers were mostly negative,[4] a number of later studies have demonstrated significant positive effects of GAs on shoot elongation in conifer,[9,10] and Ross et al.[11] have cited studies showing stimulation of shoot growth in twenty- six species within Pinaceae, Cupressaceae, and Taxodiaceae. Plants in a free growth stage have responded especially, but these examples also include plants with a fixed growth pattern. Besides GA_3, $GA_{4/7}$ mixtures have also proved to be active.

The effect of GA on conifer growth is dependent on the method of application, indicating barriers for uptake.[12] Also, the timing of application seems to be important for species with a fixed growth pattern. Considering the clear separation of the various growth processes in this pattern of growth, this requirement is well understandable. In *Pseudotsuga*, $GA_{4/7}$ enhanced shoot elongation when applied prior to bud swelling but did not have a significant influence when applied after flushing.[13,14] According to Owens et al.,[14] initiation of shoot growth in *Pseudotsuga*, until shoots have reached about 10% of their final length and begun to flush, is characterized by frequent cell divisions, while shoot elongation after flushing, accounting for almost 90% of the total shoot growth, is caused mainly by cell elongation. This could indicate that an early application of GA affects cell division. However, Owens et al.[14] suggest that the observed stimulation by an early application of GA[13] is due to GA- enhanced cell elongation. Also, in *Pinus radiata*, $GA_{4/7}$ enhanced growth of the embryonic shoot still within the bud at the time of treatment, but did not stimulate growth in already elongating internodes (according to Sweet, referred to by Ross et al.[11]).

In the case of fixed growth pattern, total shoot length is strongly correlated with the number of stem units present in the bud. There is very little information indeed on the effects of GAs on development of vegetative buds in species with predetermined growth. Owens et al.[14] observed that $GA_{4/7}$ did not influence bud formation when applied after flushing, but they point out that lack of an effect can be due to the fact that application was stopped early during the leaf initiation phase. Clearly, future application experiments on species with fixed growth should be designed to study effects on well-defined growth processes.

Growth retardants are commonly used, either alone or in combination with GAs, to obtain further information about a possible involvement of GA in a given physiological process. Studies with woody species are numerous and will not be considered here in detail. However, a general trend seems to be that these compounds inhibit shoot elongation of free growth pattern both in conifers and in hardwoods and that their effects are counteracted by GA.[15-17] In *Salix*, these compounds are able to produce a typical SD effect (cessation of elongation growth) under LD conditions.[17] Species with a fixed growth pattern are not affected by growth retardants.[11] Recent studies with paclobutrazol, an inhibitor of GA biosynthesis, have shown a significant inhibition of elongation in several conifer species and a lack of such effect on older trees.[18,19]

3 Interaction of Photoperiod and GA

Data presented above clearly indicate that plants with free growth are generally the most responsive to exogenous GA. As a rule, these plants are also sensitive to photoperiod, having short photoperiod as a primary signal for cessation of growth.[7] Short-day induced growth cessation in many deciduous tree species is effectively prevented by GA_3, indicating a connection between photoperiodic control of shoot elongation and GA metabolism.[17,20,21] Seedlings of *Salix* probably have the early-13-hydroxylation pathway for biosynthesis of GAs, and effects of five GAs from this pathway on SD-grown seedlings of *Salix* were recently studied. Gibberellins A_{20} and A_1 induced shoot elongation while the earlier members of the pathway, GA_{53}, GA_{44}, and GA_{19}, were inactive.[22] These results are consistent with the hypothesis that GA_1 is the active GA for stem elongation in *Salix* and that the step from GA_{19} to GA_{20} may be under a photoperiodic control, being inhibited by SD. Some further experiments have shown that even GA_{19} has a positive effect if the photoperiod is close

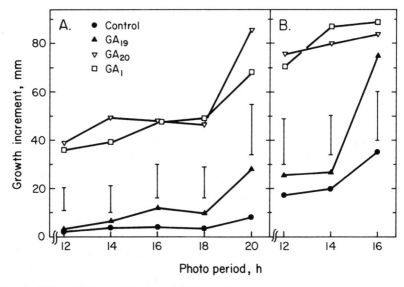

Fig. 1. Effect of GA_{19}, GA_{20}, and GA_1 on shoot elongation in seedlings of *Salix pentandra* grown under different photoperiods as indicated. (A) Ecotype from lat. 69° N. The critical photoperiod for shoot growth is about 22 h. (B) Ecotype from lat. 59° N. The critical photoperiod is about 16 h. Gibberellins were applied to the apex after 10 days at the indicated photoperiods. The figure shows total shoot growth during 2 weeks after the GA application. Temperature 18 °C. The vertical bars indicate $LSD_{0.05}$ for GA treatments

to the critical photoperiod, indicating a partial block under threshold daylength (Fig. 1). Both a northern and a southern ecotype of *Salix* gave similar responses, in spite of the difference in the critical photoperiod. Thus, in different geographic ecotypes the suggested block of GA biosynthesis occurs at different photoperiods, and this indicates that the adaptation process itself is not related to GA biosynthesis.

Also other GAs, including GA_4, GA_7, and GA_9, are able to substitute for LD in *Salix*, but it is not known whether the activity is due to these GAs per se or due to their metabolites.

To my knowledge, no careful studies on interaction of photoperiod and various GAs with regard to shoot growth in conifer seedlings, while still in the free growth stage, have been reported.

4 Interaction of Genotype and Exogenous GA

Single-gene dwarf mutants in which exogenous GA can restore the normal phenotype are known in several herbaceous species, and comprehensive studies on growth regulation in such mutants have provided a basis for our present understanding of the role of GA in regulation of stem elongation in herbaceous plants.[23] Unfortunately, such mutants have not been described in woody species. A great number of dwarfish genotypes of deciduous and coniferous species are known, but they have not been characterized genetically. However, some early studies claim that dwarf varieties of *Thuja*, *Coffea*, and *Picea* are more responsive to GA_3 than normal genotypes.[5] Also, some evidence for involvement of GAs in the dwarfing mechanism in apple has been presented.[24]

More recent application experiments with *Pseudotsuga*[25] and *Picea mariana*[26] indicate an interaction between genotype and response to GA. For example, in *Picea mariana* $GA_{4/7}$ increased shoot growth in slow-growing but not in fast-growing families.[26] On the other hand, a similar study comparing slow- and fast-growing families of *Pinus radiata*[11] did not, in my opinion, provide conclusive results.

5 Endogenous GAs in Woody Plants

Gibberellin-like activity, detected by various types of bioassays, has been reported from vegetative tissues of a great number of woody species. For example, Dunberg and Oden[27] cited, in their review, studies showing GA-like activity in thirty-one conifer species. Gibberellin A_1 was characterized from watersprouts of *Citrus*[28] as early as 1959, but even today there are only a few reports of conclusive identification of GAs from woody plants (Table 1). This is partly due to the relatively low quantities of endogenous GAs in such tissues and, until rather recently, lack of efficient methods for purification of extracts. Identification of GAs from pollen and immature or mature seeds are not included in Table 1. In addition to the cases mentioned in the table, identification by combined gas chromatography–mass spectrometry (GC–MS) of GA_3 from shoot tissue of *Pseudotsuga menziesii*, *Cupressus arizonica* and *Juniperus scopulorum* has been reported.[10]

As shown in Table 1, GA_1 has been reported from vegetative tissues of all dicotyledonous woody species analyzed so far, and it is possible that the early-13-hydroxylation pathway is the main biosynthetic pathway in many hardwoods. Quantitatively, GA_{19} and GA_{20} often dominate in these species, and GA_{19} may represent a "pool GA" for the active GA, which is probably GA_1.[30,33]

Gibberellin A_1, as well as GA_3, is also present in coniferous species, but these taxa seem to be characterized by the presence of GA_4 and GA_9. Also, conjugated GA_9, GA_9-glucosyl ester, has been identified from *Picea*.[37,40] However, too few species have been thoroughly analyzed to give a basis for suggestions about differences between conifers and dicots in their spectrum of endogenous GAs. According to Moritz et al.,[37] GA_4 and GA_9 dominated in younger plants of *Picea sitchensis*, while GA_1 and GA_3 were found in older, mature and flowering trees. This result is somewhat

Table 1. Gibberellins conclusively identified from vegetative tissues of woody plants

Species	Gibberellin	Reference
A. Deciduous species		
Citrus sinensis	GA_1	28
Citrus unshi	GA_1, iso-GA_3, GA_8^t, GA_{17}, GA_{19}, GA_{20}, GA_{29}, GA_{44}	29
Juglans regia	GA_{19}, GA_{53}	30
Malys domestica	GA_1^t, GA_9^t, GA_{19}, GA_{20}	31
Populus balsamifera x P. deltoides	GA_1, GA_{19}	32
Salix dasyclados	GA_1, GA_4^t, GA_8, GA_9, GA_{19}, GA_{20}	33
Salix pentandra	GA_1, GA_{19}, GA_{20}, GA_{29}	34
B. Coniferous species		
Picea abies	GA_1, GA_3, GA_4, GA_9	35, 36, 37
Picea sitchensis	iso-GA_9	38
Picea sitchensis	GA_1, GA_3, GA_4, GA_9, GA_{15}^h	37
Pinus radiata	GA_1, GA_3, GA_4, GA_7, GA_9, GA_{15}	39

t Tentative identification
h Detected after hydrolysis

surprising because GA_1 and GA_3 often have been associated with vegetative growth and $GA_{4/7}$ with generative growth in conifers.[11]

To date, there are no published quantitative studies on identified endogenous GAs in woody plants in relation to shoot elongation. However, a large number of investigations, based on chromatography and bioassays, have been carried out with various woody species.[21,27,41–43] These studies have focused mainly on two aspects of shoot growth: (1) budbreak and initiation of shoot elongation; and (2) cessation of shoot elongation. In general, these studies report several examples showing an increased GA-like activity prior to, or concurrent with, budbreak, and reduced activity with approaching budset, for examples, in connection with a transfer from LD to SD. Dunberg et al.[44,45] have shown results indicating a close correlation between the period of maximum shoot elongation and GA-like activity in a $GA_{1/3}$-like fraction in *Picea abies*. Also, some other studies on conifers indicate a correlation between GA_3-like activity and the rate of shoot growth (see Dunberg and Oden[27] for references). On the other hand, Lorenzi et al.[38] found that in *Picea sitchensis* the activity corresponding to iso-GA_9 correlated with the period of shoot elongation. Although these studies may indicate certain trends, their value is limited; quantification by bioassay is often unreliable and the total GA-like activity does not necessarily provide meaningful data. For example, GA_{19} is highly active in dwarf rice bioassay but is probably not biologically active per se in woody plants.[22]

Bate et al.[46] have indicated that hybrid vigour in *Populus* could be related to endogenous GA concentration. Results, based on quantification with bioassay, showed a lower total GA-like activity in bark tissue of parent clones (*P. deltoides*) than in four more rapidly growing hybrid clones. However, as a whole, the data do not show significant correlation between GA-like activity and shoot growth. The existence of relationships between endogenous GAs and inherent shoot growth potential is an interesting idea that should be explored further using reliable analytical methods and focusing on the elongating tissue itself.

6 Gibberellin Metabolism in Woody Plants

Gibberellin metabolism has recently been reviewed.[47] Only studies directly related to GA metabolism in vegetative tissues of woody plants are briefly mentioned here. Such studies are rare and provide fragmentary information.

Our studies with *Salix* have shown transformation of GA_1 to a GA_8-like compound, based on analysis by high-performance liquid chromatography (HPLC) and on-line radiocounting,[48] and of GA_{20} to GA_1.[49] In addition, recent studies with ^3H and ^2H-labeled GA_9 suggest metabolism of GA_9 to GA_{20} in seedlings of *Salix* (O. Junttila, D. Pearce, R.P. Pharis, unpublished results). Gibberellin A_9 is an endogenous compound in *Salix* (ref. 38; O. Junttila, S.B. Rood, unpublished results), and exogenous GA_9 has biological activity in *Salix*, being able to induce stem elongation under SD.[50] All these results are consistent with the hypothesis that GA_1 is the active GA for stem elongation in *Salix*. Conversion of GA_1 to biologically inactive GA_8, or to conjugates, was not affected by photoperiod, nor was the conversion of GA_{20} to GA_1.[48,49] Studies on herbaceous plants (for references, see Graebe[47]), as well as our results from application experiments with GAs from the early-13-hydroxylation pathway,[22] indicate that photoperiod may control the step from GA_{19} to GA_{20}. This possibility is currently under study.

Conversion of GA_1 to a GA_8-like compound has been shown in seedlings of *Picea*,[51] and conversion of GA_4 to GA_{34} has been reported in *Picea abies*, *Pseudotsuga*, and *Pinus* (pollen), as well as in a dicotyledonous species (*Malus*).[24,44,52,53] Also, GA_1 was detected as a metabolite of GA_4 in pollen of *Pinus* and stem tissue of *Malus*.[24,52] From a physiological point of view, information on the metabolism of GA_9 in conifers would be of great interest. One study with *Picea abies* detected a great number of metabolites, but none of them were identified.[44]

The site of synthesis of GAs in vegetative woody plants is not known for certain, although young, expanding shoot tissue is perhaps the most probable site. There are several reports showing the presence of GAs and GA-like compounds in xylem exudates and in roots of woody plants.[41] Also, GA-conjugates are often found in xylem sap, and the free GAs may be hydrolysis products of such conjugates.[30] Lavender et al.[54] have suggested that GAs transported from the roots may be involved in budbreak and initiation of shoot elongation in *Pseudotsuga*. However, there are no conclusive studies showing biosynthesis of GAs in roots of woody plants, and this possibility needs to be studied. Exact information on the site(s) of GA biosynthesis, as well as on the translocation of GAs between various organs, would be of great importance for further development of the hypothesis on GA-mediated photoperiodic control of shoot elongation.

7 Concluding Remarks

In spite of the voluminous literature on GAs and woody plants, it is quite obvious that our knowledge in this field is still in its infancy. However, the present data strongly indicate that endogenous GAs are involved in the regulation of shoot elongation in dicotyledonous species with a free growth pattern. In these cases, the regulatory mechanism may be similar to that found in stems of many herbaceous species. In *Salix*, one of the best-studied systems of such species, GA_1 probably is the active GA for stem elongation. This conclusion may apply to many, if not most, tree species with a similar type of growth. The first step toward testing this idea is to obtain conclusive identification of endogenous GAs from elongating shoot tissues of various tree species showing the typical free growth pattern. At the moment, such species also provide the best experimental systems for more in-depth studies on GA metabolism in relation to genetic and environmental control of shoot elongation. General studies on stem growth strongly suggest that the subapical meristem is the main target tissue for GA.[1] Quantitative studies of endogenous GAs should be focused, as much as possible, upon this tissue. Further, conclusive information about the site of GA biosynthesis, both at the organ and the cellular level, is greatly needed. At the whole-plant level, involvement of GAs in the interaction between root and shoot also needs to be studied. Further, experimental systems like *Salix* may be valuable for studies on the phytochrome–GA interaction. Possibilities to identify interesting developmental mutants may be more likely in species like *Salix* than in other tree species.

In 1983, Dunberg and Oden[27] concluded that "there is presently no firm evidence that GAs are major regulators of shoot elongation growth in Pinaceae species, or that they affect the cell division activity in the apical meristematic regions." Since then, various GAs, including GA_1, have been identified from vegetative tissues of Pinaceae species. Although we lack conclusive evidence, we now have reason to suggest that GAs are involved in the regulation of shoot elongation in these species. The hypothesis that free-growing stages of conifer species have regulatory systems similar to those in free-growing dicotyledonous species should be tested. Of special interest is to determine if GA_1 or some other GA is the active GA for stem growth in such conifer seedlings. Studies on GAs (qualitative and quantitative analysis, metabolism studies) in relation to fixed shoot growth should be designed carefully to focus on well-defined processes such as the bud development phase, the cell division phase, or the cell elongation phase of shoot extension. This, of course, also applies to deciduous species with a fixed growth pattern. If GA is needed for one or other phase in this type of growth, the GA action, whether it is related to the amount of the active GA present in the target cells and/or to the responsiveness of these cells to GA, must be regulated by internal mechanisms that are completely unknown.

Acknowledgments. Thanks are due to Dr. J.B. Zaerr and Dr. W.M. Proebsting for comments on the manuscript. Financial aid from the Norwegian Research Council for Sciences and Humanities is gratefully acknowledged.

References

1. Sachs RM. Stem elongation. Ann Rev Plant Physiol. 1965; 16:73–96.
2. Kozlowski TT. Growth and development of trees, Vol. II. New York: Academic Press, 1971.
3. Lanner RM. Patterns of shoot development in *Pinus* and relationship to growth potential. In: Cannell MGR, Last FT, eds. Tree physiology and improvement. New York: Academic Press, 1976: p. 223–243.
4. Westing AH. Effect of gibberellin on conifers: generally negative. J For. 1959; 57:120–122.
5. Melchior GH, Knapp R. Gibberellin-Wirkungen an Bäumen. Silvae Genet. 1962; 11:29–39.
6. Paleg LG. Physiological effects of gibberellins. Ann Rev Plant Physiol. 1965; 16:291–322.
7. Nitsch JP. Photoperiodism in woody plants. Proc Amer Soc Hort Sci. 1957; 70:526–544.
8. Junttila O. Effects of gibberellic acid and temperature on the growth of young seedlings of *Syringa vulgaris* L.J. Hort Sci. 1970; 45:315–329.
9. Pharis RP. Probable roles of plant hormones regulating shoot elongation, diameter growth and crown form of coniferous trees. In: Cannell MGR, Last FT, eds. Tree physiology and yield improvement. New York: Academic Press, 1976: p. 291–306.
10. Pharis RP, Kuo CG. Physiology of gibberellins in conifers. Can J For Res. 1977; 7:299–325.
11. Ross SD, Pharis RP, Binder WD. Growth regulators and conifers: Their physiology and potential uses in forestry. In: Nickell LG, ed. Plant growth regulating chemicals, Vol. II. Boca Raton, Florida: CRC Press, 1983: pp. 35–78.
12. Bilan MV, Kemp AK. Effect of gibberellin on height growth of one-year old seedlings of loblolly pine. J For. 1960; 58:35–37.
13. Ross SD. Enhancement of shoot elongation in Douglas-fir by gibberellin $A_{4/7}$ and its relation to the hormonal promotion of flowering. Can J For Res. 1983; 13:986–994.

14. Owens JN, Webber JE, Ross SD, et al. Interaction between gibberellin $A_{4/7}$ and root-pruning on the reproductive and vegetative processes in Douglas-fir. III. Effects on anatomy of shoot elongation and terminal bud development. Can J For Res. 1985; 15:354–364.
15. Pharis RP, Ruddat M, Phillips C, et al. Response of conifers to growth retardants. Bot Gaz. 1967; 128:105–109.
16. Dunberg A, Eliasson L. Effects of growth retardants on Norway spruce (*Picea abies*). Physiol Plant. 1972; 26:302–305.
17. Junttila O. Growth cessation and shoot tip abscission in *Salix*. Physiol Plant. 1976; 38:278–286.
18. Wheeler NC. Effect of paclobutrazol on Douglas-fir and loblolly pine. J Hort Sci. 1987; 62:101–106.
19. Rietveld W. Effect of paclobutrazol on conifer seedling morphology and field performance. In: Landis TD, ed. Proceedings of the Combined Meeting of the Western Forest Nursery Association. General Technical Report RM 167, Fort Collins, USDA Forest Service, Rocky Mountain Forest and Range Experimental Station, 1988: pp. 19–23.
20. Nitsch JP. Growth responses of woody plants to photoperiodic stimuli. Proc Amer Soc Hort Sci. 1957; 70:512–525.
21. Vince-Prue D. Photoperiod and hormones. In: Pharis RP, Reid DM, eds. Hormonal regulation of development. III. Role of environmental factors. Encyclopedia of plant physiology, New series, Vol. 11. Heidelberg: Springer-Verlag, 1985: pp. 308–364.
22. Junttila O, Jensen E. Gibberellins and photoperiodic control of shoot elongation in *Salix*. Physiol Plant. 1988; 74:371–376.
23. Phinney BO. Gibberellin A_1, dwarfism and the control of shoot elongation in higher plants. In: Crozier A, Hillman RT, eds. The biosynthesis and metabolism of plant hormones. Cambridge: Cambridge University Press, 1984: pp. 17–41.
24. Dennis R, Thompson WK, Pharis RP. The influence of dwarfing interstock on the distribution and metabolism of xylem-applied tritiated gibberellin A_4 in apple. Plant Physiol. 1986; 82:1091–1095.
25. Webber JE, Ross SD, Pharis RP, et al. Interaction between gibberellin $A_{4/7}$ and root-pruning on the reproductive and vegetative processes in Douglas-fir. II. Effects on shoot elongation and its relationship to flowering. Can J For Res. 1985; 15:348–353.
26. Williams DJ, Danick BP, Pharis RP. Early progeny testing and evaluation of controlled crosses of black spruce. Can J For Res. 1987; 17:1442–1450.
27. Dunberg A, Oden PC. Gibberellin and conifers. In: Crozier A, ed. The biochemistry and physiology of gibberellin, Vol. 2. New York: Praeger, 1983: pp. 221–295.
28. Kawarada A, Sumiki Y. The occurrence of gibberellin A_1 in watersprouts of citrus. Bull Agric Chem Soc Jpn. 1959; 23:343–344.
29. Poling SM, Mair VP. Identification of endogenous gibberellins in navel orange shoots. Plant Physiol. 1988; 88:639–642.
30. Dathe W, Sembdner G, Yamaguchi L, et al. Gibberellins and growth inhibitors in spring bleeding sap, roots and branches of *Juglans regia* L. Plant Cell Physiol. 1982; 23:115–123.

31. Koshioka M, Taylor JS, Edwards GR, et al. Identification of gibberellin A_{19} and A_{20} in vegetative apple tissue. Agric Biol Chem. 1985; 49:1223–1226.
32. Rood SB, Bate NJ, Mander LN, et al. Identification of gibberellins A_1 and A_{19} from *Populus balsamifera* × *Populus deltoides*. Phytochemistry. 1988; 27:11–14.
33. Junttila O, Abe H, Pharis RP. Endogenous gibberellin in elongating shoots of clones of *Salix dasyclados* and *Salix viminalis*. Plant Physiol. 1988; 87:781–784.
34. Davies JK, Jensen E, Junttila O, et al. Identification of endogenous gibberellin from *Salix pentandra*. Plant Physiol. 1985; 78:473–476.
35. Oden PC, Andersson B, Gref R. Identification of gibberellin A_9 in extracts of Norway spruce (*Picea abies* (L.) Karst.) by combined gas chromatography–mass spectrometry. J Chromatogr. 1982; 247:173–148.
36. Oden PC, Schwenen L, Graebe JE. Identification of gibberellins in Norway spruce (*Picea abies* (L.) Karst.) by combined gas chromatography–mass spectrometry. Plant Physiol. 1987; 84:516–519.
37. Moritz T, Philipson JJ, Oden PC. Detection and identification of gibberellins in Sitka spruce (*Picea sitchensis*) of different ages and coning ability by bioassay, radioimmunoassay and gas chromatography–mass spectrometry. Physiol Plant. 1989; 75:325–332.
38. Lorenzi R, Saunders PF, Heald JK, et al. A novel gibberellin from needles of *Picea sitchensis*. Plant Sci Lett. 1977; 8:179–182.
39. Zhang R, Pearce D, Pharis RP, et al. Identification and quantitation of endogenous gibberellins in *Pinus radiata*. In: Pharis RP, Rood SB, eds. Abstracts for the 13th International Conference on Plant Growth Substances, Calgary, 1988; Abstract No. 377.
40. Lorenzi R, Horgan R, Heald JK. Gibberellin A_9 glucosyl ester in needles of *Picea sitchensis*. Phytochemistry. 1976; 15:789–790.
41. Dathe W, Sembdner G, Kefeli VI, et al. Gibberellins, abscisic acid, and related inhibitors in branches and bleeding sap of birch (*Betula pubescens* Ehrh.). Biochem Physiol Pflanzen. 1978; 173:238–248.
42. Wareing PF, Saunders PF. Hormones and dormancy. Ann Rev Plant Physiol. 1971; 22:261–288.
43. Lavender DP, Silim SN. The role of plant growth regulators in dormancy in forest trees. In: Kossuth SV, Ross SD, eds. Hormonal control of tree growth. Dordrecht: Martinus Nijhoff Publishers, 1987: pp. 171–192.
44. Dunberg A. Dynamics of gibberellin-like substances and of indole-3-acetic acid in *Picea abies* (L.) Karst. during the period of shoot elongation. Physiol Plant. 1976; 32:186–190.
45. Dunberg A, Malmberg G, Sassa T, et al. Metabolism of tritiated gibberellin A_4 and A_9 in Norway spruce, *Picea abies* (L.) Karst. Effects of a cultural treatment known to enhance flowering. Plant Physiol. 1983; 71:257–262.
46. Bate NJ, Rood SB, Blake TJ. Gibberellins and heterosis in poplar. Can J Bot. 1988; 66:1148–1152.
47. Graebe JE. Gibberellin biosynthesis and control. Ann Rev Plant Physiol. 1987; 38:419–465.
48. Junttila O, Pharis RP. Studies on the metabolism of [^{3}H]GA$_1$ in *Salix* in relation to photoperiod. In: Schreiber K, Schutte HR, Sembdner G, eds. Conju-

gated plant hormones—structure, metabolism and function. Proceedings of the International Symposium, Gera, 1987; pp. 205–215.

49. Rood SB, Junttila O. Minor influence of photoperiod on the metabolism of gibberellin A_{20} in *Salix pentandra*. Physiol Plant. 1989; 76:506–510.
50. Junttila O. Effects of different gibberellins on elongation growth under short day conditions in seedlings of *Salix pentandra*. Physiol Plant. 1981; 53:315–318.
51. Dunberg A. Metabolism of tritiated gibberellin A_1 in seedlings of Norway spruce (*Picea abies*). Physiol Plant. 1981; 51:349–352.
52. Kamienska A, Pharis RP, Wample RL, et al. Occurrence and metabolism of gibberellin in conifers. In: Tamura S, ed. Plant growth substances 1973. Tokyo: Hirokawa Publishing Co., 1974: p. 305–313.
53. Wample RL, Durley RC, Pharis RP. Metabolism of gibberellin A_4 by vegetative shoots of Douglas fir at 3 stages of ontogeny. Physiol Plant. 1975; 35:273–278.
54. Lavender DP, Sweet GB, Zaerr JB, et al. Spring shoot growth in Douglas-fir may be initiated by gibberellins exported from the roots. Science. 1973; 182:838–839.

CHAPTER 20

The Gibberellin Control of Cell Elongation

M. Katsumi and K. Ishida

1 Introduction

The most typical physiological effect of gibberellin (GA) is the stimulation of shoot elongation, and recent studies with dwarf plants of maize, pea, rice, and others have clearly shown that GA is a key hormone controlling plant height.[1,2] However, the mechanism of GA stimulation of shoot growth has yet to be clarified.

While shoot elongation consists of the two cellular processes, cell proliferation and cell elongation (expansion), substantial contribution to the increase in plant size can be attributed to the latter process. Plant cell elongation is a dynamic and complex process of biochemical and biophysical events involving water absorption and cell wall expansion.[3–5]

Literature reviews indicate that GA decreases cell osmotic potential,[6–9] thereby increasing water uptake,[9] which may affect cell hydraulic properties. Gibberellin has also been shown to affect cell wall extensibility.[10–14]

Although Cleland et al.[15] and Katsumi and Kazama[7] were unable to show that GA changes the mechanical properties of the cell wall in cucumber hypocotyls, as determined by the Instron technique, Cosgrove and Sovonick-Dunford[14] argue, on the basis of data in pea obtained by the pressure-block technique, that GA affects not only the rate of wall relaxation but also the value of the yield threshold, which is not affected by auxin. They conclude that GA acts on the cell wall in ways different from auxin.[16]

Gibberellin has been suggested to affect cell wall architecture by changing the pattern of cellulose microfibril arrangement through its effect on cortical microtubule orientation.[17–21] The orientation of cellulose microfibrils is crucial in determining the shape of plant cells. In the wall of young growing cells, cellulose microfibrils are oriented transversely to the cell elongation axis.[22] Such cells are obviously mechanically more extensible in the longitudinal direction of the cell than those with longitudinally and/or randomly arranged cellulose micorfibrils. Therefore, the extensibility of the cell wall and hence the elongation capacity of a cell should be primarily limited by the orientation pattern of cellulose microfibrils. Cells having transversely arranged cellulose microfibrils should have a potentially higher capacity for elongation (the cell growth potential).

Ledbetter and Porter[23] first suggested that cortical microtubules control the alignment of newly synthesized cellulose microfibrils on the inner surface of the cell wall. Since then, there has been substantial evidence accumulated to support this microtubules–microfibrils relationship.[24–26] Thus, cortical microtubules are usually arranged in parallel with microfibrils that are being formed. In the above context, it can be said that the pattern of cortical microtubule arrangement determines the cell growth potential. In fact, Iwata and Hogetsu[27] have shown that in *Avena* coleoptiles and mesocotyls, transverse arrangement of cortical microtubules is correlated with cell elongation.

Gibberellin, which stimulates cell elongation, has been demonstrated to induce transverse orientation of microtubules.[17–21] Auxin, which stimulates cell elongation through cell wall loosening, seems to have the same effect on microtubule orientation.[28] In contrast to GA, kinetin[17] and ethylene,[29–31] which stimulate the expansion (thickening) of cells, have been shown to induce longitudinal orientation of microtubules.

In this report, we present evidence to support the GA control of microtubule orientation using the dwarf-5 (d_5) mutant of maize and discuss the mechanism of GA-induced cell elongation in relation to cell wall extensibility. The d_5 mutant was used because it is a single-gene mutant with a lesion in GA biosynthesis[1]; it therefore provides an excellent system for the analysis of the relation of gibberellin to cell wall extension.

2 Materials and Methods

Seeds of *Zea mays* L. segregating normal and d_5 were kindly supplied by Professor B.O. Phinney, University of California, Los Angeles. Dark-grown seedlings, 7 days after planting, were transplanted and cultured in water or 100 μM GA_3 solution for 24 h as described previously.[21] The mesocotyl length of d_5 seedlings was less than one-fifth the length of normal seedlings, and the former responded to GA_3 (Fig. 1). For both normal and d_5 seedlings, the 4-mm portion of the mesocotyl below the coleoptilar node was excised at 1-mm intervals, and the four 1-mm sections were designated as A,B,C, and D (Fig. 2).

Segments were fixed, cryosections of tissues were prepared (Fig. 3), and microtubles were observed by immunofluorescence microscopy using the method described by Hogetsu.[32] The numbers of cells with microtubules oriented transversely (60°–90°), obliquely (30°–60°), and longitudinally (0°–30°) were counted from photomicrographs in order to obtain their percentage distributions in the A–D regions.

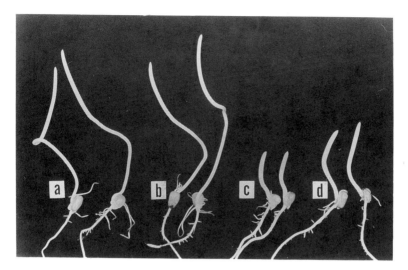

Fig. 1. Photograph showing dark-grown normal and d_5 seedlings: (a) normal control; (b) normal treated with 100 μM GA$_3$; (c) d_5 control; (d) d_5 treated with 100 μM GA$_3$. Photograph was taken 48 h after treatment. Note that the mesocotyls of d_5 seedlings responded to GA$_3$ (d)

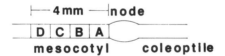

Fig. 2. Illustration showing how regional sections were excised

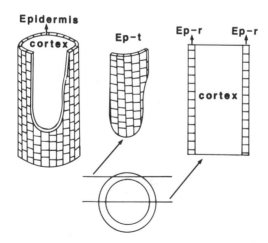

Fig. 3. Illustrations showing how tissue sections were made for microscopy. Ep-t, Tangential surface of epidermis; Ep-r, radial surface of epidermis

214

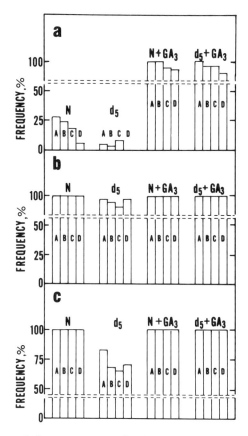

Fig. 5. Frequency of the occurrence of cells with transversely oriented microtubules in the A–D regions of the mesocotyl: (a) tangential surfaces of epidermal cells; (b) radial surfaces of epidermal cells; (c) cortex cells

3 Results and Discussion

The microtubule arrangements in the tangential surfaces of the epidermal cells were transverse, oblique, and/or longitudinal in both normal and d_5 seedlings (Fig. 4a). However, in normal seedlings the frequency of the occurrence of transversely oriented microtubules was much higher than in d_5 seedlings (Fig. 5a). In the A region of normal seedlings, about 20% of cells observed had transversely oriented microtubules, whereas in the same region of d_5 seedlings only about 5% of cells showed transversely oriented microtubules. In the lower regions of both seedlings, the percentage of cells with transversely oriented microtubules decreased, while that of cells with obliquely and/or longitudinally oriented microtubules increased. As shown clearly in Figs. 4a and 5a, in normal and d_5 GA_3-treated seedlings, more than 95% of the cells observed had transversely oriented microtubules in all the regions.

In contrast to the tangential surfaces, the radial surfaces of the epidermal cells in all the regions had transversely oriented microtubules except for less than 10% of those of d_5 seedlings, which showed obliquely oriented microtubules (Figs. 4b and 5b). After GA_3 treatment, however, all the cells had transversely oriented microtubules.

A distinct difference in the pattern of microtubule orientation was also observed between cortex cells of normal and d_5 seedlings (Figs. 4c and 5c). In normal seedlings, all the cortex cells in the four regions had transversely oriented microtubules, while in d_5 seedlings 80% or more of A region cells showed such microtubule orientation as compared to less than 70% in the lower regions (B, C, D). After GA_3 treatment, all the cortex cells of the four regions of both d_5 and normal seedlings had transversely oriented microtubules. Results from the observation of microtubule patterns in the A and D regions of normal and d_5 seedlings with and without GA_3 treatment are summarized in Fig. 6.

The present results confirm the previous report in which a similar observation was made on the tangential surface of the epidermal cells by electron microscopy.[21] In the present study, cortex cells and the radial surfaces of the epidermal cells as well as tangential surfaces were observed and compared by immunofluorescence microscopy. The present method is superior to electron microscopy in that the entire pattern of microtubule arrangement in a cell as well as the distribution of cortical microtubule patterns in the tissue can be observed.

There is a distinct difference in the distribution of microtubule orientation patterns between normal and d_5 seedlings. The difference is especially significant in the tangential surfaces of epidermal cells and in cortex cells. In d_5 seedlings, the percentage of cells with transversely oriented microtubules is apparently less than in normal seedlings, and also the occurrence of transversely oriented microtubules is restricted to a narrow region below the coleoptilar node (A region). It can be suggested, on the basis of these observations, that d_5 mesocotyls have a smaller population of cells with high growth potential as defined in Section 1. Gibberellin increases the number of cells with transversely oriented microtubules and stimulates mesocotyl elongation. Therefore, the dwarf habit of d_5 mesocotyl elongation can be attributed to the presence of a small population of cells with high growth potential due to an insufficient supply of endogenous GA. This interpretation has also been suggested in the previous paper.[21]

Endogenous GA is necessary to induce transversely oriented microtubules in both epidermal and cortex cells. However, the actual cell wall expansion, and hence cell elongation, should involve auxin action, which is associated with the cell wall matrix. Cosgrove and Sovonick-Dunford[14] have suggested that gibberellin affects the value of the "yield threshold" of the cell. At present, it is not certain what cellular element(s) determines

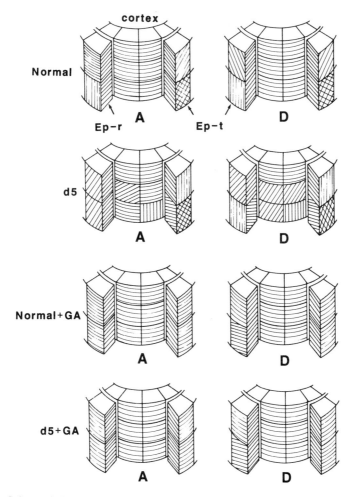

Fig. 6. Schematic illustration of patterns of the microtubule arrangement in the A and D regions of the mesocotyl. Ep-t, tangential surface of epidermis; Ep-r, radial surface of epidermis

this value. If we assume that it is the pattern of cellulose microfibril arrangement, then GA may affect the value of the yield threshold.

Acknowledgments. We wish to thank Professor N. Hara, Professor K. Syono, and Dr. T. Hogetsu, who kindly gave us permission to use their research facilities at the University of Tokyo, Komaba. This study was supported by a Grant-in-Aid for Special Project Research (No. 63110007) to M. Katsumi from the Ministry of Education, Science and Culture, Japan. The data presented here will be included in Ms. Ishida's Master's thesis.

References

1. Phinney BO. Gibberellin A_1, dwarfism and the control of shoot elongation in higher plants. In: Crozier A, Hillman J, eds. The biosynthesis and metabolism of plant hormones. Soc. Exp. Biol. Seminar Ser. 23. Cambridge: Cambridge University Press, 1984: pp. 17–41.
2. MacMillan J. Gibberellin deficient mutants of maize and pea and molecular basis of gibberellin action. In: Hoad GV, Lenton JR, Jackson MB, et al, eds. Hormone action in plant development—a critical appraisal. London: Butterworths, 1987: pp. 73–88.
3. Taiz L. Plant cell expansion: Regulation of cell wall mechanical properties. Annu Rev Plant Physiol. 1984; 35:585–657.
4. Cosgrove DJ. Biophysical control of plant cell growth. Annu Rev Plant Physiol. 1986; 37:377–405.
5. Ray PM. Principles of plant cell growth. In: Cosgrove DJ, Knievel DJ, eds. Physiology of cell expansion during plant growth. Symposium in Plant Physiology, Pennsylvania State University. Rockville, Maryland: American Society of Plant Physiologists, 1987: pp. 1–17.
6. Ende J, Koornneef P. Gibberellic acid and osmotic pressure. Nature. 1968; 219:510–511.
7. Katsumi M, Kazama H. Gibberellin control of cell elongation in cucumber hypocotyl sections. Bot Mag Tokyo. 1978; Special Issue 1:141–158.
8. Katsumi M, Kazama H, Kawamura N. Osmotic potential of the epidermal cells of cucumber hypocotyls as affected by gibberellin and cotyledons. Plant Cell Physiol. 1980; 21:933–937.
9. Kazama H, Katsumi M. Gibberellin-induced changes in the water absorption, osmotic potential and starch content of cucumber hypocotyls. Plant Cell Physiol. 1983; 24:1209–1216.
10. Adams PA, Montague MJ, Tepfer M, et al. Effect of gibberellic acid on the plasticity and elasticity of *Avena* stem segments. Plant Physiol. 1975; 56:757–760.
11. Kawamura H, Kamisaka S, Masuda Y. Regulation of lettuce hypocotyl elongation by gibberellic acid. Correlation between cell elongation, stress-relaxation properties of the cell wall and wall polysaccharide content. Plant Cell Physiol. 1976; 17:23–34.
12. Nakamura T, Sekine S, Arai K, et al. Effects of gibberellic acid and indole-3-acetic acid on stress-relaxation properties of pea hook cell wall. Plant Cell Physiol. 1975; 16:127–138.
13. Stuart DA, Jones RL. Roles of extensibility and turgor in gibberellin- and dark-stimulated growth. Plant Physiol. 1977; 59:61–68.
14. Cosgrove DJ, Sovonick-Dunford SA. Mechanism of gibberellin-dependent stem elongation in pea. Plant Physiol. 1989; 89:184–191.
15. Cleland R, Thompson ML, et al. Differences in effects of gibberellins and auxins on wall extensibility of cucumber hypocotyls. Nature. 1968; 219:510–511.
16. Cleland RE. The mechanism of wall loosening and wall extension. In: Cosgrove DJ, Knievel DJ, eds. Physiology of cell expansion during plant growth. Symposium in Plant Physiology, Pennsylvaria State University. Rockville, Maryland: American Society of Plant Physiologists. 1987: pp. 18–27.

17. Shibaoka H. Involvement of wall microtubules in gibberellin promotion and kinetin inhibition of stem elongation. Plant Cell Physiol. 1974; 15:255–263.
18. Shibaoka H, Hogetsu T. Effects of ethyl N-phenylcarbamate on wall microtubules and on gibberellin- and kinetin-controlled cell expansion. Bot Mag Tokyo 1977; 90:317–321.
19. Mita T, Shibaoka H. Gibberellin stabilizes microtubules in onion leaf sheath cells. Protoplasma. 1984; 119:100–109.
20. Mita T, Shibaoka H. Effects of S-3307, an inhibitor of gibberellin biosynthesis, on swelling of leaf sheath cells and on the arrangement of cortical microtubules in onion seedlings. Plant Cell Physiol. 1984; 25:1531–1539.
21. Mita T, Katsumi M. Gibberellin control of microtubule arrangement in the mesocotyl epidermal cells of the d_5 mutant of *Zea mays* L. Plant Cell Physiol. 1986; 27:651–659.
22. Preston RD. The physical biology of plant cell walls. London: Chapman and Hall, 1974: pp. 383–409.
23. Ledbetter MC, Porter KP. A "microtubule" in plant cell fine structure. J Cell Biol. 1963; 19:239–250.
24. Gunning BES, Hardham AR. Microtubules. Annu Rev Plant Physiol. 1982; 33: 651–698.
25. Robinson DG, Quader H. The microtubule–microfibril syndrome. In: Lloyed CW, ed. The cytoskelton in plant growth and development. London: Academic Press, 1982: pp. 109–126.
26. Lloyed CW. Toward a dynamic helical model for the influence of microtubules on wall patterns in plants. Int Rev Cytol. 1984; 86:1–51.
27. Iwata K, Hogetsu T. Arrangement of cortical microtubules in *Avena* coleoptiles and mesocotyls and *Pisum* epicotyls. Plant Cell Physiol. 1988; 29:807–815.
28. Iwata K, Hogetsu T. The effects of light irradiation on microtubule orientation in seedlings of *Avena sativa* L. and *Pisum sativum* L. Plant Cell Physiol. 1989; 30:1011–1016.
29. Steen DA, Chadwick AV. Ethylene effects in pea stem tissue: Evidence of microtubule mediation. Plant Physiol. 1981; 67:460–466.
30. Lang JM, Eisinger WR, Green PB. Effects of ethylene on the orientation of microtubules and cellulose microfibrils of pea epicotyl cells with polylamellate walls. Protoplasma. 1982; 110:5–14.
31. Roberts IN, Lloyed CW, Roberts K. Ethylene- induced microtubule reorientations: Mediation by the helical arrays. Planta. 1985; 116:439–447.
32. Hogetsu T. Immunofluorescence microscopy of microtubule arrangement in root cells of *Pisum sativum* L. var. Alaska. Plant Cell Physiol. 1986; 27:939–945.

CHAPTER 21

The Role of Gibberellin in the Formation of Onion Bulbs

H. Shibaoka

1 Introduction

Onion plants form bulbs in response to the stimulus of long-day conditions. More than forty years ago, Heath[1] reported that the bulb is formed by the swelling of leaf-sheath cells and that cell division is not involved in this phenomenon. Under short-day conditions, leaf-sheath cells elongate longitudinally, whereas under long-day conditions they expand laterally. This means that leaf-sheath cells respond to a long-day stimulus by a change in direction of cell expansion. The direction of cell expansion depends largely on the orientation of the cellulose microfibrils in the cell wall.[2] The orientation of cellulose microfibrils, in turn, is considered to be controlled by cortical microtubules.[3] Since changes in the direction of cell expansion are part of bulb development, the cortical microtubules may be involved in the regulation of this development. Therefore, we examined the behavior of cortical microtubules in the leaf-sheath cells of onion plants during bulb development. We have reported previously that the suppression by gibberellin A_3 (GA_3) of the lateral expansion of cells in azuki bean epicotyls is correlated with an increased number of transversely oriented cortical microtubules.[4] This is followed by the accumulation of transversely oriented cellulose microfibrils in the cell wall.[5] Thus, in the work reported here, we examined the effects of GA_3 on the arrangement of microtubules during the development of onion bulbs.

2 Changes in Microtubules During Bulb Development

Onion plants (a late-growing cultivar of *Allium cepa* L.) were grown in a field near Osaka City. Small pieces of tissue were cut from the leaf sheath of plants from the time before initiation of bulb development (April 23) to that of visible bulb development (May 21). The tissues were fixed and embedded in Spurr's resin.[6] Transverse and longitudinal sections were obtianed and used for the measurment of cross-sectional areas of the cells and for the examination of the orientation of cortical microtubules.

Table 1. Changes in the arrangement of microtubules adjacent to the outer tangential wall of cells, just beneath the outer epidermis of leaf sheaths from onion plants, during bulb development. From Mita and Shibaoka[6]

Sampling date[a]	Position and morphology of leaf	Length of leaf sheath (mm)	Position of cells[b] (mm)	Microtubule arrangement[c] Density[d]	Microtubule arrangement[c] Orientation	Cross-sectional area of cell[e] (μm^2)
April 23 (13:19)	Emerged leaf	40–45	5	++	transverse	704 ± 25
	Emerged leaf	20–25	15	+	transverse	762 ± 49
			5	++	transverse	526 ± 49
			15	+	transverse	700 ± 36
	Innermost emerged leaf	8–9	5	++	transverse	426 ± 16
May 1 (13:35)	Innermost emerged leaf	25–30	5	+	transverse	603 ± 19
			15	++	transverse	647 ± 31
	First scale	10	5	+	random	452 ± 31
May 14 (13:57)	Innermost emerged leaf	25–30	5	++	random	1081 ± 46
			15	+	random	1214 ± 61
	First scale	20–25	5	+	random	727 ± 38
			15	+	random	634 ± 27
	Second scale	6	5	+	random	382 ± 22
May 21 (14:07)	Innermost emerged leaf	90	5	±	random	2619 ± 109
			15	±	random	3544 ± 167
	Innermost emerged leaf	70	5	±	random	1031 ± 43
			15	±	random	1174 ± 63
	First scale	40–50	5	±	random	2101 ± 118
			15	−	—	3927 ± 200
	First scale	20–25	5	+	random	569 ± 26
			15	±	random	861 ± 41
	Second scale	8	5	+	random	445 ± 19

[a] In parentheses, hours and minutes from sunrise to sunset
[b] Distance of cells from the basal end of the leaf sheath
[c] At least 40 cells from 4 different plants were examined
[d] ++, Dense; +, less dense; ±, sparse; −, none
[e] Average of results from 40 cells with standard error of the mean. Expressed as (radial width) × (tangential width) in μm^2

Swellings were evident in the cortical cells, not the epidermal cells, of the leaf sheath.[6] Therefore, the orientation of microtubules was examined in cortical cells just beneath the epidermis. Transversely oriented microtubules were present near the face of the outer tangential walls of cells in tissues sampled at a time when leaf-sheath swelling had not yet begun (April 23) (Table 1). Cells were sampled at a time when leaf-sheath swelling had just begun (May 14) and at a time when leaf-sheath swelling was evident (May 21); the orientation of microtubules in these samples was examined. Transversely oriented microtubules were replaced by scattered, mostly nonoriented microtubules (Table 1). Microtubule number decreased gradually, being lower in cells from tissues sampled on May 21 than in those sampled on May 14. Microtubule number was lower in cells 15 mm above the basal end of the leaf sheath than in cells 5 mm above the basal end. No microtubules were observed in cells 15 mm above the basal end of scales with a 45-mm-long leaf sheath, on May 21 (Table 1). In general, no microtubules were visible near the faces of the walls of the cells taken from swollen scales. This swelling of the cells may be caused by the absence of the transversely oriented cortical microtubules necessary for the deposition of transversely oriented cellulose microfibrils.

3 Effects of Microtubule-Disrupting Agents

Heath and Holdsworth[7] suggested that onion leaf blades produced a "bulbing hormone" in response to the stimulus of long day. The observation that formation of onion bulbs under long-day conditions was accompanied by a decrease in the number of cortical microtubules suggests that the "bulbing hormone" could be some kind of a microtubule-disrupting agent. Thus, we examined the effects of two microtubule-disrupting agents, colchicine and cremart,[8] on the swelling of leaf-sheath cells.

Two- and three-leafed young onion seedlings were grown under short-day conditions (12 h light/12 h dark) and treated either with colchicine ($3 \times 10^{-4}\,M$) or cremart ($2 \times 10^{-5}\,M$). Both colchicine and cremart resulted in a leaf-sheath swelling and increased the cross-sectional areas of both epidermal and cortical cells. Both colchicine- and cremart-treated plants had a lowered number of microtubules in the leaf-sheath cells. It appears that the presence of transversely oriented cortical microtubules prevents leaf-sheath swelling.

4 Effects of Root Removal

Under continuous light, no swelling of leaf sheaths of seedlings at the two- or three-leaf stage were observed. However, leaf-sheath swelling was observed following removal of the roots from seedlings grown in continuous light, suggesting that roots are the source of an "antibulbing hormone" and supply it to leaf sheaths. Histological observations also revealed that the swelling of root-removed seedlings was due to the lateral expansion of both cortical and epidermal leaf-sheath cells.[9] Root removal also resulted in changes in the arrangement of microtubules in the leaf-sheath cells.[9] Transversely or nearly transversely oriented microtubules were present near the face of the outer tangential walls of leaf-sheath cells sampled from seedlings with intact roots. By contrast, microtubules from root-removed seedlings were oriented parallel or obliquely to the cell axis. Changes in the orientation of microtubules were evident 2 days after the root removal and before visible evidence of swelling. Changes in the orientation of microtubules were not accompanied by a decrease in the number of microtubules. Abundant microtubules were observed in cells sampled at a time when swelling of cells had just begun (3 days after root removal).

When seedlings were grown under short day, removal of the roots caused neither lateral expansion of leaf-sheath cells nor changes in the arrangement of microtubules. Transversely or nearly transversely oriented microtubules were present in leaf-sheath cells of both intact and root-removed seedlings. These observations indicate that the swelling is not solely caused by removal of the supply of the "antibulbing hormone," but that swelling also requires a presumptive "bulbing hormone" that appears to be produced in leaf blades in response to the stimulus of long day.[7]

5 Effects of S-3307, a GA Biosynthesis Inhibitor

Gibberellin is a candidate for the "antibulbing hormone" coming from the root, since GA has been observed to cause the accumulation of transversely oriented cortical microtubules, especially in azuki bean epicotyls,[4] in dwarf maize mesocotyls,[10] and in dwarf pea epicotyls.[11] Therefore, we have hypothesized that roots suppress the swelling of the leaf sheath in onions by supplying the leaf sheath with endogenous GA. To support this supposition, we treated the seedlings with the GA biosynthesis inhibitor S-3307[12] to see if it affects the swelling of leaf sheaths and the arrangement of cortical microtubules in these leaf sheaths.

Treatment with S-3307 (10^{-4} M) caused swelling of leaf sheaths in two- and three-leafed young onion seedlings grown under long-day conditions (Fig. 1). The swelling was brought about by the lateral expansion of leaf-sheath cells; no increases in the number of cells were observed. The treatment resulted in the lateral expansion of both epidermal and cortical cells.

Treatment with S-3307 (10^{-4} M) also resulted in changes in the arrangement of the microtubules. Cortical microtubules in the leaf-sheath cells were oriented longitudinally or obliquely in seedlings treated with S-3307,[13] whereas they were oriented transversely to the cell axis in un-

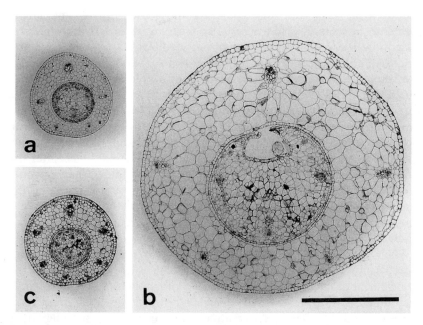

Fig. 1. Antagonism between GA_3 and S-3307 in the swelling of the basal parts of onion seedlings grown under long-day conditions. Transverse sections were cut 2 mm above the basal ends of the second leaves. (a) Pretreated with water for 2 days and again with water for 10 days. (b) Pretreated with water for 2 days and then with S-3307 (10^{-4} M) for 10 days. (c) Pretreated with GA_3 (10^{-4} M) for 2 days and then with GA_3 (10^{-4} M) plus S-3307 (10^{-4} M) for 10 days. Bar = 1 mm. From Mita and Shibaoka[13]

treated seedlings. Longitudinally or obliquely oriented microtubules were observed in cells sampled at a time when the swelling was not yet evident (2 days after the initiation of treatment). The changes in the orientation of microtubules were not accompanied by a change in the number of microtubules. Abundant microtubules were observed in cells sampled at a time when cell swelling had just begun (3 days after the initiation of treatment). It is concluded that endogenous GA plays an important role in the arrangement of cortical microtubules in the leaf sheaths of the onion.

The lateral expansion of cells resulting from S-3307 (10^{-4} M) treatment was greatly reduced by the simultaneous or prior application of GA_3 (10^{-4} M). In some experiments, simultaneous or prior application of GA_3 only partially reversed the effects of S-3307; however, a combination of the two treatments always resulted in complete reversal of the S-3307 effect. Seedlings pretreated with GA_3 (10^{-4} M) for 2 days and then with GA_3 (10^{-4} M) plus S-3307 (10^{-4} M) showed little or no swelling (Fig. 1).

When seedlings were grown in short day, S-3307 (10^{-4} M) did not cause noticeable swelling of the seedling leaf sheaths. As previously mentioned, treatment with S-3307 caused changes in the orientation of cortical microtubules in seedlings grown under long day. However, it did not cause any changes in the orientation of cortical microtubules in seedlings grown under short day.

The hypothesis that the "antibulbing hormone" from the root is GA and that root removal results in swelling via a reduction in levels of GA in leaf sheaths seems tenable.

6 Antagonism Between GA and Microtubule-Disrupting Agents

If the "bulbing hormone" is some kind of a microtubule-disrupting agent and if the "antibulbing hormone is GA, exogenous GA should have the ability to reverse the effects of the microtubule-disrupting agents. Thus, we examined whether or not GA_3 could act as an antagonist to the microtubule-disrupting agent with respect to both the swelling of leaf-sheath cells and the disruption of microtubules.

Gibberellin A_3 has been shown to reverse the stimulation of cell swelling in onion by both colchicine and cremart.[14] The basal parts of leaf sheaths of onion seedlings did not become bulbous when seedlings were treated with colchicine (3×10^{-4} M) plus GA_3 (10^{-4} M) or with cremart (2×10^{-5} M) plus GA_3 (10^{-4} M). In contrast, seedlings treated with colchicine (3×10^{-4} M) alone or cremart (2×10^{-5} M) alone became bulbous.[14] Gibberellin A_3 (10^{-4} M) does not promote the elongation of leaf sheaths in onion seedlings. Gibberellin A_3 (10^{-4} M) alone slightly promoted the lateral expansion of cells. Gibberellin A_3 (10^{-4} M) almost completely reversed the effect of colchicine (3×10^{-4} M) on the lateral expansion of cells; GA_3 (10^{-4} M) only partially reversed the effect of cremart (2×10^{-5} M) (Table 2).

Microtubules were oriented predominantly transversely or nearly transversely to the cell axis in cells from onion seedlings treated with GA_3 (10^{-4} M). The majority of microtubules were oriented transversely or nearly transversely to the cell axis also in cells from water-treated seedlings. However, detailed examination revealed that there was a difference in the arrangement of microtubules between cells from GA_3-treated seedlings and cells from the water-treated seedlings (Fig. 2). In the GA_3-treated seedlings, about 80% of microtubules were oriented within the angular ranges of 80°–90° to the cell axis. In the water-treated seedlings, microtubules were oriented in a relatively wide range of directions, although about 65% of microtubules were oriented within the angular ranges of 60°–90° to the cell axis (Fig. 2).

We note that GA_3 prevents the microtubule disruption by colchicine and

Table 2. Antagonism between microtubule-disrupting agents and GA_3 in the lateral expansion of onion leaf-sheath cells. From Mita and Shibaoka[14]

Days after the start of treatment	Treatments	Cross-sectional area of cell (μm^2)[a]	
		Epidermal cell	2nd layer cell[b]
2	Water	224 ± 5	300 ± 21
	Colchicine ($3 \times 10^{-4}\ M$)	222 ± 5	335 ± 27
	Cremart ($2 \times 10^{-5}\ M$)	229 ± 5	359 ± 23
9	Water	263 ± 6	385 ± 25
	GA_3 ($10^{-4}\ M$)	310 ± 7	489 ± 29
	Colchicine ($3 \times 10^{-4}\ M$)	2343 ± 128	3343 ± 270
	Colchicine + GA_3 ($10^{-4}\ M$)	360 ± 6	457 ± 29
	Cremart ($2 \times 10^{-5}\ M$)	3116 ± 202	4629 ± 475
	Cremart + GA_3 ($10^{-4}\ M$)	1078 ± 66	1035 ± 66

[a] Cells 2 mm above the basal end of the second leaf were examined.
 Average of results from 45 cells with standard error of the mean.
 Expressed as (radial width) × (tangential width) in μm^2
[b] Cell just beneath the epidermis

cremart. Transversely oriented microtubules were abundant in cells from seedlings treated with colchicine ($3 \times 10^{-4}\ M$) plus GA_3 ($10^{-4}\ M$) or with cremart ($2 \times 10^{-5}\ M$) plus GA_3 ($10^{-4}\ M$).[14] Gibberellin A_3 ($10^{-4}\ M$) prevented the microtubule disruption normally resulting from exposure to low temperature. Abundant microtubules were observed in cells from seedlings pretreated with GA_3 ($10^{-4}\ M$) for 2 days and then kept at 1 °C for 2.5 h. No microtubules or only C-shaped microtubules were observed in cells from seedlings that were pretreated with water for 2 days and then kept at 1 °C for 2.5 h.[14]

7 Conclusions

The formation of onion bulbs seems to be controlled by two hormones—the "bulbing hormone," which originates from the leaf blade kept under long-day conditions, and the "antibulbing hormone," which originates from the root. The "bulbing hormone" probably induces bulb formation by the disruption of the transverse orientation of cortical microtubules in the leaf-sheath cells. Transversely oriented microtubules are necessary for the construction of transversely arranged cellulose microfibrils, which, in turn, presumably suppress the swelling of the cells. The "antibulbing hormone" is probably GA; it suppresses bulb formation by controlling the disruption of the transverse orientation of cortical microtubules by the "bulbing hormone." It also suppresses the swelling of leaf sheaths by causing the cortical microtubules to be oriented transversely to the cell axis.

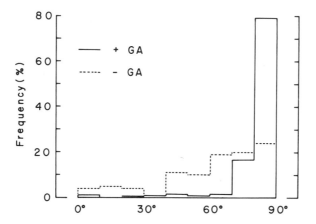

Fig. 2. Effects of GA_3 on the arrangement of microtubules adjacent to the outer tangential walls of onion leaf-sheath epidermal cells. Cells 2 mm above the basal end of the second leaf sampled 2 days after the start of treatment were examined. The histogram shows the percentage of microtubules oriented within the angular ranges of 0°–10°, 10°–20°, etc., to the cell axis. From Mita and Shibaoka [14]

References

1. Heath OVS. Formative effects of environmental factors as exemplified in the development of the onion plant. Nature. 1945; 155:623–626.
2. Preston RD. The physical biology of plant cell walls. London: Chapman and Hall, 1974: pp. 385–409.
3. Robinson DG, Quader H. The microtubule–microfibril syndrome. In: Lloyd CW (ed) The cytoskeleton in plant growth and development. London: Academic Press, 1982: pp. 109–126.
4. Shibaoka H. Involvement of wall microtubules in gibberellin promotion and kinetin inhibition of stem elongation. Plant Cell Physiol. 1974; 15:255–263.
5. Takeda K, Shibaoka H. Effects of gibberellin and colchicine on microfibril arrangement in epidermal cell walls of *Vigna angularis* Ohwi et Ohashi epicotyls. Planta. 1981; 151:393–398.
6. Mita T, Shibaoka H. Changes in microtubules in onion leaf sheath cells during bulb development. Plant Cell Physiol. 1983; 24:109–117.
7. Heath OVS, Holdsworth M. Morphogenic factors as exemplified by the onion plant. Soc Exp Biol Symp. 1948; 2:326–350.
8. Sumida S, Ueda M. Effect of *O*-ethyl-*O*-(3-methyl-6-nitrophenyl)-*N-sec*-butylphosphorothioamidate (S-2846), an experimental herbicide, on mitosis in *Allium cepa*. Plant Cell Physiol. 1976; 17:1351–1354.
9. Mita T, Shibaoka H. Effects of root excision on swelling of leaf sheath cells and on the arrangement of cortical microtubules in onion seedlings. Plant Cell Physiol. 1984; 25:1521–1529.
10. Mita T, Katsumi M. Gibberellin control of microtubule arrangement in the mesocotyl epidermal cells of the d_5 mutant of *Zea mays* L. Plant Cell Physiol. 1986; 27:651–659.

11. Akashi T, Shibaoka H. Effects of gibberellin on the arrangement and the cold stability of cortical microtubules in epidermal cells of pea internodes. Plant Cell Physiol. 1987; 28:339–348.
12. Izumi K, Yamaguchi I, Wada A, et al. Effects of a new plant growth retardant (E)-1-(4-chlorophenyl)-4,4-dimethyl-2-(1,2,4-triazol-1-yl)-1-penten-3-ol (S-3307) on the growth and gibberellin content of rice plants. Plant Cell Physiol. 1984; 25:611–617.
13. Mita T, Shibaoka H. Effects of S-3307, an inhibitor of gibberellin biosynthesis, on swelling of leaf sheath cells and on the arrangement of cortical microtubules in onion seedlings. Plant Cell Physiol. 1984; 25:1531–1539.
14. Mita T, Shibaoka H. Gibberellin stabilizes microtubules in onion leaf sheath cells. Protoplasma. 1984; 119:100–109.

CHAPTER 22

Gibberellin Requirement for the Normal Growth of Roots

E. Tanimoto

1 Introduction

Gibberellin (GA) strongly promotes shoot growth whereas it shows little effect on root elongation in GA-deficient dwarf plants (cf. Fig. 1) and in rosette plants.[1] Although roots of dwarf maize and lettuce respond to exogenous GA in some experimental conditions,[2,3] roots of these plants elongate normally without GA application. These phenomena suggest that roots do not require GA or require less GA than shoots. Thus, studies on the role of GA in root growth have been limited as compared with those in shoots.[4-6] In order to evaluate GA requirements for root growth, the effects of the interaction of ancymidol[7] and GA_3 on the elongation growth of roots and shoots have been studied in lettuce seedlings and in dwarf and normal pea plants.

2 Materials and Methods

Seedling roots of *Lactuca sativa* cv. Grand Rapids and *Pisum sativum* cv. Alaska and cv. Little Marvel were prepared as described elsewhere.[1,8] Pea seeds were germinated on a frame in a humid box[8] in the dark at 23° to prepare straight roots. Lettuce seeds were germinated on a vertical filter paper in a humid cup in fluorescent light at 23 °C. All experiments were carried out in fluorescent light of ca. 300 lux at plant level at 23 °C.

Ancymidol was applied to roots by hydroculture using growth solution containing inorganic nutrients with or without the growth regulators.[1,8] Gibberellin A_3 was applied to roots and/or to shoots by hydroculture or by dipping whole seedlings in GA_3 solution or by spraying the shoots with GA_3 solution.

2.1 Application of GA_3 by the Dipping Method

Seedling roots were pretreated with 3 μM ancymidol for 2 days by hydroculture and transferred to the rhizometer measuring box,[9] which automati-

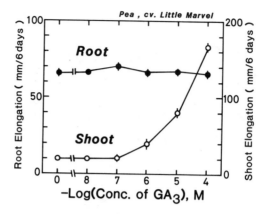

Fig. 1. Effect of root-applied GA_3 on root and shoot elongation of dwarf peas. Root length of 2-day-old seedlings was recorded, and the seedling roots were dipped in hydroculture solution containing different concentrations of GA_3. Lengths of roots and shoots were measured after a 6-day culture period in fluorescent light. Means of 20 seedlings are shown with SE (vertical bars)

cally records root elongation during hydroculture. Whole seedlings were automatically dipped in GA_3 solution every 30 min for 4 days by a computer-controlled system designed for the rhizometer.

2.2 Application of GA_3 by the Spray Method

Seedling roots were pretreated with 3 μM ancymidol for 2 days by hydroculture. Shoots of the seedlings were sprayed with GA_3 solution containing 0.05% Tween-20 every day. Roots were then thoroughly washed with distilled water after every spray and transferred to new hydroculture solution containing ancymidol. The sugar composition of the cell wall was analyzed by gas chromatography after hydrolysis of the cell wall using trifluoroacetic acid as described elsewhere.[8]

3 Results

Root-applied GA_3 showed no effect on root elongation of dwarf peas whereas it strongly promoted shoot elongation (Fig. 1). The same result was obtained in lettuce seedlings.[1] Root elongation of lettuce was suppressed by higher concentrations of ancymidol than those required for suppression of hypocotyl elongation (Fig. 2). When ancymidol was applied to roots in the hydroculture growth solution, hypocotyl elongation was significantly inhibited by 0.12 μM ancymidol while root elongation was inhibited at 4 μM. When GA_3 was applied to roots with 4 μM ancymidol (Fig. 3), the roots overcame ancymidol inhibition at 10 nM while the hypo-

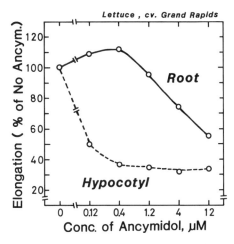

Fig. 2. Effect of ancymidol concentrations on root and hypocotyl elongation of lettuce seedlings. Vertically grown seedling roots were dipped in hydroculture solution containing different concentrations of ancymidol. Root and hypocotyl lengths were measured after a 2-day incubation period in fluorescent light. Elongation is indicated as percent of control (elongation without ancymidol)

cotyl elongation was stimulated at 10 μM. Concentration-dependent promotion of root elongation is indicated in Fig. 4 for experiments in which lettuce seedlings were pretreated with 1 μM ancymidol for 2 days and then transferred to new growth solutions containing 1 μM ancymidol with or without GA_3. Root elongation was stimulated at 1–10 nM GA_3 whereas stimulation of hypocotyl elongation required higher concentrations (1–10 μM). Similar results were obtained in normal pea plants.[8]

The interaction of ancymidol and GA_3 on the seedlings of dwarf peas is shown in Fig. 5. Dwarf peas were further dwarfed by ancymidol. Lengths of shoot and main root were significantly decreased by 10 μM ancymidol, and the development of lateral roots was also inhibited at 10 μM. Ancymidol not only inhibited root elongation but also caused thickening of the apical portion of the roots. Gibberellin A_3 applied with 10 μM ancymidol completely eliminated the ancymidol inhibition. Quantitative data for GA_3 promotion were obtained from experiments in which dwarf pea seedlings were pretreated with ancymidol for 2 days and then a different concentration of GA_3 was applied to roots together with ancymidol. Gibberellin A_3 showed clear promotioin of root elongation at concentrations higher than 10 nM whereas shoot elongation was promoted at 10 μM (Fig. 6). Although GA is known to move smoothly from roots to shoots, root-applied GA_3 may not have reached the shoots if roots preferentially consumed root-applied GA_3. In the next experiment, GA_3 was applied equally to roots and shoots by dipping whole seedlings repeatedly in GA_3 solution using the rhizometer. At a concentration of 10 nM, GA_3 promoted root elongation but failed to increase shoot length, whereas 1 μM GA_3 pro-

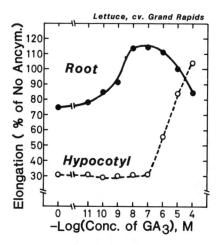

Fig. 3. Effect of GA_3 concentrations on root and hypocotyl elongation of lettuce seedlings in the presence of 4 μM ancymidol. Seedling roots were dipped in hydroculture solution with or without 4 μM ancymidol and containing different concentrations of GA_3. Lengths of roots and hypocotyls were measured after a 2-day incubation period in fluorescent light. Elongation is indicated as percent of control (elongation without ancymidol)

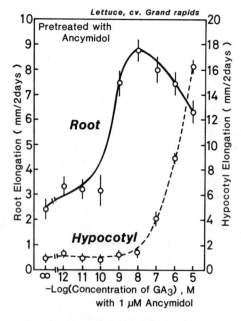

Fig. 4. Effect of GA_3 concentrations on root and hypocotyl elongation in lettuce seedlings pretreated with 1 μM ancymidol. Seedling roots were dipped in hydroculture solution with 1 μM ancymidol for 2 days and then treated with new solutions containing different concentrations of GA_3 in the presence of 1 μM ancymidol for 2 days. Means of 20–30 seedlings are indicated with SE (vertical bars)

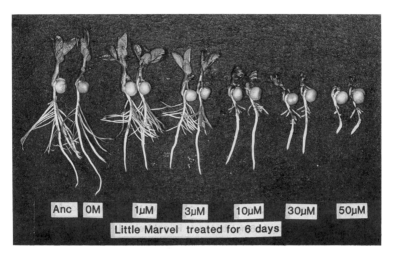

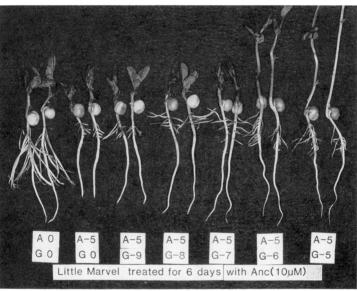

Fig. 5. Photographs of dwarf pea seedlings, cv. Little Marvel, grown by hydroculture in fluorescent light. Two-day old seedling roots were dipped in hydroculture solution containing growth regulators and cultured for 6 days at 23 °C. A–5, 10^{-5} M (10 μM) ancymidol; A0, without ancymidol; G-X, 10^{-x} M GA$_3$; G0, without GA$_3$

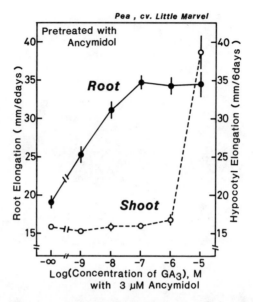

Fig. 6. Effect of GA_3 on root and shoot elongation of dwarf pea seedlings pretreated with 3 μM ancymidol for 2 days. Seedling roots of 2-day-old dwarf peas were dipped in hydroculture solution containing 3 μM ancymidol for 2 days and then seedling roots were treated with 3 μM ancymidol with different concentrations of GA_3 for 6 days. Means of 20 seedlings are indicated with SE (vertical bars)

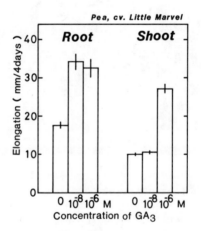

Fig. 7. Effect of GA_3 on root and shoot elongation of dwarf pea seedlings. Seedling roots were pretreated with 3 μM ancymidol for 2 days and then transferred to the rhizometer measuring box.[9] Gibberelin A_3 solution was applied to both roots and shoots equally by dipping whole seedlings in culture solution containing 3 μM ancymidol with or without GA_3. Final lengths of roots and shoots were measured after 4 days during which seedlings were dipped in hydroculture solution every 30 min (192 times per 4 days). Means of 13–15 plants with SE (vertical bars) are indicated

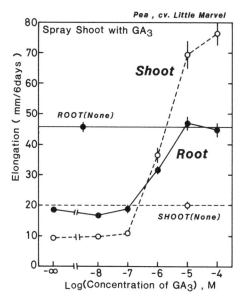

Fig. 8. Effect of spraying shoot with GA_3 on the elongation of roots and shoots of dwarf peas. Two-day old seedling roots were dipped in 3 μM ancymidol for 2 days and then shoots were sprayed every day with GA_3 solution containing 0.05% Tween-20 for 4 days. Roots were dipped in hydroculture solution containing 3 μM ancymidol throughout the experiments. Elongation was measured 6 days after the first spray. Horizontal solid line (None) and broken line (None) indicate the elongation of roots and shoots, respectively, without application of the growth regulators. Means of 20 seedlings are indicated with SE (vertical bars)

moted both (Fig. 7). The effect of varying the site of GA application was further tested by spraying the shoots with GA_3 solution. Shoot-applied GA_3 promoted root elongation at concentrations in the same range as those which promoted shoot elongation (Fig. 8).

Ancymidol led to thickening of the elongation zone of the roots (Fig. 5) whereas GA completely canceled the ancymidol effect on dwarf peas. The same effect of ancymidol was observed in Alaska peas.[8] Morphological change of root cells in Alaska peas is shown in Fig. 9. Microscopic sections were taken from the elongation zone of the root. Ancymidol led to expansion of cortical cells while GA_3 completely canceled the ancymidol effect.

The composition of cell wall neutral sugars of GA_3-treated slender roots was compared with that of ancymidol-treated thick roots of Alaska peas. Sugar composition changed remarkably along the root axis from the tip to the base (Fig. 10). The arabinose content was highest in the tip and rapidly decreased toward the base, whereas galactose complementarily increased toward the base. The thickened zone of ancymidol-treated roots had a higher galactose content than GA_3-treated slender roots (Fig. 11).

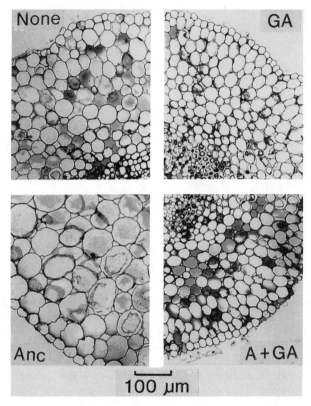

Fig. 9. Effect of ancymidol and GA_3 on the cross section of root cells. Seedling roots of Alaska peas were treated with or without 10 μM ancymidol (A, Anc) and/or 0.1 μM GA_3 (GA) for 2 days and the apical portion of the root was excised and fixed for microscopic observation. Photographs were taken from the sections at 4 mm behind the root tip

4 Discussion

Roots of GA-deficient dwarf plants elongate normally without exogenous application of GA. The concept that roots require less GA than shoots has been proposed in lettuce[1] and in normal peas.[8,10] Physiological and genetic studies on GA-deficient dwarf plants suggested that these plants are leaky mutants.[11-19] It is thus highly probable that GA produced through the leak is sufficient for root growth but is insufficient for shoot growth in GA-deficient dwarf plants. Present results for dwarf peas are consistent with this concept. Ancymidol inhibition of root elongation was completely prevented by GA_3 at concentrations significantly lower than those required for shoot elongation (Fig. 6). When GA_3 was applied with ancymidol, a minimal concentration of GA_3 to give maximal growth of ancymidol-

22. GA Requirements for Root Growth   237

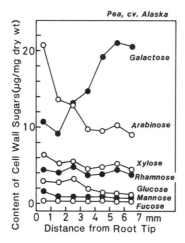

Fig. 10. Changes in neutral sugars in the cell wall of Alaska pea roots along the root axis. Seedling roots, 20 ± 3 mm long, were consecutively sectioned at 1-mm intervals from the tip to the base of the root. Contents of neutral sugars per dry weight of 1-mm segments are indicated as a function of the distance from the root tip. The sugars were released from the cell wall by trifluoroacetic acid hydrolysis

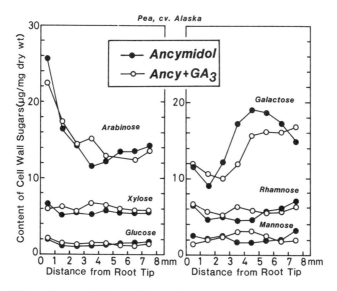

Fig. 11. Effect of ancymidol and GA_3 on the contents of neutral sugars in the cell wall. Seedling roots were treated with or without 0.1 μM GA_3 in the presence of 10 μM ancymidol for 2 days. Neutral sugar contents per dry weight of 1-mm segments are indicated as a function of the distance from root tip

suppressed roots was found to be 1–10 nM in lettuce (Figs. 3 and 4) and normal[8] and dwarf peas (Fig. 6).[10] Thus, roots of lettuce and dwarf peas, cv. Little Marvel, are as sensitive to GA as those of normal peas, cv. Alaska. These concentrations of GA_3 failed to promote shoot growth (Figs. 3, 4, 6, and 7) even if the seedling shoots were dipped in GA_3 solution many times for 4 days (Fig. 7). In addition, shoot-applied GA_3 promoted root elongation at concentrations in the same range as those which promoted shoot growth (Fig. 8). These results support the concept that a low level of GA regulates the root elongation also in rosette and dwarf plants. The concept may explain why roots of rosette and dwarf plants are not dwarf.

Ancymidol not only suppressed elongation growth but also induced expansion of the elongation zone of the root (Figs. 5 and 9).[1,8] A similar phenomenon of swelling has been found in grapevine roots[20] and in onion leaf sheaths[21] with the growth retardants CCC and S-3307, respectively. These phenomena coincided with the change in sugar composition of the cell wall (Fig. 11). Gibberellin A_3 completely reversed ancymidol-induced expansion and ancymidol-induced change in the cell wall of root cells (Figs. 9 and 11).

The specificity of ancymidol and other growth retardants must be carefully considered to connect the growth inhibition and GA biosynthesis.[18,22] A minimum condition that must be met is that any inhibition be completely canceled by exogenous GA. This condition was met in the results obtained in the present study. Ancymidol effects on the elongation growth and the morphological changes in root cells were completely overcome by exogenously applied GA_3.

Although GA transport from roots to shoots was not measured in the present study, GA has been shown to move from roots to shoots and vice versa.[23-28] Thus, the present results indicating different sensitivity of roots and shoots to exogenous GA_3 support the concept that "root organs" require smaller amounts of GA than "shoot organs" for normal elongation growth. This differential sensitivity may be related causally to the fact that the total volume of elongating root cells is far less than that of elongating stem cells in young seedlings. The concept does not necessarily indicate that "root cells" require less GA than "shoot cells." The organ-dependent difference in GA requirement may play an important role in supporting root growth of young rosette plants and GA-deficient dwarf plants. The chemically induced dwarf roots will be helpful for further study on the mechanism of action of GA in the root system.

5 Conclusions

Growth responses of roots and shoots to GA_3 and ancymidol were compared in a rosette plant, *Lactuca sativa* L. cv. Grand Rapids, and in liana plants, *Pisum sativum* L. cv. Alaska and cv. Little Marvel. Root elongation was inhibited by higher concentrations of ancymidol than those required for the inhibition of shoot elongation. Gibberellin A_3 applied to the roots completely prevented the ancymidol inhibition, with the concentration required for roots (1–10 nM) being lower than that for shoots (1–10 μM). When a low concentration of GA_3 (1–10 nM) was applied to both roots and shoots, it failed to promote shoot elongation whereas it enhanced root elongation. Shoot-applied GA_3 reversed the ancymidol inhibition of root elongation in the same concentration range as that which promoted shoot elongation. Ancymidol led to thickening of the elongation zone of the roots, while GA_3 completely canceled this. Gibberellin A_3 and ancymidol affected the sugar composition of the cell wall in peas. Ancymidol-treated thick roots of cv. Alaska had a higher galactose content than those of GA-treated slender roots.

These results support the concept that roots require smaller amounts of GA than shoots for normal elongation growth and that ancymidol-induced and GA-reversed morphological changes in root cells are accompanied by a change in cell wall components. These findings may explain why roots of rosette plants and of dwarf plants are not dwarf.

References

1. Tanimoto E. Gibberellin-dependent root elongation in *Lactuca sativa*: Recovery from growth retardant-suppressed elongation with thickening by low concentration of GA_3. Plant Cell Physiol. 1987; 28:963–973.
2. Mertz D. Hormonal control of root growth I. Plant Cell Physiol. 1966; 7:125–135.
3. Paleg L, Aspinall D, Coombe B, et al. Physiological effects of gibberellic acid. IV. Other gibberellins in three test systems. Plant Physiol. 1964; 39:286–290.
4. Scott TK. Auxins and roots. Ann Rev Plant Physiol. 1972; 23:235–258.
5. Torrey JG. Root hormones and plant growth. Ann Rev Plant Physiol. 1976; 27:435–459.
6. Feldman LJ. Regulation of root development. Ann Rev Plant Physiol. 1984; 35:223–242.
7. Coolbaugh RC, Hamilton R. Inhibition of *ent*-kaurene oxidation and growth by α-cyclopropyl-α-(p-methoxyphenyl)-5-pyrimidine methyl alcohol. Plant Physiol. 1976; 57:245–248.
8. Tanimoto E. Gibberellin regulation of root growth with change in galactose content of cell walls in *Pisum sativum*. Plant Cell Physiol. 1988; 29:269–280.
9. Tanimoto E, Watanabe J. Automated recording of lettuce root elongation as affected by indole-3-acetic acid and acid pH by a new rhizometer with minimum mechanical contact to root. Plant Cell Physiol. 1986; 27:1475–1487.
10. Tanimoto E. Roots require less gibberellin than shoots for normal elongation: An explanation for why roots of dwarf plants are not dwarf. In: Abstracts for the 13th International Conference on Plant Growth Substances, Calgary, 1988; Abstract No. 370.
11. Shive JB, Sisler HD. Effect of ancymidol (a growth retardant) and triarimol (a

fungicide) on the growth, sterols, and gibberellins of *Phaseolus vulgaris* (L.). Plant Physiol. 1976; 57:640–644.
12. Suge H. Inhibition of flowering and growth in *Pharbitis nil* by the growth retardant ancymidol. Plant Cell Physiol. 1980; 21:1187–1192.
13. Wada K, Imai T. Effect of 1-*n*-decylimidazole on gibberellin biosynthesis in Tan-ginbouzu, a dwarf variety of rice. Agric Biol Chem. 1980; 44:2511–2512.
14. Potts WC, Reid JB, Murfet IC. Internode length in *Pisum*. I. The effect of the *Le/le* gene difference on endogenous gibberellin-like substances. Physiol Plant. 1982; 55:323–328.
15. Phinney BO. Gibberellin A_1, dwarfism and the control of shoot elongation in higher plants. In: Crozier A, Hillman JR, eds. The biosynthesis and metabolism of plant hormones. Cambridge: Cambridge University Press, 1984: pp. 17–41.
16. Spray CR, Yamane H, Phinney BO, et al. Endogenous gibberellins (GAs) & GA-like substances from the vegetative shoots of normal, dwarf-1 & dwarf-5 maize (*Zea mays*). Plant Physiol. 1984; 75:S94.
17. Potts WC, Reid JB, Murfet IC. Internode length in *Pisum*. Gibberellin and the slender phenotype. Physiol Plant. 1985; 63:357–364.
18. Britz SJ, Saftner RA. Inhibition of growth by ancymidol and tetcyclacis in the gibberellin-deficient *dwarf-5* mutant of *Zea mays* L. and its prevention by exogenous gibberellin. J Plant Growth Regul. 1987; 6:215–219.
19. Ross JJ, Reid JB, Gaskin P, et al. Internode length in *Pisum*. Estimation of GA_1 levels in genotypes *Le*, *le* and *le*^{d}. Physiol Plant. 1989; 76:173–176.
20. Skene KGM, Mullins MG. Effect of CCC on the growth of roots of *Vitis vinifera* L. Planta. 1967; 77:157–163.
21. Mita T, Shibaoka H. Effects of S-3307, an inhibitor of gibberellin biosynthesis, on swelling of leaf sheath cells and on the arrangement of cortical microtubules in onion seedlings. Plant Cell Physiol. 1984; 25:1531–1539.
22. Lang A. Gibberellins: Structure and metabolism. Ann Rev Plant Physiol. 1970; 21:537–570.
23. Carr DJ, Reid DM, Skene KGM. The supply of gibberellins from the root to the shoot. Planta. 1964; 63:382–392.
24. Reid DM, Crozier A, Harvey, BMR. The effects of flooding on the export of gibberellins from the root to the shoot. Planta. 1969; 89:376–379.
25. Crozier A, Reid DM. Do roots synthesize gibberellins? Can J Bot. 1971; 49:967–975.
26. Reid DM, Crozier A. Effects of waterlogging on the gibberellin content and growth of tomato plants. J Exp Bot. 1971; 22:39–48.
27. Prochazka S. Translocation of growth regulators from roots in relation to the stem spical dominance in pea (*Pisum sativum* L.) seedlings. In: Brouwer R, Gasparikova O, Kolek J, Loughman BC, eds. Structure and function of plant roots. The Hague: Martinus Nijhoff/Dr W. Junk Publishers, 1981: p. 407–409.
28. Kaldewey H. Transport and other modes of movement of hormones (mainly auxins). In: Scott TK, ed. Hormonal regulation of development II. Encyclopedia of plant physiology, New series, Vol. 10. Berlin: Springer-Verlag, 1984: pp. 80–148.

CHAPTER 23

Effects of Gibberellin A_3 on Growth and Tropane Alkaloid Synthesis in Ri Transformed Plants of *Datura innoxia*

H. Kamada, T. Ogasawara, and H. Harada

1 Introduction

Infection of dicotyledonous plants with *Agrobacterium rhizogenes* results in adventitious root formation at the infection sites.[1] These adventitious roots, called hairy roots, grow well and form numerous lateral roots in hormone-free medium after elimination of the bacterium. Recent work clearly demonstrates that a large plasmid (pRi) present in the bacterium is a causal factor in the root formation.[2,3] A portion of the Ri plasmid, containing the two distinct regions TL and TR T-DNA (together called the T-DNA), is integrated into the plant genomic DNA and stably maintained in the transformed plant cells. Genes present in the T-DNA are transcribed as polyadenylated mRNA and translated in the transformed cells. The right-hand side of the T-DNA region (TR-DNA) of the Ri plasmid contains genes synthesizing indole-3-acetic acid (IAA) and agropine. The functional genes concerned with the induction and morphology of hairy roots are located in the TL-DNA region of agropine-type Ri plasmids and are designated as rol A, B, C, and D.[4] However, the physiological functions of these genes in hairy root systems have not been clarified.

Recently, researchers have examined effects of plant growth regulators such as auxin and cytokinin on the growth and morphology of hairy roots. Treatment with anti-auxins, such as 2,3,5-triiodobenzoic acid, 2,4,6-trichlorophenoxyacetic acid, or *p*-chloroisobutyric acid, inhibited the growth of hairy roots in potato.[5] These results indicate that auxin plays an important role in hairy root growth. However, both mannopine-type Ri plasmids (in which no IAA-synthesizing gene is present on the T-DNA) and mutated agropine-type Ri plasmids (in which TR-DNA has been deleted) are able to induce hairy roots in several plant species which grow well on hormone-free medium.[2,6] Thus, it is likely that some factors other than IAA regulate the growth of hairy roots.

Regenerated plants from hairy roots exhibit dwarf and wide leaf morphologies in many plant species.[7] However, the physiological roles of T-

DNA genes in the induction of the abnormal morphology in the regenerated plants remain to be clarified.

It is well known that tropane alkaloids, such as l-hyoscyamine, 6β-hydroxyhyoscyamine, and scopolamine, are synthesized mainly in roots of several solanaceous plants.[8] Axenic cultures of hairy roots of *Datura innoxia*, *Atropa belladonna*, *Hyoscyamus niger*, and others produce these alkaloids,[9-11] but the factor(s) that controls the tropane alkaloid synthesis remains unclear.

We induced hairy roots in several plant species and examined effects of plant growth regulators on the phenomena described above. During the experiments, we observed stimulating effects of gibberellin (GA) on growth[12] and tropane alkaloid synthesis[13] by the hairy roots. In this paper, we describe results obtained with *Datura innoxia* and some other solanaceous plants and discuss the physiological roles of GA_3 on the growth and tropane alkaloid synthesis in hairy roots.

2 Materials and Methods

2.1 Induction of Hairy and Normal Roots

Hairy roots of *Datura innoxia*, *Atropa belladonna*, and *Hyoscyamus niger* were induced by inoculating axenic plants or leaf disks with *Agrobacterium* harboring Ri plasmids (pRiA4b). Axenic hairy roots were obtained by culturing on hormone-free Murashige and Skoog medium[14] (hereafter referred to as MSHF medium) containing the antibiotic claforan (0.5 mg/ml). Normal root cultures of *Datura innoxia* originated from root tips of axenic plants were cultured on MSHF medium containing 0.01 mg of 1-naphthaleneacetic acid (NAA) per liter (hereafter referred to as MSNAA medium). The axenic cultures of hairy and normal roots were subcultured in MSHF and MSNAA media, respectively, in the dark at 25 °C with gyratory shaking at 75 rpm.

2.2 Growth Analyses of Hairy Roots and Normal Roots

The axenic hairy roots (ca. 750 mg) grown in liquid MSHF medium were cultured in MSHF medium containing GA_3 and/or a specific inhibitor of gibberellin biosynthesis, S-3307 [(E)-1-(4-chlorophenyl)-4,4-dimethyl-2-(1,2,4-triazol-1-yl)-1-penten-3-ol], under the culture conditions described above. The fresh weight was measured before and after 9 days of culture. The normal roots (ca. 0.65 g) grown in liquid MSNAA medium were cultured in MSNAA medium containing GA_3 and/or S-3307. The fresh weight was measured before and after 12 days of culture. Growth rate was expressed as percent of the control value (no addition of test chemicals). For measuring the elongation and the lateral root branching, root tips

(1 cm) were cultured in test media in the dark at 25 °C with gyratory shaking at 40 rpm. The number of lateral roots was counted by using 2-cm-long hairy root segments, excised between 5 and 7 cm from the root tips after 20 days of culture. The data are expressed as percent of the control value (no addition of test chemicals).

2.3 Analyses of Tropane Alkaloids

Nine-and 12-day-old cultures of hairy and normal roots, respectively, were harvested, lyophilized, and ground. The powdered sample was weighed and the alkaloids were extracted. Extraction was performed according to the method described earlier.[10] The identification and quantification of each alkaloid was carried out by high-performance liquid chromatography (HPLC) analysis.[10]

3 Effects of GA on the Growth of Hairy and Normal Roots

Exogenously applied GA_3 stimulated the elongation of hairy roots of *Datura innoxia*, *Atropa belladonna*, and *Hyoscyamus niger* (Table 1). In *Datura innoxia*, treatment of the hairy roots with GA_3 also stimulated both fresh weight increase and lateral root branching; the maximum stimulative effects were obtained at a concentration of 0.1 mg/liter (Table 2). These stimulatory effects of exogenously applied GA_3 were also observed in hairy roots of other plant species such as *Solanum nigrum*, *Daucus carota*, and *Pharbitis nil*.[12] The treatment of normal root cultures of *Datura innoxia* with GA_3 stimulated the fresh weight increase even at a concentration of 1.0 mg/liter (Table 2). It was previously reported that exogenously applied GA_3 inhibited initiation and growth of normal roots in many plant species, but promoted it in others.[15] Recently, Tanimoto[16] clearly demonstrated that endogenous GA played an important role in the elongation of normal roots in *Pisum sativum* and *Lactuca sativa*.

Table 1. Effects of GA_3 on the elongation of hairy roots in some solanaceous plants. Root tips (1 cm) were cultured in liquid MSHF medium with GA_3 at the concentrations indicated. Data are expressed as the length (cm) of the hairy roots with standard error, measured after 9 days of culture in *Atropa* and 8 days in *Datura* and *Hyoscyamus*

Plant species	GA_3 (mg/liter)		
	0.0	0.01	0.1
Datura innoxia	7.1 ± 0.7	9.0 ± 0.9	9.8 ± 1.0
Atropa belladonna	5.9 ± 0.6	8.0 ± 0.8	8.6 ± 0.8
Hyoscyamus niger	5.8 ± 1.6	9.5 ± 0.6	9.8 ± 1.6

Table 2. Effects of GA_3 on the growth of hairy roots and normal roots of *Datura innoxia*. Hairy and normal roots were cultured in liquid MSHF and MSNAA media, respectively, with GA_3 at the concentrations indicated. Fresh weight was measured before and after 9 days of culture for the hairy roots and 12 days for the normal roots. Hairy root elongation was measured at 2-day intervals, and elongation rate was calculated by using linear regression. The number of lateral roots of the hairy roots was counted after 20 days of culture. Data are expressed as percentages of the values for the control (no addition of GA_3)

Concentration of GA_3 (mg/liter)	Fresh weight		Elongation rate of HR	No. of lateral roots of HR
	NR	HR		
0.0	100	100	100	100
0.001	—	137	130	109
0.01	195	129	163	129
0.1	220	149	167	138
1.0	317	127	141	111

NR, Normal roots; HR, hairy roots; —, not determined

Table 3. Effects of S-3307 with or without GA_3 on the growth of hairy roots of *Datura innoxia*. Hairy roots were cultured in MSHF medium containing S-3307 (0.01 mg/liter) with or without GA_3 (0.01 mg/liter). Data were obtained and are expressed as described in Table 2

Treatment	Fresh weight	Elongation rate	No. of lateral roots
Control[a]	100	100	100
S-3307	72	46	64
S-3307 + GA_3	133	168	128

[a] Control: No addition of S-3307 and GA_3

When the hairy roots of *Datura innoxia* were treated with an inhibitor of gibberellin biosynthesis, S-3307, at a concentration of 0.01 mg/liter, the fresh weight increase was 70% of the control value (no addition of S-3307 and GA_3) (Table 3). However, when S-3307 and GA_3 were applied simultaneously, both at 0.01 mg/liter, GA_3 counteracted the inhibitory effect of S-3307 and restored the growth to a level higher than that of the control (Table 3). Similar effects of S-3307 and GA_3 were also observed on the elongation rate and the number of lateral roots in the hairy roots of *Datura innoxia* (Table 3). Thus, endogenous GA must exert its action on the growth of both normal and hairy roots.

4 Effects of GA on Tropane Alkaloid Synthesis

In the hairy roots of *Datura innoxia*, treatment with GA_3 lowered the content of *l*-hyoscyamine and enhanced the content of 6β-hydroxy-

23. Growth and Alkaloid Synthesis in Ri Transformed Plants 245

Table 4. Effects of GA_3 on tropane alkaloid contents in hairy and normal roots of *Datura innoxia*. Hairy and normal roots were cultured in liquid MSHF and MSNAA media, respectively, containing GA_3 at the concentrations indicated. Alkaloid contents were determined in the hairy roots after 10 days of culture and in the normal roots after 20 days. The alkaloid contents are expressed as mg/g dry weight

Concentration of GA_3 (mg/liter)	Hairy roots			Normal roots		
	HYO	6B-HYO	SCO	HYO	6B-HYO	SCO
0.0	4.97	0.11	0.08	0.61	0.39	2.69
0.001	4.55	0.48	0.23	—[a]	—	—
0.01	3.43	0.71	0.43	0.43	0.25	2.41
0.1	3.25	0.58	0.32	0.17	0.05	1.84
1.0	2.41	0.21	0.19	0.13	trace[b]	0.87

HYO, *l*-Hyoscyamine; 6B-HYO, 6β-hydroxyhyoscyamine; SCO, scopolamine
[a]—, Not determined
[b]Trace amount

Table 5. Effects of S-3307 with or without GA_3 on tropane alkaloid contents in hairy roots of *Datura innoxia*. Hairy roots were cultured in liquid MSHF medium with S-3307 (0.01 mg/liter) and/or GA_3 (0.01 mg/liter) for 10 days. Alkaloid contents are expressed as mg/g dry weight

Treatment	HYO	6B-HYO	SCO
Control	4.34	0.36	0.09
GA_3	2.75	0.51	0.18
S-3307	4.81	0.13	0.06
S-3307 + GA_3	2.74	0.53	0.22

HYO, *l*-Hyoscyamine; 6B-HYO, 6β-hydroxyhyoscyamine; SCO, scopolamine
[a]Control: no addition of S-3307 and GA_3

hyoscyamine and scopolamine (Table 4). The content of *l*-hyoscyamine was decreased by the treatment with GA_3 at all the concentrations tested (Table 4). On the other hand, maximum values of content of 6β-hydroxyhyoscyamine and scopolamine were observed in the treatment with GA_3 at a concentration of 0.01 mg/liter (Table 4). When the hairy roots of *Datura innoxia* were treated with S-3307 (0.01 mg/liter), the content of 6β-hydroxyhyoscyamine and scopolamine decreased slightly (Table 5). However, when the hairy roots were treated with S-3307 (0.01 mg/liter) and GA_3 (0.01 mg/liter), the content of 6β-hydroxyhyoscyamine and scopolamine increased and that of *l*-hyoscyamine decreased to values comparable to those obtained for the treatment with GA_3 (0.01 mg/liter) (Table 5).

Activity of *l*-hyoscyamine 6β-hydroxylase, which catalyzes the conversion of *l*-hyoscyamine to 6β-hydroxyhyoscyamine, was about 7-fold higher than that in the control (no addition of GA_3) when the hairy roots of *Datura innoxia* were treated with GA_3 (0.01 or 0.1 mg/liter) (our unpublished data).

In adventitious buds regenerated from hairy roots of *Datura innoxia*, none of the alkaloids, such as l-hyoscyamine, 6β-hydroxyhyoscyamine, and scopolamine, could be detected. When the buds were cultured in MSHF medium with GA_3 (1 mg/liter) and l-hyoscyamine (1 g/liter) for 4 days, the conversion rate of l-hyoscyamine to scopolamine via 6β-hydroxyhyoscyamine was 2-fold higher than that in the control (no addition of GA_3), and the content of scopolamine increased compared to that for the control (no addition of GA_3) (our unpublished data).

These results indicate that GA influences tropane alkaloid synthesis in hairy roots. However, l-hyoscyamine content and 6β-hydroxyhyoscyamine and scopolamine content in normal roots of *Datura innoxia* were very low and high, respectively, compared to those in the hairy roots (Table 4). In normal roots of *Datura innoxia*, the content of l-hyoscyamine, 6β-hydroxyhyoscyamine, and scopolamine was decreased by treatment with GA_3 at concentrations higher than 0.01 mg/liter (Table 4). On the other hand, exogenous application of GA_3 did not alter the content of tropane alkaloids in normal root cultures of *Hyoscyamus niger* and *H. albus*.[17] The reason why different results were obtained for hairy and normal roots is not clear. It may be that NAA, used for normal root cultures, affected the GA_3 action on tropane alkaloid synthesis.

5 Effects of GA on Morphology of Plants Regenerated from Hairy Roots

In various species, plants transformed with Ri plasmid exhibited abnormal morphology such as wrinkling leaves, shortened stems, and increased width/length ratio of leaves.[2,7] Plants regenerated from hairy roots of *Atropa belladonna* differentiated numerous roots and exhibited abnormal morphology (transformed phenotype) such as shortened stems and increased width/length ratio of leaves compared to normal (not transformed) plants (normal phenotype). Progenies obtained by self-pollination of the regenerated plants could be divided into two groups, each exhibiting transformed or normal phenotypes. When shoot tips of both plants were cultured on MSHF medium with GA_3 (1 mg/liter), the length of internodes increased and the width/length ratio of leaves decreased compared to those in the control (no addition of GA_3) (Table 6). The width/length ratio of the leaves of transformed phenotype was comparable to that of normal phenotype, when the plants were treated with GA_3. Similar effects of GA_3 on plant morphology were reported in Ri-transformed plants of *Datura innoxia*.[13] Our preliminary results indicate that the quality and quantity of endogenous GA in the transformed plants of *Atropa belladonna* are comparable to those in the normal plants (data not shown). Thus, it seems

Table 6. Effects of GA_3 on the morphology (width/length ratio in leaves and length of internodes) of Ri transformed plants of *Atropa belladonna*. The materials used were progenies exhibiting transformed phenotype (transformant) or normal phenotype (normal) after self-pollination of plants regenerated from hairy roots transformed with Ri plasmid (pRi 15834). They were cultured on MSHF medium containing GA_3 (1 mg/liter) for 3 weeks. Data are expressed as ratio of width/length in the leaves (A) and length (cm) of the internodes (B) with standard errors

	Plant material	$-GA_3$	$+GA_3$
(A)	Transformant-1	0.56 ± 0.11	0.44 ± 0.12
	Transformant-2	0.58 ± 0.06	0.48 ± 0.11
	Normal-1	0.49 ± 0.05	0.46 ± 0.05
	Normal-2	0.53 ± 0.09	0.42 ± 0.07
(B)	Transformant-1	0.35 ± 0.13	1.73 ± 0.79
	Transformant-2	0.67 ± 0.32	1.30 ± 0.24
	Normal-1	1.37 ± 0.40	2.77 ± 0.70
	Normal-2	1.06 ± 0.37	2.24 ± 0.64

likely that some of the T-DNA genes on the Ri plasmid produce one or more factors that interact with GA.

6 Conclusion

The effect of GA on growth and tropane alkaloid synthesis in roots of some solanaceous plants has been discussed. The evidence presented here supports stimulative effects of GA on root growth (fresh weight, elongation, and lateral root branching) of *Datura innoxia*, *Atropa belladonna*, and *Hyoscyamus niger* and on conversion of *l*-hyoscyamine to scopolamine via 6β-hydroxyhyoscyamine in the hairy roots of *Datura innoxia*. Normal morphology was reestablished in those transgenic plants with abnormal morphology, such as dwarfism and increased width/length ratio, by exogenously applied GA_3. Further work is needed to clarify this response.

Acknowledgments. This research was supported in part by a Grant-in-Aid for Scientific Research on Priority Areas and a Grant-in-Aid for Scientific Research from the Ministry of Education, Science and Culture of Japan to H. Kamada and H. Harada. The authors are grateful to Dr. K. Izumi of Sumito Chemical Co., Ltd., for the supply of S-3307 and to Dr. A.D. Powell for critical reading of the manuscript.

References

1. Riker AJ, Banfield WM, Wright WH, et al. Studies on infectious hairy-root of nursery apple trees. J Agric Res. 1930; 41:507–540.

2. White FF, Sinkar VP. Molecular analysis of root induction by *Agrobacterium rhizogenes*. In: Hohn TH, Schell JS, eds. Plant DNA infectious agents. Vienna: Springer-Verlag, 1987: pp. 149–177.
3. Zambryski P, Tempé J, Schell J. Transfer and function of T-DNA genes from *Agrobacterium* Ti and Ri plasmids in plants. Cell. 1989; 56:193–201.
4. White FF, Taylor BH, Huffman GA, et al. Molecular and genetic analysis of the transferred DNA regions of the root-inducing plasmid of *Agrobacterium rhizogenes*. J Bacteriol. 1985; 164:33–44.
5. Quattrocchio F, Benvenuto E, Tavazza R, et al. A study of the possible role of auxin in the root-inducing plasmid of the agropine type *Agrobacterium rhizogenes* 1855. J Plant Physiol. 1986; 123:143–149.
6. Combred A, Baucher M-F. A common organization of the T-DNA genes expressed in plant hairy roots induced by different plasmids of *Agrobacterium rhizogenes*. Plant Mol Biol. 1988; 10:499–509.
7. Tepfer D. Transformation of several plant species of higher plants by *Agrobacterium rhizogenes*: Sexual transmission of the transformed genotype and phenotype. Cell. 1984; 37:959–967.
8. West FR Jr, Mika ES. Synthesis of atropine by isolated roots and root-callus cultures of belladonna. Bot Gaz. 1957; 119:50–54.
9. Flores HE, Filner P. Metabolic relationships of putrescine, GABA and alkaloids in cell and root cultures of Solanaceae. In: Neumann K-H, Barz W, Reinhard E, eds. Primary and secondary metabolism of plant cell cultures. Berlin: Springer-Verlag, 1985: pp. 174–185.
10. Kamada H, Okamura N, Satake M, et al. Alkaloid production by hairy root cultures in *Atropa belladonna*. Plant Cell Rep. 1986; 5:239–242.
11. Hamill JD, Parr AJ, Rhodes JC, et al. New routes to plant secondary products. Biotechnology 1987; 5:800–804.
12. Ohkawa H, Kamada H, Sudo H, et al. Effects of gibberellic acid on hairy root growth in *Datura innoxia*. J Plant Physiol. 1989; 134:633–636.
13. Kamada H, Ohkawa H, Harada H, et al. Effects of GA_3 on growth and alkaloid production by hairy roots of *Datura innoxia*. In: Proceedings of the 14th Annual Plant Growth Regulator Society of America Meeting. 1987: pp. 227–232.
14. Murashige T, Skoog F. A revised medium for rapid growth and bioassays with tobacco tissue cultures. Physiol Plant. 1962; 15:473–497.
15. Goodwin PB. Phytohormones and growth and development of organs of the vegetative plant. In: Letham DS, Goodwin PB, Higgins TJV, eds. Phytohormones and related compounds—a comprehensive treatise. Vol. II. Phytohormones and the development of higher plants. Amsterdam Elsevier/North-Holland Biochemical Press, 1978: pp. 31–173.
16. Tanimoto E. Gibberellin-dependent root elongation in *Lactuca sativa*: Recovery from growth retardant-suppressed elongation with thickening by low concentrations of GA_3. Plant Cell Physiol. 1987; 28:963–973.
17. Hashimoto T, Yamada Y, Yukimune Y. Tropane alkaloid production in *Hyoscyamus* root cultures. J Plant Physiol. 1986; 124:61–75.

CHAPTER 24

Biochemical and Physiological Aspects of Gibberellin Conjugation

G. Sembdner, W. Schliemann, and G. Schneider

1 Introduction

Conjugation represents part of the metabolism of all groups of plant hormones. The process of conjugation is characterized by coupling of a plant hormone molecule to another low-molecular-weight component by covalent binding.[1,2] Contrary to biosynthetic and other metabolic pathways, conjugation does not exclude reversibility. Plant hormone conjugates differ from the corresponding free hormones with respect to their physical, chemical, and biological properties.[3] Though the physiological role of conjugation is not fully understood, it is considered to be involved in the regulation of the biologically active hormone levels as well as in transport and even compartmentation within the cell.

The first gibberellin (GA) conjugate was isolated from fruits of *Phaseolus coccineus* and structurally elucidated to be GA_8-2-O-glucoside.[4,5] Several more GA-O-glucosides (GA-O-Glc) as well as GA glucosyl esters (GA-Glc esters) and some other types of GA conjugates have been subsequently detected.[3,6]

2 Naturally Occurring GA Conjugates

Of the GA conjugates isolated from higher plants, the glucosyl conjugates represent the major group, comprising the GA-Glc esters and GA-O-Glc. The GA-Glc ester type is characterized by the connection of the glucosyl moiety to the C-7 carboxy group of the GA molecule, resulting in a neutral polar GA conjugate (Fig. 1).

In the case of GA-O-Glc, the conjugating sugar moiety is attached to one of the hydroxy groups of the GA, resulting in a series of isomeric GA-O-Glc (Fig. 1). Most of the isolated GA-O-Glc belong to the group of GA-2-O-Glc. In terms of physiological relevance, the group of GA-3-O-Glc seems be the most important. Table 1 summarizes the isolated and structurally elucidated GA glucosyl conjugates.

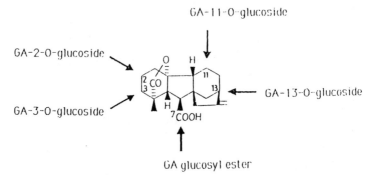

Fig. 1. Schematic structures of GA glucosyl conjugates

Table 1. Endogenously occurring GA glucosyl conjugates

Conjugate	Plant source	References
GA_1-glucosyl ester	*Phaseolus vulgaris*	7
GA_1-3-*O*-glucoside	*Dolichos lablab*	8
GA_3-3-*O*-glucoside	*Pharbitis nil*	9
	Quamoclit pennata	10
GA_4-glucosyl ester	*Phaseolus vulgaris*	7
GA_5-glucosyl ester	*Pharbitis purpurea*	11
GA_8-2-*O*-glucoside	*Phaseolus coccineus*	4
	Phaseolus vulgaris	7
	Althaea rosea	12
	Pharbitis nil	9
GA_9-glucosyl ester	*Picea sitchensis*	13
GA_{26}-2-*O*-glucoside	*Pharbitis nil*	9
GA_{27}-2-*O*-glucoside	*Pharbitis nil*	9
GA_{29}-2-*O*-glucoside	*Pharbitis nil*	9
GA_{35}-11-*O*-glucoside	*Cytisus scoparius*	14
GA_{37}-glucosyl ester	*Phaseolus vulgaris*	7
GA_{38}-glucosyl ester	*Phaseolus vulgaris*	7
GA_{44}-glucosyl ester	*Pharbitis purpurea*	11

There are numerous indications in the literature that many further GA conjugates may occur. Their identity remains obscure since only the free parent GA has been identified after hydrolysis. Besides the GA glucosyl conjugates, GA acyl derivatives as well as GA alkyl esters have been found in plants.[3,6]

3 Synthesis and Analysis of GA Glucosyl Conjugates

Partial syntheses of GA glucosyl conjugates have been performed in order to confirm the structure of isolated conjugates and to synthesize potentially native conjugates not yet isolated. The synthetic specimens are also re-

Table 2. Chemically synthesized GA glucosyl conjugates

Ring A-glucosides	Reference	Ring C-glucosides	Reference	Glucosyl esters	Reference
GA_1-3-O-Glc	15	GA_1-13-O-Glc	15	GA_1-Glc ester	16
3-epi-GA_1-3-O-Glc	17	3-epi-GA_1-13-O-Glc	17		
16,17-H_2-GA_1-3-O-Glc	17	16,17-H_2-GA_1-13-O-Glc	17		
GA_3-3-O-Glc	18	GA_3-13-O-Glc	18	GA_3-Glc ester	16
iso-GA_3-3-O-Glc	19	iso-GA_3-13-O-Glc	19		
GA_4-3-O-Glc	17			GA_4-Glc ester	16
		GA_5-13-O-Glc	19	GA_5-Glc ester	19
GA_7-3-O-Glc	17			GA_7-Glc ester	20
GA_8-2-O-Glc	20	GA_8-13-O-Glc	20	GA_8-Glc ester	20
				GA_9-Glc ester	13
		GA_{20}-13-O-Glc	20	GA_{20}-Glc ester	20
GA_{29}-2-O-Glc	21	GA_{29}-13-O-Glc	21		
				GA_{37}-Glc ester	16
		GA_3-3,13-di-O-Glc	18	GA_{38}-Glc ester	16
		13-O-Glc-GA_5-Glc ester	22		

quired for analytical purposes as well as for biological and metabolic studies.[3] In Table 2 the GA glucosyl conjugates that have been synthesised to date are compiled.

In view of the high polarity and the structural similarities of GA glucosyl derivatives, their isolation, purification, and identification require specifically dedicated analytical methods. For the prepurification and in order to separate neutral compounds (GA-Glc esters) from acidic components (GA-O-Glc), ion exchange chromatography [DEAE-Sephadex, Nucleosil N(CH$_3$)$_2$] is recommended.[3,6,23] Reverse-phase high-performance liquid chromatography (HPLC) is another useful prepurification step that also serves to separate the individual GA glucosyl conjugates.[6,23,24]

With purified samples, gel permeation chromatography (e.g., Biobeads SX-4) can be used to separate the group of GA glucosyl conjugates from free GAs.[25] For final qualitative and quantitative estimation of GA-O-Glc, combined gas chromatography–mass spectrometry (GC–MS) of their permethyl derivatives[26] should be used. On the basis of comprehensive GC–MS investigations with permethylated GA-O-Glc, accurate identification can be achieved.[27]

4 Metabolic Formation of GA-O-Glc in Maturing Fruits of Runner Beans (*Phaseolus coccineus*)

In maturing fruits of *P. coccineus*, the only structurally elucidate GA-O-Glc is GA$_8$-2-O-β-D-glucoside,[4,5] although evidence has been given for the occurrence of conjugates of GA$_1$, GA$_{17}$, GA$_{20}$, GA$_{28}$, and GA$_{34}$ by means of hydrolysis experiments[28,29] and GC–MS.[26]

On the basis of improved analytical conditions, we reinvestigated the formation of GA-O-Glc after injecting labeled as well as unlabeled GA$_3$ into immature runner bean fruits in vivo. After the usual workup, the *n*-butanol fraction, which contained the GA$_3$ glucosyl conjugates, was purified by DEAE-Sephadex A-25 chromatography[30] and HPLC. By comparison with synthetic standard compounds in HPLC and by capillary GC–MS of the permethylated derivatives, the glucosidic metabolites were identified as GA$_3$-3-O-Glc, GA$_3$-13-O-Glc, and iso-GA$_3$-3-O-Glc in the ratio 9:1:1 (Fig. 2).[31] As in other GA$_3$ feeding experiments, the iso-GA$_3$-3-O-Glc is regarded as an artifact of workup derived from GA$_3$-3-O-Glc. The preferential formation of GA$_3$-3-O-Glc in vivo corresponds with the stereospecificity of the GA$_3$ glucosylating enzyme activity detected in the same plant material in vitro (cf. Section 5).

Following application of [1β,2β-^{3}H$_2$]GA$_1$ (specific activity, 1.39 TBq/mmol) to immature fruits of *P. coccineus*, the metabolically formed conjugates were isolated and identified in the same way by HPLC and capillary GC–MS of the permethylated derivatives as [1β-^{3}H]GA$_8$-2-O-Glc and

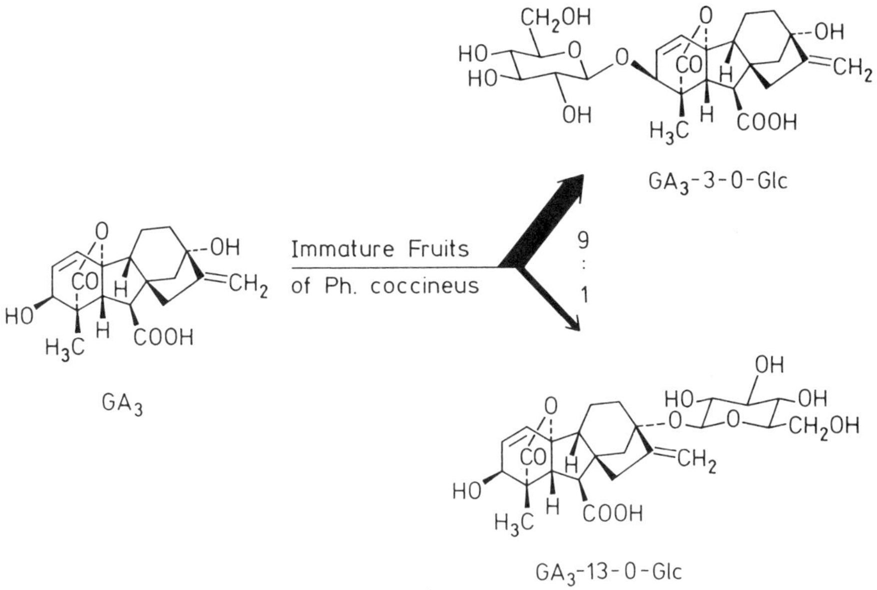

24. GA Conjugation   253

Fig. 2. Glucosidic metabolites of GA_3 in immature fruits of *Phaseolus coccineus* L.

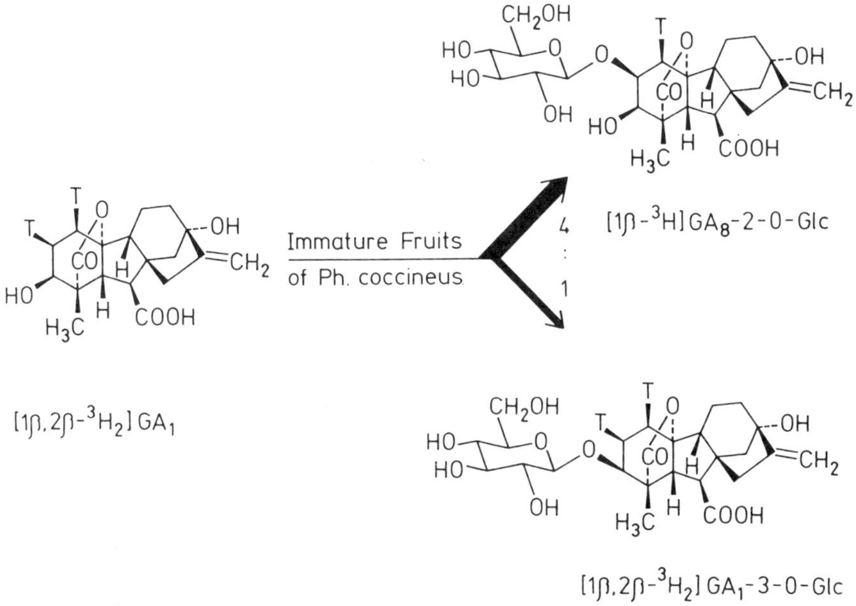

Fig. 3. Glucosidic metabolites of $[1\beta, 2\beta,-^3H_2]GA_1$ in immature fruits of *Phaseolus coccineus* L.

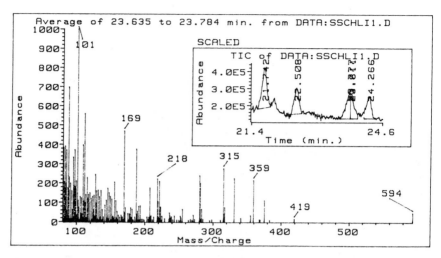

Fig. 4. Mass spectrum of GA_1-3-O-Glc isolated from *Phaseolus coccineus* fruits (permethyl derivative, 70 eV, electron impact-MS)

$[1\beta,2\beta\text{-}^3H_2]GA_1$-3-$O$-Glc in the ratio 4:1 (Fig. 3).[31–33] Furthermore, by this metabolic labeling of the pool of endogenous GA glucosides, the natural occurrence of GA_1-3-O-Glc in this plant material could be detected by HPLC and was quantified by comparison with authentic standards to be about 10 µg/100 g fresh weight. Final structural identification of the supposed GA_1 conjugate[28,29] was achieved by GC–MS of the permethylated derivative (Fig. 4), which resulted in a fragmentation pattern characteristic of GA_1-3-O-β-D-glucoside.[27] The endogenous occurrence of the 3-O-glucoside of the highly biologically active GA_1 in runner bean fruits indicates that further experiments are necessary to examine the possible physiological function of this GA glucoside in seed development and germination.

5 Enzymatic Formation of GA_3-3-O-Glc

In order to study enzymatic GA-O-Glc formation, a screening for GA glucosylating activities was undertaken, including mainly immature legume fruits. Homogenates from various plant materials were centrifuged at 20,000 × g. The sediments were immediately incubated with GA_3 and UDP-glucose, whereas supernatants were first purified by gel filtration on Sephadex G-25 and checked afterwards. Among the legume species tested, the supernatants from fruits of *P. coccineus* and *P. angularis* showed considerable glucosylating activities. The *P. coccineus* system was chosen for further investigations.[34,35] The enzyme was found to be a soluble protein that is localized in the pericarp. The product formed has been structurally identified as GA_3-3-O-Glc. Further investigations of the *P. coccineus* enzyme were focused on its substrate and cosubstrate specificity and showed that UDP-glucose exclusively served as a donor for the glucosyl moiety.

The aglycon specificity of the glucosyltransferase was studied by testing more than a dozen naturally occurring and chemically modified GAs. Only GA_7 and GA_{30} were found to mimic GA_3 in forming 3-O-glucosides, but to a smaller extent. Thus, the enzyme from fruits of *P. coccineus* was characterized to be a UDP-glucose: GA_3 3-O-glucosyltransferase.[34,35]

6 Enzymatic Hydrolysis of GA-O-Glc

Because of unreliable results, bioassay is an unsuitable method for the analysis of intact GA glucosyl conjugates (cf. Table 6). On the other hand, chemical hydrolysis of GA glucosyl conjugates mostly results in the degradation of the biologically active GA. Hence, enzymes are recommended for the liberation of free GAs from crude GA conjugate fractions.[36] With β-glucosidase preparations of different origins, a remarkable dependence of the hydrolysis rates on the structure of the GA-O-Glc was observed.[37]

A more detailed study of the relationships between the stereochemistry of the glucosyl linkage in the GA-O-Glc and their enzymatic cleavage by a β-glucosidase-containing cellulase preparation has been carried out. A set of sixteen synthetically prepared GA-O-Glc was used, and their hydrolyses were measured with the glucose oxidase–peroxidase reaction. In the group of GA-3-O-Glc, an unequivocal increase of the hydrolysis rates was found for the transition from axial to quasiequatorial and especially from axial to equatorially arranged GA-O-Glc. The highest hydrolysis rate was measured with GA_8-2-O-Glc having an equatorially arranged glucosyl moiety, and, surprisingly, this rate was identical to the rates obtained with the spatially hindered tertiary GA-13-O-Glc.*

For the characterization of plant β-glucosidases with GA glucosyl conjugate hydrolyzing activity, the radiolabeled GA-O-Glc obtained in the metabolic studies (Section 4) served as highly sensitive enzyme substrates.

In the dwarf rice bioassay (via leaf application), GA_3-3-O-Glc was nearly inactive whereas the biological activity of GA-Glc esters and GA_8-2-O-Glc was almost the same as that of the corresponding free GAs.[16,40,41] From shoots of 4-day-old etiolated dwarf rice seedlings, an enzyme fraction (CM 4) was obtained by CM-Sephadex C-50 chromatography which exhibited a 200 times faster hydrolysis of GA_8-2-O-Glc as compared to GA_3-3-O-Glc. Furthermore, as this fraction and the main β-glucosidase com-

---

*References 3, 31, 38, and 39.

Table 3. GA glucose conjugate hydrolyzing activity of soluble β-glucosidases from shoots of dwarf rice seedlings (4-day-old, etiolated, *Oryza sativa* L. cv. Tanginbozu)

Enzyme fraction	[³H]GA-O-Glc hydrolysis rate		β-Glucosidase (pNP-β-Glc)[a] specific activity (nkat/mg)
	[³H]GA$_8$-2-O-Glc [Bq/(mg·h)]	[³H]GA$_3$-3-O-Glc [Bq/(mg·h)]	
CM 4	14,570	18	4.55
CM 5	740	29	54.72

	Hydrolysis rate [nmol GA-O-Glc/(mg protein·h)]			
	GA$_8$-2-O-Glc	GA$_3$-3-O-Glc	GA$_3$-Glc ester	GA$_5$-13-O-Glc
CM 4	359	1.7	47.9	0.96
CM 5	n.d.[b]	n.d.	57.5	n.d.

[a] pNP-β-Glc: *p*-Nitrophenyl-β-D-glucopyranoside
[b] n.d., Not determined

Table 4. GA glucose conjugate hydrolyzing β-glucosidases from seeds of Japanese morning glory (*Pharbitis nil* (L.) Choisy cv. Violet)

Enzyme fraction	[³H]GA-O-Glc hydrolysis rate		β-Glucosidase (pNP-β-Glc)[a] specific activity (nkat/mg)
	[³H]GA$_8$-2-O-Glc [Bq/(mg·h)]	[³H]GA$_3$-3-O-Glc [Bq/(mg·h)]	
D 1	157	15.5	2.226
D 2	23	2.6	0.127
D 3	12	1.1	0.066
CM 1	395	46	9.780

	Hydrolysis rate [nmol GA-O-Glc/(mg protein·h)]			
	GA$_8$-2-O-Glc	GA$_3$-3-O-Glc	GA$_3$-Glc ester	GA$_5$-13-O-Glc
CM 1	7.3	1.9	2.9	0.8

[a] pNP-β-Glc: *p*-Nitrophenyl-β-D-glucopyranoside

ponent (CM 5) showed remarkable hydrolyzing activity toward GA$_3$-Glc ester, the bioassay data for GA-O-Glc and GA-Glc esters in dwarf rice bioassay are explainable on the basis of our results[31,42,43] (Table 3).

With the same methodology, the β-glucosidases from seeds of *Pharbitis nil* were partly characterized. The results (Table 4) indicate that β-glucosidases with hydrolyzing activity toward the endogenous GA$_8$-2-O-Glc and GA$_3$-3-O-Glc were present in the seed. Especially the GA$_3$-3-O-Glc hydrolyzing activity might be important for the liberation of biologically active GAs during seed germination.[31,43,44]

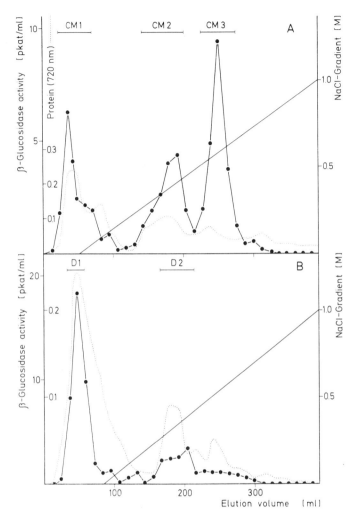

Fig. 5. Separation of soluble β-glucosidases from the pericarp of immature runner bean fruits (*Phaseolus coccineus* L.): (A) CM-Sephadex C-50; (B) DEAE-Sephadex A-50. ●——●, β-Glucosidase activity (pNP-β-Glc)

In enzyme extracts from maturing fruits of *P. coccineus*, a β-glucosidase with high hydrolyzing activity toward the endogenous GA_8-2-O-Glc was detected and found to be predominantly localized in the pericarp as a soluble enzyme. The GA_8-2-O-Glc hydrolase (pH optimum, 5.0) could be separated from two other soluble β-glucosidases (Fig. 5) and enriched 650-fold (Table 5). In the study of this activity during pod development, a decrease in the later stages was observed,[43,45] and this may be functionally related to the increase in GA glucosylating activity in the same tissue.[34,35] The physiological function of the GA_8-2-O-Glc hydrolase cannot be the

Table 5. GA_8-2-O-Glc hydrolyzing activity of soluble β-glucosidases from the pericarp of immature runner bean fruits (*Phaseolus coccineus* L. cv. Prizewinner)

Enzyme fraction	Specific activity	
	[^{3}H]GA_8-2-O-Glc hydrolase (pkat/mg protein)	β-Glucosidase (pNP-β-Glc)[a] (nkat/mg protein)
Crude extract	0.002	0.031
CM 1	1.460	0.022
CM 2	0.012	0.057
CM 3	0	0.180
Crude extract	0.004	0.014
D 1	0.020	0.048
D 2	2.598	0.013

[a] pNP-β-Glc: *p*-Nitrophenyl-β-D-glucopyranoside

liberation of a biologically active GA because GA_8 itself is already deactivated, but this enzyme may catalyze an important step of GA_8 glucose conjugate catabolism finally leading to the formation of GA_8 catabolite.

7 Biological Activity and Possible Physiological Function of GA Glucosyl Conjugates

The biological activity of GA glucosyl conjugates has been investigated in different bioassays such as dwarf rice, dwarf maize, dwarf pea, lettuce hypocotyl, and wheat endosperm bioassays. In each case, the activity of a GA conjugate has to be compared with that of the parent GA, and this "relative activity" can be calculated from lognormal dose–response curves of the conjugate and its aglycon by comparing parallel segments of the straight curves.[23,46] The relative activities of some GA-O-Glc and GA-Glc esters[3,47] are summarized in Table 6. These data demonstrate that the bioassays respond differently to the various conjugates. In general, the GA-Glc esters are more active than the GA-O-Glc. The highest responses are obtained in the nonsterile dwarf rice bioassay (via root application). Studies were done in order to assess whether the response of bioassays originates from the applied GA conjugate per se or from free GA released in vivo by enzymatic hydrolysis (cf. Section 6). The results demonstrated that rates of hydrolysis parallel the observed bioactivities.* These data have led to the view[2] that GA conjugates per se are biologically inactive and that the responses they induce are dependent upon the degree of hydrolysis and the structure of the parent GA.

---

* References 3, 16, 48, and 49.

Table 6. Relative biological activity of some GA glucosyl esters and GA-*O*-glucosides (% of activity of parent GA)[3,47]

Conjugate	Tan-ginbozu dwarf rice bioassay		Dwarf maize mutant bioassay		Dwarf pea bioassay	Lettuce hypocotyl bioassay	Wheat half-seed α-amylase bioassay
	Nonsterile	Sterile	d1	d5			
GA_1-Glc ester	60–100	80–100	5–10	5–10	10–50	20	1
GA_1-3-*O*-Glc	30	10–20		1	1	5–10	1
GA_1-13-*O*-Glc	50	20–30		2–5	2–5	10–20	
GA_3-Glc ester	80–100	60–70	30–60	40–50	10–50	20	10
GA_3-3-*O*-Glc	50–90	1–20	1	1–10	1	5	2
GA_3-13-*O*-Glc	50–100			1	1		1
GA_5-Glc ester	80–100			5–10		5–10	
GA_5-13-*O*-Glc	100			1			1
GA_8-2-*O*-Glc	100	50	100	100	70–100		0
GA_9-Glc ester	50–70				30–40	30–40	
GA_{20}-Glc ester	100			20			
GA_{20}-13-*O*-Glc	100			1–3			

With regard to the possible physiological functions of GA glucosyl conjugates, which have already been reviewed and extensively discussed,[3,47,50] special attention has to be drawn to the structural differences of the GA glucosyl conjugates resulting in different properties and, consequently, different functions. Thus, the GA-Glc esters are likely to be favored storage forms and preferentially involved in rapid changes in the levels of physiologically active GAs, because of their accumulation in ripening seeds and the ease of hydrolysis. It may also be that formation of GA-Glc esters is a rapid metabolic action that is followed by other metabolic or catabolic events. Furthermore, GA-Glc esters might be translocation forms with high xylem mobility.

The 3-O-glucosides of bioactive GAs such as GA_1 and GA_3 seem to be highly qualified to play a regulatory role in GA metabolism. This is supported by the high specificity of the enzymes catalyzing conjugate formation and hydrolysis.

In contrast, the formation of 2-O-glucosides of, for example, GA_8, GA_{26}, GA_{27}, and GA_{29} has no significance in regulation of the active GA level, because the parent GAs are already deactivated irreversibly by 2β-hydroxylation. However, the glucose moiety might be important for transport or compartmentation. Glucosylation and oxidation of 2β-hydroxy-GAs are assumed to be alternative catabolic processes, realized to different extents in different plant species. The formation of GA-2-O-glucosides, however, might also be an intermediate step in GA catabolism followed by further catabolic reactions, for example, oxidation to 2-keto GAs after hydrolysis of the corresponding glucose conjugate.

References

1. Sembdner G. Conjugation of plant hormones. In: Schreiber K, Schütte H-R, Sembdner G, eds. Biochemistry and chemistry of plant growth regulators. Halle/Saale: Inst Plant Biochem Acad Sci GDR, 1974: pp. 283–302.
2. Sembdner G, Gross D, Liebisch H-W, et al. Biosynthesis and metabolism of plant hormones. In: MacMillan J, ed. Hormonal regulation of development. I. Molecular aspects of plant hormones. Encyclopedia of plant physiology, New series, Vol. 9. Berlin: Springer-Verlag, 1980: pp. 281–444.
3. Schneider G. Gibberellin conjugates. In: Crozier A, ed. The biochemistry and physiology of gibberellins, Vol. 1. New York: Praeger, 1983: pp. 389–456.
4. Schreiber K, Weiland J, Sembdner G. Isolierung und Struktur eines Gibberellinglucosides. Tetrahedron Lett. 1967; 4285–4288.
5. Schreiber K, Weiland J, Sembdner G. Isolierung von Gibberellin-A_8-O(3)-β-D-glucopyranosid aus Früchten von *Phaseolus coccineus*. Phytochemistry. 1970; 9:189–198.
6. Takahashi N, Yamaguchi I, Yamane H. Gibberellins. In: Takahashi N, ed. Chemistry of plant hormones. Boca Raton, Florida: CRC Press, 1986: pp. 51–151.
7. Hiraga K, Yokota T, Murofushi N, et al. Isolation and characterization of

gibberellins in mature seeds of *Phaseolus vulgaris*. Agric Biol Chem. 1974; 38:2511–2520.
8. Yokota T, Kobayashi S, Yamane H, et al. Isolation of a novel gibberellin glucoside, 3-*O*-β-D-glucopyranosyl gibberellin A_1 from *Dolichos lablab* seed. Agric Biol Chem. 1978; 42:1811–1812.
9. Yokota T, Murofushi N, Takahashi N, et al. Gibberellins in immature seeds of *Pharbitis nil*. Part III. Isolation and structures of gibberellin glucosides. Agric Biol Chem. 1971; 35:583–595.
10. Yamaguchi I, Yokota T, Yoshida, S, et al. High pressure liquid chromatography of conjugated gibberellins. Phytochemistry. 1979; 18:1699–1702.
11. Yamaguchi I, Kobayashi M, Takahashi N. Isolation and characterization of glucosyl esters of gibberellin A_5 and A_{44} from immature seeds of *Pharbitis purpurea*. Agric Biol Chem. 1980; 44:1975–1977.
12. Harada H, Yokota T. Isolation of gibberellin A_8 glucoside from shoot apices of *Althaea rosea*. Planta. 1970; 92:100–104.
13. Lorenzi R, Horgan R, Heald JK. Gibberellin A_9 glucosyl ester in needles of *Picea sitchensis*. Phytochemistry. 1976; 15:789–790.
14. Yamane H, Yamaguchi I, Murofushi N, et al. Isolation and structure of gibberellin A_{35} and its glucoside from immature seeds of *Cytisus scoparius*. Agric Biol Chem. 1974; 35:1144–1146.
15. Schneider G, Sembdner G, Schreiber K. Synthese von *O*(3)- und *O*(13)-glucosylierten Gibberellinen. Tetrahedron, 1977; 33:1391–1397.
16. Hiraga K, Yokota T, Takahashi N. Biological activity of some synthetic gibberellin glucosyl esters. Phytochemistry. 1974; 13:2371–2376.
17. Schneider G. Über strukturelle Einflüsse bei der Glucosylierung von Gibberellinen. Tetrahedron. 1980; 37:545–549.
18. Schneider G, Sembdner G, Schreiber K. Zur Synthese von Gibberellin-A_3-β-D-glucopyranosiden. Z Chem. 1974; 14:474–475.
19. Schneider G. Synthese von Gibberellinglucosiden. Thesis, Inst Plant Biochem Acad Sci GDR Halle/Saale, 1981.
20. Schneider G, Sembdner G, Schreiber K, et al. Partial synthesis of some physiologically relevant gibberellin glucosyl conjugates. Tetrahedron. 1989; 45: 1355–1364.
21. Schneider G, Schreiber K, Jensen E, et al. Synthesis of gibberellin A_{29} β-D-glucosides and β-D-glucosyl derivatives of [17-^{13}C, T_2] gibberellin A_5, A_{20}, and A_{29}. Liebigs Ann Chem. 1990, 491–494.
22. Schneider G, Miersch O, Liebisch H-W. Synthese von *O*-β-D-Glucopyranosylgibberellin-*O*-β-D-glucopyranosylestern. Tetrahedron. 1977; 405–406.
23. Sembdner G, Schneider G, Schreiber K. Methoden zur Pflanzenhormonanalyse. Berlin: Springer-Verlag, 1987.
24. Jensen E, Crozier A, Monteiro AM. Analysis of gibberellins and gibberellin conjugates by ion-suppression reversed-phase high-performance liquid chromatography. J Chromatogr. 1986; 367:377–384.
25. Turnbull CGN, Crozier A, Schneider G. HPLC-based methods for the identification of gibberellin conjugates: metabolism of [^{3}H]gibberellin A_4 in seedlings of *Phaseolus coccineus*. Phytochemistry. 1986; 25:1823–1828.
26. Rivier L, Gaskin P, Albone KS, et al. GC–MS identification of endogenous gibberellins and gibberellin conjugates as their permethyl derivatives. Phytochemistry. 1981; 20:687–692.

27. Schmidt J, Schneider G, Jensen E. GC–MS investigations of permethylated gibberellin glucosides. Biomed Environ Mass Spectrom. 1988; 17:7–13.
28. Gaskin P, MacMillan J. Polyoxygenated *ent*-kauranes and water-soluble conjugates in seed of *Phaseolus coccineus*. Phytochemistry. 1975; 14:1575–1578.
29. Albone KS, Gaskin P, MacMillan J, et al. Identification and localization of gibberellins in maturing seeds of the cucurbit *Sechium edule*, and a comparison between this cucurbit and the legume *Phaseolus coccineus*. Planta. 1984; 162:560–565.
30. Gräbner R, Schneider G, Sembdner G. Fraktionierung von Gibberellinen, Gibberellinkonjugaten und anderen Phytohormonen durch DEAE-Sephadex-Chromatographie. J Chromatogr. 1976; 121:110–115.
31. Schliemann W. Untersuchungen zur enzymatischen Hydrolyse von Gibberellinglucosekonjugaten. Thesis, 1989, Inst Plant Biochem Acad Sci GDR, Hallesaale, 4990.
32. Schliemann W. Partial characterization of butanol-insoluble metabolites of [^{3}H]GA$_1$ in maturing fruits of *Phaseolus coccineus* L. Biochem Physiol Pflanzen. 1987; 182:153–163.
33. Schliemann W, Schneider G. Metabolic formation and occurrence of gibberellin A$_1$-3-*O*-β-D-glucopyranoside in immature fruits of *Phaseolus coccineus* L. Plant Growth Regul. 1989; 8:85–90.
34. Knöfel H-D, Schwarzkopf E, Müller P, et al. Enzymic glucosylation of gibberellins. J Plant Growth Regul. 1984; 3:127–140.
35. Sembdner G, Knöfel H-D, Schwarzkopf E, et al. In vitro glucosylation of gibberellins. Biol Plant. 1985; 27:231–236.
36. Schliemann W, Liebisch H-W. Enzymic and chemical hydrolysis of plant hormone conjugates. Biochem Physiol Pflanzen. 1984; 179:533–552.
37. Müller P, Knöfel H-D, Liebisch H-W, et al. Untersuchungen zur Spaltung von Gibberellinglucosiden. Biochem Physiol Pflanzen. 1978; 173:396–409.
38. Schliemann W, Schneider G. Untersuchungen zur enzymatischen Hydrolyse von Gibberellin-*O*-glucosiden I. Hydrolysegeschwindigkeiten von Gibberellin-13-*O*-glucosiden. Biochem Physiol Pflanzen. 1979; 174:738–745.
39. Schneider G, Schliemann W. Untersuchungen zur enzymatischen Hydrolyse von Gibberellin-*O*-glucosiden. II. Hydrolysegeschwindigkeiten von Gibberellin-2-*O*- und Gibberellin-3-*O*-glucosiden. Biochem Physiol Pflanzen. 1979; 174:746–751.
40. Crozier A, Kuo CC, Durley RC, et al. The biological activities of 26 gibberellins in nine plant bioassays. Can J Bot. 1970; 48:867–877.
41. Yokota T, Murofushi N, Takahashi N, et al. Gibberellins in immature seeds of *Pharbitis nil*. IV. Biological activities of gibberellins and their glucosides in *Pharbitis nil*. Phytochemistry. 1971; 10:2943–2949.
42. Schliemann W. Hydrolysis of conjugated gibberellins by β-glucosidases of dwarf rice (*Oryza sativa* L., cv. "Tan-ginbozu"). J Plant Physiol. 1984; 116:123–132.
43. Schliemann W. Enzymic hydrolysis of gibberellin conjugates. In: Schreiber K, Schütte H-R, Sembdner G, eds. Conjugated plant hormones. Structure, metabolism and function. Berlin: Deutscher Verlag der Wissenschaften, 1987: pp. 191–198.
44. Schliemann W. Hydrolyse von Gibberellin-*O*-glucosiden durch β-Glucosidasen aus *Pharbitis nil*. Biochem Physiol Pflanzen. 1983; 178:359–372.

45. Schliemann W. β-Glucosidase with gibberellin A_8-2-O-glucoside hydrolysing activity from pods of runner beans. Phytochemistry. 1988; 27:689–692.
46. Sembdner G, Borgmann E, Schneider G, et al. Biological activity of some conjugated gibberellins. Planta. 1976; 132:249–257.
47. Lehmann H, Sembdner G. Plant hormone conjugates. In: Purohit SS, ed. Hormonal regulation of plant growth and development, Vol. 3. Bikaner: Agro Botanical Publishers, 1986: pp. 245–310.
48. Liebisch H-W. Uptake, translocation and metabolism of labelled GA_3 glucosyl ester. In: Schreiber K, Schütte H-R, Sembdner G, eds. Biochemistry and chemistry of plant growth regulators. Halle/Saale: Inst Plant Biochem Acad Sci GDR, 1974: pp. 109–113.
49. Sembdner G, Weiland J, Schneider G, et al. Recent advances in the metabolism of gibberellins. In: Carr D, ed. Plant growth substances 1970. Berlin: Springer-Verlag, 1972: pp. 143–150.
50. Rood SB, Pharis RP. Evidence for reversible conjugation of gibberellins in higher plants. In: Schreiber K, Schütte H-R, Sembdner G, eds. Conjugated plant hormones. Structure, metabolism and function. Berlin: Deutscher Verlag der Wissenschaften, 1987: pp. 183–190.

CHAPTER 25

Metabolism of [^{3}H]Gibberellin A$_4$ and [^{2}H]Gibberellin A$_4$ in Cell Suspension Cultures of Rice, *Oryza sativa* cv. Nihonbare

M. Koshioka, E. Minami, H. Saka, R. P. Pharis, and L. N. Mander

1 Introduction

A number of gibberellins (GAs), namely, GA_1, GA_4, GA_9, GA_{12}, GA_{17}, GA_{19}, GA_{20}, GA_{24}, GA_{29}, GA_{34}, GA_{44}, GA_{51}, and GA_{53}, have been characterized from rice cv. Nihonbare (Japonica).[1-3] In metabolic studies of [^{3}H]GA$_4$ in dwarf rice seedlings, Durley and Pharis[4] have reported the conversion of [^{3}H]GA$_4$ to [^{3}H]GA$_1$, [^{3}H]GA$_2$, and [^{3}H]GA$_{34}$ based on the results of packed-column gas chromatography–radiochromatogram scanning, and Koshioka et al.[5] have also reported taht [^{3}H]GA$_4$ was metabolized to [^{3}H]GA$_1$, [^{3}H]GA$_2$, [^{3}H]GA$_{34}$, their glucosides (GA-G), and [^{3}H]GA$_8$-G and that the main metabolite of [^{3}H]GA$_4$ was an unidentified glucoside. Hence, as part of an examination of GA metabolic sequences in rice plants, we have analyzed the metabolism of [^{3}H]GA$_4$ and [^{2}H]GA$_4$ in cell suspension cultures of *Oryza sativa* cv. Nihonbare.

2 Experimental

Cell suspension cultures were derived from seeds of rice (*Oryza sativa* L.) of cultivar Nihonbare, on Murashige and Skoog (MS) medium and maintained by weekly passages in R2 medium.[6] Filter-sterilized [^{3}H]GA$_4$ [0.185 MBq, 1.24 TBq/mmol, from Amersham Japan Co., dissolved in 100 µl of 95% ethanol (EtOH)] and [^{2}H]GA$_4$ (prepared by a modification of the Takai procedure[7]; 200 µg dissolved in 100 µl of 95% EtOH) were each applied to separate 40-ml volumes of the rice cultures (2-day-old inoculum). After 2 days of incubation, the cells and their medium were separated and freeze-dried. Freeze-dried cells (0.48 g for [^{3}H]GA$_4$ feeds and 0.35 g for [^{2}H]GA$_4$ feeds) were extracted with 40 ml of aqueous 80% methanol (MeOH). The extracts were passed through small C_{18} reverse-phase columns to remove pigments and nonpolar compounds[8] and then dried. The dried extracts were dissolved in 20 ml of distilled water, then adjusted to pH 3.0, and shaken 3 times with equal to volumes of ethyl acetate (EtOAc), followed with n-butanol (n-BuOH). The freeze-dried medium was partitioned in the same manner. The acidic EtOAc extracts of the cells and the medium were subjected to chromatography on separate small silica-gel (Si-gel) partition columns in order to separate most of the free GAs from putative GA glucosyl conjugates.[9] The free GA fractions (n-hexane:EtOAc-soluble) from the Si-gel columns and the combined MeOH-soluble fractions (eluted subsequent to n-hexane: EtOAc) from the Si-gel columns and acidic BuOH extracts were further purified on gradient-eluted C_{18} reverse-phase high-performance liquid chromotography (HPLC) columns.[10] The C_{18} HPLC fractions of free GAs were further chromatographed on Nucleosil $N(CH_3)_2$ HPLC columns.[11] For the C_{18} HPLC fractions of putative GA glucosyl conjugates, an additional isocratic-eluted C_{18} HPLC was used.[12] A part of each putative GA conjugate fraction was further separated on a DEAE-Sephadex A-25 column designed to separate GA glucosyl ester (GA-GE) from GA glucosyl ether (GA-G).[13] The other part was hydrolyzed with β-glucosidase prior to additional HPLC analysis. The HPLC fractions of the free GAs from the [^{2}H]GA$_4$ feeds were further analyzed by combined gas chromatography–mass spectrometry–selected ion monitoring (GC–MS–SIM) as the methyl ester trimethylsilyl ether (MeTMSi) derivatives.

3 Results and Discussion

3.1 Separation and Identification of Metabolites of [^{3}H]GA$_4$

Because of the high specific radioactivity of [^{3}H]GA$_4$, tentative identifications from the [^{3}H]GA$_4$ feeds were based only on a series of sequential chromatography steps, with comparison of retention times (t_R) to those of authentic standards. However, the reader is referred to results (Table 1) from the [^{2}H]GA$_4$ feeds for more definitive analyses of certain of the metabolites of [^{2}H]GA$_4$. As shown in the radioactive profiles obtained from gradient-eluted C_{18} HPLC of free GA fractions from the cells and the medium (Fig. 1), [^{3}H]GA$_1$-like (at t_R 27–28 min), [^{3}H]GA$_2$-like (at t_R 30–31 min), and [^{3}H]GA$_{34}$-like (at t_R 36–37 min) compounds were found; two additional peaks (at t_R 24–25 min and t_R 33–34 min, respectively) were of unknown origin. From the combined MeOH-soluble fraction from the short Si-gel column and the BuOH fraction, at least six peaks were observed in the radioactive profile as shown in Fig. 2. After several analytical procedures, [^{3}H]GA$_8$-G-like (at t_R 11–12 min), [^{3}H]GA$_2$-G-like (at t_R 27–28 min), [^{3}H]GA$_{34}$-G-like (at t_R 33–34 min), and [^{3}H]GA$_4$-GE-like and [^{3}H]GA$_4$-G-like (at t_R 37–39 min) compounds were tentatively identified, leaving two unidentified peaks (compounds C1 and C2) in the radioactive profile, at t_R 23–24 min and t_R 30–31 min, respectively. A hydrolysate of compound C1 was eluted at t_R 27–28 min on gradient-eluted C_{18}

Table 1. Kovats retention indices and relative intensities of characteristic m/z ions from GC–MS–SIM for extracts of cell suspension cultures or rice cv. Nihonbare fed [²H₂]GA₄

Sample	Gradient-eluted C$_{18}$ HPLC t_R (min)	KRI	m/z (% intensity)	Identity
Fractions[a]	27–28	2672	508 (100), 450 (13.4), 379 (10.1), 315 (9.9)	
GA$_1$	27–28	2673	506 (100), 448 (17.4), 377 (13.8), 313 (9.2)	
Fractions[a]	30–31	2775	510 (29.1), 495 (15.0), 420 (15.3), 291 (26.3), 132 (100)	[²H]GA$_2$
GA$_2$	30–31	2776	508 (20.2), 493 (16.7), 418 (12.6), 289 (23.6), 130 (100)	
Fractions[a]	39–40	2550	420 (18.9), 388 (17.6), 291 (41.3), 286 (100), 227 (76.4), 226 (81.0)	
[²H₂]GA$_4$	39–40	2550	420 (18.5), 388 (17.8), 291 (41.6), 286 (100), 227 (80.3), 226 (84.3)	[²H]GA$_4$

[a]Free GA fractions

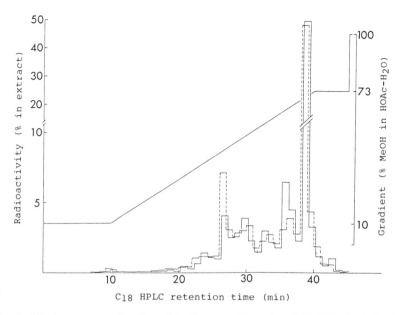

Fig. 1. Elution pattern of radioactivity from gradient-eluted C_{18} HPLC of the free GA fraction from an extract of cell suspension cultures of *Oryza sativa* cv. Nihonbare fed [^{3}H]GA$_4$. ———, Cell extract; — — —, medium extract

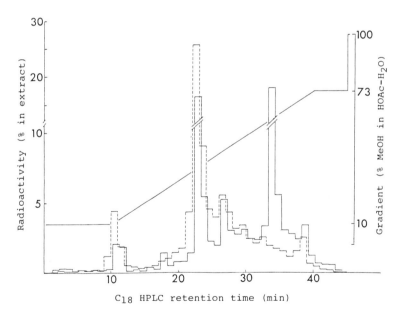

Fig. 2. Elution pattern of radioactivity from gradient-eluted C_{18} HPLC of the putative GA glucosyl conjugate fraction from an extract of cell suspension cultures of *Oryza sativa* cv. Nihonbare fed [^{3}H]GA$_4$. ———, Cell extract; — — —, medium extract

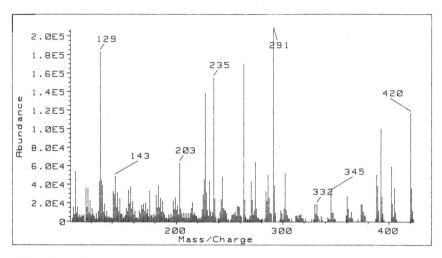

Fig. 3. Mass spectrum of methyl ester trimethylsilyl ether of the unknown free acid compound eluted at t_R 27–28 min on C_{18} HPLC and at t_R 25–26 min on Nucleosil $N(CH_3)_2$ HPLC[5]

Table 2. Kovats retention indices for the hydrolysate of the unknown putative glucosyl ether [t_R 24–25 min on gradient-eluted C_{18} HPLC], 16,17-dihydro GA_4 epimers, and C/D-ring-rearranged GA_4 epimers[5]

Compound	KRI
Hydrolysate of the unknown compound	2692
3α-OH, 16α-CH_3-dihydro-GA_4[a]	2658
3α-OH, 16β-CH_3-dihydro-GA_4[a]	2707
3β-OH, 16α-CH_3-dihydro-GA_4[a]	2524
3β-OH, 16β-CH_3-dihydro-GA_4[a]	2570
3α-OH C/D-ring-rearranged GA_4[a]	2616
3β-OH C/D-ring-rearranged GA_4[a]	2485

[a] Synthesized by L.N. Mander (unpublished results)

HPLC and at t_R 25–26 min on Nucleosil $N(CH_3)_2$ HPLC. These t_R values coincided with those of an as yet uncharacterized GA-like compound from a feed of [^{3}H]GA_4 of low specific activity to dwarf rice cv. Tan-ginbozu seedlings. The MeTMSi derivative of the unknown compound in Tanginbozu seedlings had prominent ions of m/z 420 (M$^+$), 405 [(M-15)$^+$], 402 [(M-18)$^+$], 392 [(M-28)$^+$], 388 [(M-32)$^+$], 361 [(M-59)$^+$], 360 [(M-60)$^+$], 332 [(M-88)$^+$], 330 [(M-90)$^+$], and 302 [(M-118)$^+$], as shown in Fig. 3, and the fragmentation pattern was quite similar to that of 16,17-dihydro-GA_4.[5] However, the Kovats retention index (KRI) of the unknown compound did not coincide with any of the KRI values of dihydro-GA_4 epimers and

C/D-ring-rearranged GA_4 epimers as shown in Table 2.5. Thus, the metabolites of [^{3}H]GA_4 in cell suspension cultures cv. Nihonbare were the same as those found in dwarf rice cv. Tan-ginbozu seedlings.

3.2 Separation and Identification of Metabolites of [^{2}H]GA_4

The sequential analyses for [^{2}H]GA_4 feeds were done in essentially the same way as described above for the [^{3}H]GA_4 feeds. [^{2}H]GA_4 metabolites were collected from HPLC column fractions corresponding to the t_R values of [^{3}H]GA_4 metabolites and then derivatized (MeTMSi) prior to GC–MS–SIM analyses. As shown in Table 1, [^{2}H]GA_1 was found at t_R 27–28 min from gradient-eluted C_{18} HPLC, corresponding to a radioactive peak at the same t_R from the [^{3}H]GA_4 feeds (see Fig. 1); [^{2}H]GA_2 and [^{2}H]GA_4 were also found at t_R 30–31 min and t_R 39–40 min, respectively. All identifications were confirmed by GC–SIM, using four to six characteristic ions and KRI values. From the C_{18} HPLC fraction at t_R 36–37 min, only one definitive mass peak of m/z 508 was found at the same KRI (2679) as that of GA_{34}.

3.3 Metabolism of [^{3}H]GA_4

In the cell suspension cultures of rice, most of the applied radioactivity (99.6%) was recovered in the extracts, as shown in Table 3. The distribution of radioactivity was 14.6% in the cells, 85.0% in the medium, and 0.4% in the cell residue. Within the cultures, 31.2% of total radioactivity was found in free GA fractions other than [^{3}H]GA_4, and 22.0% in the putative GA conjugate fraction (see Table 4). About 35% of total radioactivity remained as free [^{3}H]GA_4, with 1.4% as its putative glucosyl conju-

Table 3. Recovery of radioactivity from cell suspension cultures of rice cv. Nihonbare fed [^{3}H]GA_4

Fraction or residue	Radioactivity recovered (% of total applied [^{3}H]GA_4)		
	Cells	Medium	Total
Acidic EtOAc-soluble fraction (from Si-gel partition column)			
Free GA fraction (n-hexane: EtOAc)	6.0	59.8	65.8
Subsequent MeOH wash fraction	0.9	10.2	11.1
Acidic BuOH-soluble fraction	4.6	6.3	10.9
Aqueous residue	3.1	8.7	11.8
Subtotal	14.6	85.0	99.6
Cell residue			0.4
Total			100.0

Table 4. Radioactivity in free GA fraction and putative GA conjugate fraction (as a percentage of the total radioactivity) from cell suspension cultures of rice cv. Nihonbare incubated with [^{3}H]GA$_4$ for 48 h

Tentative identity of GAs	Gradient-eluted C$_{18}$ HPLC fraction t_R (min)	Percentage of radioactivity		
		Cell extract	Medium extract	Total
Free GA fraction				
Unknown (F1)	24–25	0.2	1.4	1.6
GA$_1$	27–28	0.4	6.1	6.5
GA$_2$	30–31	0.4	3.8	4.2
Unknown (F2)	33–34	0.3	2.9	3.2
GA$_{34}$	36–37	0.7	4.1	4.8
GA$_4$	39–40	3.2	31.4	34.6
Others		0.8	10.1	10.9
Subtotal		6.0	59.8	65.8
Conjugated GA fraction				
GA$_8$-G-like	11–12	0.2	1.3	1.5
Unknown-G (C1)	23–24	1.3	6.5	7.8
GA$_2$-G-like	27–28	0.5	2.3	2.8
Unknown-G (C2)	30–31	0.3	1.0	1.3
GA$_{34}$-G-like	33–34	1.3	0.8	2.1
GA$_4$-G/GE-like	37–39	0.5	0.9	1.4
Others		1.4	3.7	5.1
Subtotal		5.5	16.5	22.0
Total		11.5	76.3	87.8

gates. The major metabolites of [^{3}H]GA$_4$ were [^{3}H]GA$_1$ at 6.5% of the total radioactivity, and the ^{3}H-containing unknown putative glucosyl compound C1 at 7.8% of the total radioactivity. Although compound C1 was found in a relatively high amount, its free acid was not found. This suggests that the rate of conjugation is very rapid and/or that other putative [^{3}H]GA glucosyl conjugates may be directly converted to compound C1. The presence of [^{3}H]GA$_8$-G and the apparent absence of [^{3}H]GA$_8$ also suggest either a very rapid conjugation of free [^{3}H]GA$_8$ or the existence of metabolic pathways from [^{3}H]GA$_1$-G and/or [^{3}H]GA$_{34}$-G to [^{3}H]GA$_8$-G (Fig. 4). It is noteworthy that a relatively high amount of [^{3}H]GA$_2$ (3.8%) was found along with a relatively high amount of its putative glucoside (2.8%). Although [^{3}H]GA$_2$ has been suggested as an artifact product from [^{3}H]GA$_4$,[4,14,15] we consider it to be a normal metabolite of [^{3}H]GA$_4$, since no [^{3}H]GA$_2$ was found in feedings of [^{3}H]GA$_4$ to anise[12] and carrot[8] extracts that were extracted by the same method[9]. As noted by Durley and Pharis,[4] [^{3}H]GA$_{34}$ was also found as a metabolite of [^{3}H]GA$_4$, with

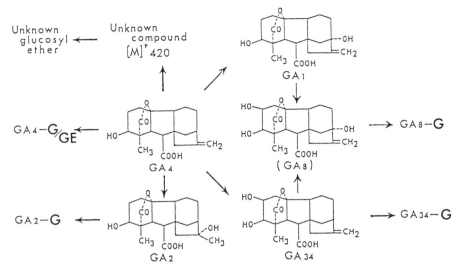

Fig. 4. Possible metabolic pathways of GA_4 in cell suspension cultures of *Oryza sativa* cv. Nihonbare

[^{3}H]GA_{34}-G. In comparison with the level of unknown compound C1 (50% of extracted radioactivity) in [^{3}H]GA_4-fed dwarf rice cv. Tanginbozu seedlings,[5] the level of compound C1 in this study was fairly low (7.8%), although compound C1 was the major metabolite in both of these systems.

3.4 Metabolic Pathways of GA_4

Possible metabolic pathways of GA_4 in cell suspension cultures of *Oryza sativa* cv. Nihonbare are shown in Fig. 4. The pattern of GA_4 metabolism in these cultures is thus similar to that found in the dwarf rice seedlings.

References

1. Kobayashi M, Yamaguchi I, Murofushi N, et al. Endogenous gibberellins in immature seeds and flowering ears of rice. Agric Biol Chem. 1984; 48:2725–2729.
2. Kobayashi M, Yamaguchi I, Murofushi N, et al. Fluctuation and localization of endogenous gibberellins in rice. Agric Biol Chem. 1988; 52:1189–1194.
3. Kurogochi S, Murofushi N, Ota Y, et al. Identification of gibberellin in the rice plant and quantitative changes of gibberellin A_{19} throughout its life cycle. Planta. 1979; 146:185–191.
4. Durley RC, Pharis RP. Interconversion of gibberellin A_4 to gibberellin A_1 and A_{34} by dwarf rice, cultivar Tan-ginbozu. Planta. 1973; 109:357–361.

5. Koshioka M, Beall FD, Takeno K, et al. Metabolism of tritiated gibberellin A_4 in seedlings of dwarf rice cv. Tan-ginbozu. Plant Physiol. 1989; submitted.
6. Ohira K, Ojima K, Fujiwara A. Studies on the nutrition of rice cell culture I. A simple, defined medium for rapid growth in suspension culture. Plant Cell Physiol. 1973; 14:1113–1121.
7. Lombardo L. Methylenation of carbonyl compounds with $Zn-CH_2-TiCl_4$, application to gibberellins. Tetrahedron Lett. 1982; 4293–4296.
8. Koshioka M, Jones A, Koshioka MN, et al. Metabolism of [^{3}H]gibberellin A_4 in somatic suspension cell cultures of carrot. Phytochemistry. 1983; 22: 1585–1590.
9. Koshioka M, Takeno K, Beall FD, et al. Purification and separation of plant gibberellins from their precursors and glucosyl conjugates. Plant Physiol. 1983; 73:398–406.
10. Koshioka M, Harada J, Takeno K, et al. Reversed-phase C_{18} high-performance liquid chromatography of acidic and conjugated gibberellins. J Chromatogr. 1983; 256:101–115.
11. Koshioka M, Pharis RP, Matsuta N, et al. Metabolism of [^{3}H]gibberellin A_5 and [^{2}H]gibberellin A_5 in cell suspension cultures of *Prunus persica*. Phytochemistry. 1988; 27:3799–3805.
12. Koshioka M, Douglas TJ, Ernst D, et al. Metabolism of [^{3}H]gibberellin A_4 in somatic suspension cultures of anise. Phytochemistry. 1983; 22:1577–1584.
13. Turnbull CGN, Crozier A, Schneider G. HPLC-based methods for the identification of gibberellins: Metabolism of [^{3}H]gibberellin A_4 in seedlings of *Phaseolus coccineus*. Phytochemistry. 1986; 25:1823–1828.
14. Reeve DR, Crozier A, Durley RC, et al. Metabolism of ^{3}H-gibberellin A_1 and ^{3}H-gibberellin A_4 by *Phaseolus coccineus* seedlings. Plant Physiol. 1975; 55:42–44.
15. Wample RL, Durley RC, Pharis RP. Metabolism of gibberellin A_4 by vegetative shoots of Douglas fir at three stages of ontogeny. Physiol Plant. 1975; 35:273–278.

CHAPTER 26

Stem Growth and Gibberellin Metabolism in Spinach in Relation to Photoperiod

J.A.D. Zeevaart, M. Talon, and T.M. Wilson

1 Introduction

Long-day plants (LDP) with a rosette growth habit under short days (SD) can be induced to flower and elongate their stems by exposure to long-day (LD) conditions. As has been discussed previously,[1,2] stem elongation and flower formation are two separate, albeit concomitant, developmental events, the former being controlled by gibberellins (GAs), the latter by an as yet unidentified floral stimulus.

In this paper, recent advances in our knowledge of the GAs in spinach are reviewed as well as how the photoperiod controls certain steps in the latter part of the GA biosynthetic pathway.

2 Photoperiodic After-Effect in Long-Day Plants

By definition, all LDP respond with stem elongation and flower formation after transfer from SD to LD conditions. However, there are marked differences among LDP in the so-called photoperiodic after-effect, that is, the response exhibited after the plants are returned to SD following exposure to a certain number of LD. Two extreme cases are represented by spinach and *Silene armeria*. In spinach, the effect of LD is direct; in other words, there is no photoperiodic after-effect. When plants were exposed to 15 or 18 LD, and then returned to SD, stem elongation ceased (Fig. 1) and the inflorescences did not develop further. Following transfer to SD, the LD growth habit changed to that of plants grown permanently in SD. On the other hand, in *Silene armeria* the effect of LD is inductive, since inflorescence development and stem elongation continued when the plants were returned to SD.[3,4] Another LDP, *Agrostemma githago*, is intermediate between spinach and *Silene*. When *Agrostemma* plants were transferred from LD to SD, the stems continued to elongate at a reduced rate for some time, but their ultimate height never approached that of plants in continuous LD.[5]

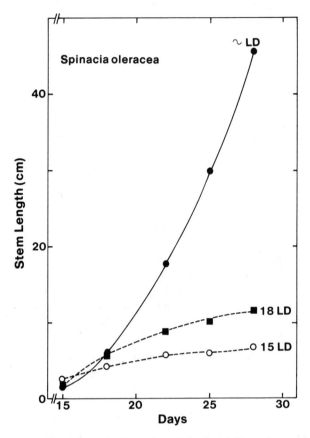

Fig. 1. Stem growth of spinach plants in continuous LD, and cessation of stem elongation when returned to SD after exposure to 15 or 18 LD

3 Gibberellins and Growth Responses

Exposure of spinach plants to LD induces the setting of the leaves in a vertical position and promotes petiole and stem growth. These responses to LD can also be elicited in SD by treatment with GA_3. On the other hand, the same responses can be suppressed in LD by treatment with the growth retardant AMO-1618, which inhibits GA biosynthesis. The effects of this retardant can be fully overcome by simultaneous applications of GA_3.[1] These observations indicate that petiole and stem growth in spinach are regulated by GA and that stem growth induced in LD is mediated by GAs.

4 Endogenous Gibberellins in Spinach

In earlier work, several members of the early-13-hydroxylation pathway, viz., GA_{53}, GA_{44}, GA_{19}, GA_{17}, GA_{20}, and GA_{29}, were identified as endogenous GAs in spinach.[1] More recently, GA_{12}, GA_5, GA_1, 3-*epi*-GA_1, GA_8, and GA_{29}-catabolite have also been detected. In addition, several GAs lacking a C-13 hydroxyl, viz., GA_{15}, GA_{24}, GA_9, and GA_{51}, have been identified. Furthermore, the following 3-hydroxylated GAs have been found in spinach: GA_4, GA_{34}, GA_7, and iso-GA_7. Spinach extracts also contained several uncharacterized GAs. One of them had a mass spectrum indicative of monohydroxy-GA_{53} that has been tentatively assigned the structure 2β-hydroxy-GA_{53}. All these GAs are present in plants grown under SD as well as under LD conditions, although the levels vary greatly, depending on the photoperiod under which the plants had been grown prior to harvest (M. Talon, unpublished results).

Gibberellins lacking a C-13 hydroxyl are present in relatively low levels in spinach. The early-13-hydroxylation pathway is apparently the main GA biosynthetic pathway in this plant.

The photoperiod has a pronounced effect on the levels of GAs in spinach. In earlier work, it was found that following transfer from SD to LD the content of GA_{19} declined fivefold, whereas the levels of GA_{20} and GA_{29} increased almost sevenfold during the same period.[1] These observations led to the idea that in spinach the conversion of GA_{19} to GA_{20} is under photoperiodic control. In recent work, it has been found that the levels of GA_1 and GA_8 also increase substantially during LD treatment, whereas levels of 2β-hydroxy-GA_{53} decrease (M. Talon, unpublished results).

5 In Vivo Gibberellin Metabolism

In metabolic studies, emphasis has been on the early-13-hydroxylation pathway. As shown in Table 1, [^{14}C]GA_{19} applied to four groups of plants under different light–dark conditions was only metabolized to a significant extent when the plants were in continuous light (CL). When plants in CL were kept in darkness (D) for 8 h prior to [^{14}C]GA_{19} application, only a small amount was converted to [^{14}C]GA_{20}, indicating that most GA_{19}-oxidase activity had been lost during the dark period. A similar experiment with [^{14}C]GA_{44} gave quite different results (Table 2). In all treatments, a substantial amount was converted to [^{14}C]GA_{19}, but further metabolism to [^{14}C]GA_{20} and [^{14}C]GA_{29} took place only in CL. Thus, the conversion of GA_{44} to GA_{19} is not regulated by light. It should be noticed that [^{14}C]GA_{44} in the lactone form is metabolized by spinach in vivo as well as by cell-free extracts,[6] whereas in a cell-free system from pea embryos it was only metabolized when in the hydroxy acid form.[7]

Table 1. Metabolism of [^{14}C]GA$_{19}$ by spinach plants[a] kept under different light–dark conditions

Light–dark conditions	% Radioactivity recovered in:			
	GA$_{19}$	GA$_{20}$	GA$_{29}$	Polar[b]
8 h L–8 h D ↓ 8 h D–8 h L	99.4	0.2	0.3	0
8 h L–8 h D ↓ 16 h D	98.5	1.5	0	0
CL ↓ 16 h L	48.2	22.5	14.6	14.4
CL–8 h D ↓ 16 h D	97.1	2.7	0	0

[a] [^{14}C]GA$_{19}$ (125,000 dpm = 0.5 nmol) applied to young expanding leaves of two plants in each treatment
[b] On reverse-phase high-performance liquid chromatography eluting prior to GA$_{29}$
↓ Time of [^{14}C]GA$_{19}$ application

Table 2. Metabolism of [^{14}C]GA$_{44}$ by spinach plants[a] kept under different light–dark conditions

Light–dark conditions	% Radioactivity recovered in:				
	GA$_{44}$	GA$_{19}$	GA$_{20}$	GA$_{29}$	Polar[b]
8 h L–8 h D ↓ 8 h D–8 h L	47.9	48.9	0.3	1.4	0.5
8 h L–8 h D ↓ 16 h D	55.0	43.0	0	1.0	0.8
CL ↓ 16 h L	39.4	32.7	2.6	10.0	14.7
CL–8 h D ↓ 16 h D	58.1	39.4	0.9	1.0	0.4

[a] [^{14}C]GA$_{44}$ (125,000 dpm = 0.5 nmol) applied to young expanding leaves of two plants in each treatment
[b] On reverse-phase high-performance liquid chromatography eluting prior to GA$_{29}$
↓ Time of [^{14}C]GA$_{44}$ application

Metabolism of [^{14}C]GA$_{53}$ via the early-13-hydroxylation pathway in spinach leaves also occurred preferentially in CL, but the situation was complicated by the finding that a large portion of the substrate was converted to 2β-hydroxy-GA$_{53}$ (data not shown).

Metabolic studies with [^{14}C]GA$_{20}$ have failed to show any conversion to [^{14}C]GA$_1$ and [^{14}C]GA$_8$, although GA$_{20}$ is assumed to be the precursor of GA$_1$ in spinach. A possible explanation of this observation is that the 3β-hydroxylase is localized in a compartment that is not directly accessible to applied radioactive GA$_{20}$.

6 Gibberellin Conversions in Cell-Free Extracts

Cell-free extracts capable of converting ^{14}C-labeled GAs have been obtained from spinach leaves.[6] In extracts from leaves under LD conditions, the following pathway was established: GA$_{12}$ → GA$_{53}$ → GA$_{44}$ →

Table 3. Conversions of [^{14}C]GA$_{53}$, [^{14}C]GA$_{44}$, and [^{14}C]GA$_{19}$ in cell-free extracts from spinach leaves, as affected by light–dark conditions under which the plants were grown prior to harvest[a]

Light–dark conditions	% Conversion to products[b] of:		
	[^{14}C]GA$_{53}$	[^{14}C]GA$_{44}$	[^{14}C]GA$_{19}$
D–8 h L	69.0	100.0	93.4
D–8 h L–8 h D	1.0	95.3	8.5
CL	32.1	99.1	59.7
CL–8 h D	4.2	100.0	7.7

[a] Extracts prepared as described in ref. 6, except HEPES buffer was replaced by potassium phosphate. The fraction precipitating between 40 and 55% (NH$_4$)$_2$SO$_4$ saturation was purified on a hydroxyapatite column with a gradient of phosphate buffer. The samples were dialyzed prior to assaying
[b] Purified extract of 10 g of spinach leaves (8 mg of protein) assayed for 60 min at 30°C in 50 mM potassium phosphate, 5 mM dithiothreitol, 5 mM 2-oxoglutarate, 5 mM ascorbate, and 0.5 mM FeSO$_4$ at pH 7.0, with each substrate at approximately 1 μM.

GA$_{19}$ → GA$_{20}$. So far, no 3β-hydroxylation has been detected in cell-free systems from spinach leaves, and 2β-hydroxylation has been detected only for the conversion of GA$_{12}$ to 2β-hydroxy-GA$_{12}$.[6]

The enzymes oxidizing GA$_{53}$ and GA$_{19}$ were increased in leaves under LD, whereas the enzyme oxidizing GA$_{44}$ was not affected by the photoperiod under which the plants had been grown.[6] Results obtained with purified cell-free preparations support these observations. As shown in Table 3, there was always high GA$_{44}$-oxidase activity regardless of the light–dark conditions, which is in agreement with the in vivo metabolic studies conducted with [^{14}C]GA$_{44}$ (Table 2).

The activities of both GA$_{53}$-oxidase and GA$_{19}$-oxidase were high only when the leaves had been exposed to L prior to harvest. After an 8-h D treatment, the activities of both enzymes were low (Table 3). Thus, instead of referring to SD as done previously,[6] it is more appropriate to indicate the time of harvest with respect to lights "on" or "off," since at the end of an 8-h L period (= SD) the activities of GA$_{53}$- and GA$_{19}$-oxidases are high (see also Section 7).

The results obtained with cell-free preparations and with in vivo metabolism of [^{14}C]GA$_{19}$ are in general agreement. The high GA$_{53}$- and GA$_{19}$-oxidase activities following 8-h L (Table 3) compared to insignificant [^{14}C]GA$_{19}$ conversion in vivo in a similar treatment (Table 1) can be explained as follows. The enzyme activities in Table 3 were measured after the plants had been in L for 8 h. On the other hand, in vivo feeding of [^{14}C]GA$_{19}$ was performed after an 8-h D period when GA$_{19}$-oxidase activity was low; this was followed by an additional 8-h D period. By the time GA$_{19}$-oxidase had increased during the following 8-h L period, the radioactive substrate was probably compartmentalized.

The regulation of GA_{19}-oxidase by light offers an explanation for the accumulation of GA_{19} under SD.[1] However, a similar accumulation of GA_{53} under SD conditions has not been observed. This can be explained by the alternative pathway for GA_{53} metabolism via 2β-hydroxylation to the putative 2β-hydroxy-GA_{53} which accumulates in SD.

Both GA_{53}-oxidase and GA_{19}-oxidase are regulated by L and D. These enzymes have also very similar physicochemical properties.[8] So far, it has not been possible to separate the two enzyme activities by five different purification procedures (anion exchange, hydroxyapatite, gel filtration, chromatofocusing, and hydrophobic chromatography) (T.M. Wilson, unpublished results),[8] so that the possibility remains that both catalytic sites reside in the same protein.

7 Concluding Remarks

Our results indicate that under SD conditions (8 h L:16 h D) the latter part of the GA biosynthetic pathway operates at a low rate or not at all at the end of the 16-h night. Enzyme activities increase during the following 8-h light period and then decline again during the subsequent night (Fig. 2). Thus, the amounts of GA_{20} and GA_1 produced in spinach under SD are low. On the other hand, in LD conditions (CL), enzyme activities increase to a constant level (Fig. 2), which explains the much higher GA_{20} and GA_1 content in LD than in SD. It is this high level of bioactive GA which causes stem elongation.

The question, how does light regulate enzyme activities for GA conversions in spinach, can ultimately only be resolved at the molecular level. As a first step in that direction, preparation of antibodies against GA oxidases

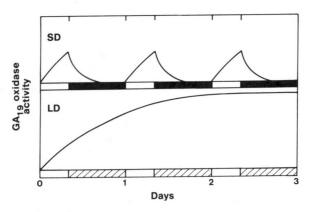

Fig. 2. Schematic presentation of GA_{19} oxidase activity in spinach leaves of plants grown in 8-h L and 16-h D periods (SD) or with the main light period supplemented with weak light from incandescent lamps (LD)

will make it possible to measure enzyme levels in extracts from plants grown under different conditions and to determine whether in D these enzymes are merely inactivated or completely degraded. Once the gene for GA_{19}-oxidase has been isolated, the effect of light on transcription can be studied.

Finally, it should be mentioned that stem growth in rosette plants is not solely regulated by the level of the bioactive GA. There is considerable evidence that LD, in addition to enhancing GA biosynthesis, also increases the responsiveness of the plants to GA.[9,10] It is this latter effect of LD treatment that appears to be at least in part responsible for the persistent effect of LD in *Silene* after return to SD.[4] The nature of increased sensitivity to GA in plants under LD is unknown, but it is thought to be related to receptors for GA and/or to the subsequent transduction chain that leads to cell division and elongation in the subapical meristem.[11]

Acknowledgments. The work of the authors was supported by the U.S. Department of Energy under Contract No. DE-AC02-76ER01338 and by the U.S. Department of Agriculture through Grant No. 88-37261-3434.

References

1. Metzger JD, Zeevaart JAD. *Spinacia oleracea*. In: Halevy AH, ed. Handbook of flowering, Vol. IV. Boca Raton, Florida: CRC Press, 1985: pp. 384–392.
2. Zeevaart JAD. Gibberellins and flowering. In: Crozier A, ed. The biochemistry and physiology of gibberellins, Vol 2. New York: Praeger, 1983: pp. 333–374.
3. Wellensiek SJ. *Silene armeria*. In: Halevy AH, ed. Handbook of flowering, Vol. IV. Boca Raton, Florida: CRC Press, 1989: pp. 320–330.
4. Talon M, Zeevaart JAD. Gibberellins and stem growth as related to photoperiod in *Silene armeria*. Plant Physiol. 1990; 92: 1094–1100.
5. Zeevaart JAD. *Agrostemma githago*. In: Halevy AH, ed. Handbook of flowering, Vol. VI. Boca Raton, Florida: CRC Press, 1989: pp. 15–21.
6. Gilmour SJ, Zeevaart JAD, Schwenen L, Graebe JE. Gibberellin metabolism in cell-free extracts from spinach leaves in relation to photoperiod. Plant Physiol. 1986; 82:190–195.
7. Kamiya Y, Graebe JE. The biosynthesis of all major pea gibberellins in a cell-free system from *Pisum sativum*. Phytochemistry. 1983; 22:681–689.
8. Gilmour SJ, Bleecker AB, Zeevaart JAD. Partial purification of gibberellin oxidases from spinach leaves. Plant Physiol. 1987; 85:87–90.
9. Jones MG, Zeevaart JAD. Gibberellins and the photoperiodic control of stem elongation in the long-day plant *Agrostemma githago* L. Planta. 1980; 149:269–273.
10. Zeevaart JAD. Effects of photoperiod on growth rate and endogenous gibberellins in the long-day plant spinach. Plant Physiol. 1971; 47:821–827.
11. Sachs RM, Bretz CF, Lang A. Shoot histogenesis: The early effects of gibberellin upon stem elongation in two rosette plants. Amer J Bot. 1959; 46:376–384.

CHAPTER 27

Phytochrome Mediation of Gibberellin Metabolism and Epicotyl Elongation in Cowpea, Vigna sinensis L.

N. Fang, B.A. Bonner, and L. Rappaport

1 Introduction

Stem elongation is typically very responsive to a variety of genetic, environmental and physiological factors which interact to determine the final internode length. While the potential roles of these factors are well established in isolation, the means by which they form a coordinated system of control is not well understood. One cluster of factors that seems to be coordinated includes: (a) the maintenance of an effective level of a growth-active gibberellin (GA), (b) the inhibition of elongation by phytochrome P_{FR}, and (c) the regulatory effects of leaves on elongation of neighboring internodes.

There is convincing evidence that the concentration of GA_1 is a major controlling factor in maize and peas.[1-3] There is much evidence that light, perceived by phytochrome, acts to inhibit elongation in a manner counter to the promotion mediated by gibberellins and that light perception may reside in both stem and leaf.[4,5] Some evidence suggests that P_{FR} action alters GA metabolism by blocking the formation of active GAs or by enhancing inactivation steps involving 2β-hydroxylation.[6-8]

This paper presents data concerning the endogenous levels of several key GAs, two of which were previously identified in cowpeas (Vigna sinensis L.),[5] and phytochrome mediation of their metabolism. They are components of the early-13-hydroxylation pathway.

Gibberellins produced in leaves were reported to regulate first internode extension.[9-11] Later, García-Martínez et al.[5] found that the cowpea internode tissue is a site of perception of red (R) and far-red (FR) light and that phytochrome mediates both GA metabolism and stem growth. The report was restricted to the responses of debladed, derooted, and decapitated explants. Therefore, we undertook studies of light-induced changes in concentration in leaves, in the active region of stem elongation, and in the basal, more mature stem. The light regime utilized elicits the end-of-day far-red promotion of elongation.[4]

2 Materials and Methods

2.1 Plant Material

Cowpea (*Vigna sinensis* L. cv. Blackeye pea No.5, Foundation Seeds and Plant Materials Services, University of California, Davis) seeds were sown in vermiculite in plastic dishpans, irrigated with half-strength Hoagland's solution, and transferred to growth chambers maintained at a constant temperature of 27 °C. An 8-h photoperiod was provided [225–250 $\mu M/$(m^2·s)] by mixed fluorescent/incandescent lamps.

2.2 Growth Studies

In studies of growth responses to GAs and light, 5-day-old seedlings were transferred to vermiculite in 110-ml plastic vials, irrigated with half-strength Hoagland's solution containing 10^{-5} M paclobutrazol, and then returned to the growth chamber for 2 days. At the end of day 7, the apices were excised and gibberellins A_1 and A_{20} (1 μg per leaf) were applied. Five-minute R or FR light treatments were made, and the plants were kept in the dark for 3 additional days and then they were measured. Red and far-red sources were as described in ref. 6 and provided 2 $\mu M/$(m^2·s) red and 5.6 $\mu M/$(m^2·s) far-red between 725–735 nm.

For determination of GA content, plants were growth as above for 7 days without paclobutrazol, irradiated for 10 min with either red or far-red light, and then held 16 h in darkness. At harvest, under a green safelight, the plants were divided into three samples each for the R and FR light treatments: (1) the primary leaves with petioles, (2) the upper 2 cm of stem, and (3) the remainder (base) of the stem above the cotyledonary node. The samples were immediately frozen and kept at −20 °C until they were extracted.

2.3 Extraction and Purification

Gibberellins were extracted and purified according to the procedure outlined in the flow chart (Fig. 1). Four regions were collected from the first silica adsorption high-performance liquid chromatography (HPLC) fractionation, and these were separately chromatographed on a reverse-phase C_{18} column using trace quantities of tritiated standards to locate peaks. Full-scan mass spectra were obtained by capillary combined gas chromatography–mass spectrometry (GC–MS) using a Trio-2 quadrupole mass spectrometer (VG Masslab, Manchester, U.K.). A Hewlett-Packard gas chromatograph and mass-selective detector was used for identification by selected ion monitoring (SIM) and isotope dilution determinations with deuterated samples.[12] Both instruments used a DB-1 fused silica capillary column.

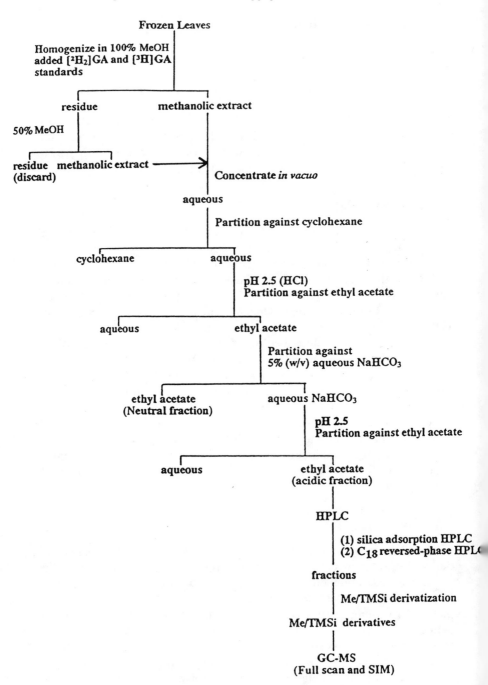

Fig. 1. General scheme of GA purification procedure

Table 1. The effect of gibberellins A_1 and A_{20} (1 μg/plant) and 5-min red or far-red light on stem elongation of decapitated cowpeas (mean of four experiments)

Treatment	Δ Stem elongation (mm)		Treated − Control	
	Red	Far-red	Red	Far-red
No GA				
Leaves	0.9	1.6	—	—
Stems	1.4	1.8	—	—
GA_{20} (1μg)				
Leaves	2.1	19.6	1.2	18.0
Stems	10.7	17.9	9.3	16.1
GA_1 (1 μg)				
Leaves	3.6	20.4	2.7	18.8
Stems	11.0	21.6	9.6	19.8

3 Results and Discussion

A series of experiments was conducted using different protocols to study the relationships between light and applied GAs. Table 1 is an example wherein 5-day-old light-grown plants were treated with 10^{-5} M paclobutrazol to inhibit GA biosynthesis. By day 7, elongation had essentially ceased. The terminal bud was excised and GAs were applied to the leaves or injected into the stems. The plants were then given red or far-red light treatment and kept in darkness for 3 days. Further elongation was dependent upon the state of phytochrome. If P_{FR} was removed by far-red irradiation, elongation was maximized, but there was essentially no difference in effectiveness of GAs between stem and leaf application. However, when the plants were red irradiated, the site of application of GAs was critical. After red, when GA_1 was applied to leaves, the GA response was 2.7 mm versus 18.8 mm after far-red. If applied to stems, the GAs were much more effective at overcoming the red inhibition.

The results suggest that P_{FR} is able to diminish the effectiveness of GA applied to the leaves in a way or to an extent not found when GA is applied to the stems. These and related experiments* led us to continue our characterization of the endogenous GAs and of phytochrome mediation of the quantities and distribution of GAs among organs of the cowpea seedling.

The previous experiment was designed to make the plants dependent upon exogenous GA. We returned to plants with no paclobutrazol treatment, differing only in red or far-red treatment at the end of the 7th day prior to the dark period. Plants were harvested under a safelight, separated into leaves, top 2 cm of the stem, and stem bases to the cotyledons. At the

---

*References 6, 5, 11, and 13.

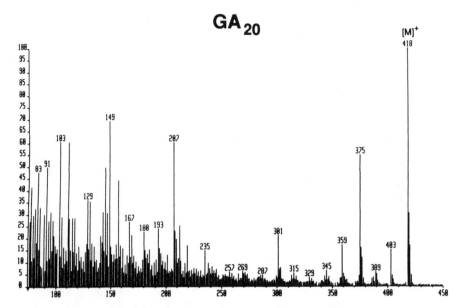

Fig. 2. Mass spectrum of the MeTMSi derivative of GA_{20} from red-treated tissues

time of homogenization, deuterated standards of GA_{19}, GA_{20}, and GA_1 were added in amounts to closely match the expected concentrations determined by a previous experiment. Tritiated standards of GA_1, GA_3, GA_5, GA_8, GA_9, and GA_{20} were added in trace amounts to facilitate tracing through purification. Fractions from the second HPLC separation were subjected to SIM of appropriate masses for identification and quanitation of isotope ratios. Confirmation of identification was obtained by full-scan mass spectra of consolidated leaf and stem samples from one treatment. A representative mass spectrum from the GA_{20} peak is shown in Fig. 2. The sample scan includes a portion of the deuterated internal standard. Selected ion monitoring determination for GA_{20} (Fig. 3) shows a ratio of endogenous and deuterated GA_{20} close to one. This allows quantitation when areas under the molecular ion for endogenous and +2 peaks are measured and compared to standard curves using known mixtures of labeled and unlabeled GA_{20}. Other intermediates of the early-13-hydroxylation pathway, GA_{53} and GA_{44}, as well as GA_8 were also identified (data to be presented elsewhere).

Table 2 lists the concentrations as well as the total amounts of GA_1, GA_{19}, and GA_{20} determined for the separated leaves and stem regions of 200 plants. It is useful to consider both concentration and total amounts resident in the different locations for insights as to available pools and possible metabolic and translocation flux.

The two light treatments resulted in significant differences in concentrations of key GAs of the early-13-hydroxylation pathway. These changes

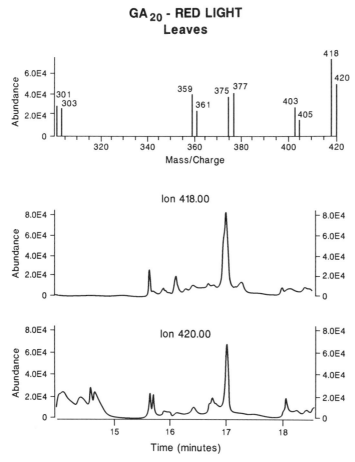

Fig. 3. Selected ion monitoring of a purified sample containing $[^2H_2]GA_{20}$ and endogenous GA_{20}. Specific ion chromatograms are shown for MeTMSi derivative of endogenous and deuterated (+2) GA_{20} in extracts of plants that had been irradiated with red light

are consistent with the role of GA_1 concentrations in regulating growth. While GA_{19} concentration varied from 2.9 to 5.5 ng/g fr wt among the different treatments and locations and GA_{20} varied from 1.2 to 5.5 ng/g fr wt, the responses of GA_1 and GA_{20} treated plants to far-red treatment were quite different. Gibberellin A_{19} concentration either did not change at all or slightly increased. Gibberellin A_{20} on the other hand, decreased both in the leaf (by 55%) and the growing region (by 70%) while remaining essentially constant in the lower stem. Gibberellin A_1 fell in the leaves and rose in the growing region; even more accumulated in the lower stem. The GA_1 concentration in the growing region compared with that of leaves was 9-fold higher under red and 23-fold higher under far-red.

Table 2. The effect of irradiation with red and far-red light on quantity and concentration of GA_s in cowpeas leaves and stems

GA	Total (ng)		Concn (ng/g)	
	Far-red	Red	Far-red	Red
Leaves, 180 g				
GA_{19}	580	576	3.22	3.20
GA_{20}	216	477	1.20	2.65
GA_1	77	126	0.43	0.70
GA_8[a]			3×10^5	3×10^5
Stems (0–2 cm), 30 g				
GA_{19}	166	136	5.53	4.52
GA_{20}	50	164	1.65	5.48
GA_1	303	201	10.10	6.70
GA_8[a]			111×10^5	6×10^5
Bases, 155 g				
GA_{19}	467	443	3.01	2.86
GA_{20}	315	330	2.03	2.13
GA_1	1054	186	6.80	1.20
GA_8[a]			55×10^5	16×10^5

[a] Endogenous GAs; concentration based on area of molecular ion

The light-mediated concentration changes suggest that the conversion of GA_{20} to GA_1 may be enhanced by far-red removal of P_{FR} in both stems and leaves, that is, that P_{FR} decreases conversion of GA_{20} to GA_1. Consideration of the total amounts involved shows a loss of 424 ng of GA_{20} in the leaves and growing region plus GA_1 in the leaves while the stem tissues gain 970 ng of GA_1. The pool size of earlier intermediates and GA_8 and subsequent metabolites may change, but the data suggest the possibility of a flux from the leaves to the stems with phytochrome influencing the rate of interconversion and, either directly or indirectly, the transport process.

Although the concentrations of the two precursors to GA_1 are slightly lower in the leaves than in the growing region of the stem, the 6-fold greater tissue mass of the leaves contains a significantly larger pool of the two precursors. One might expect that control of one or two reactions could manipulate the level of GA_1 in the growing region to accomplish short-term growth modulation.

4 Conclusions

End-of-day far-red treatments to GA-deficient (paclobutrazol-treated) decapitated cowpea seedlings strongly promoted growth only when exogenous GAs were provided. Applications to leaf or stem were equally effective after far-red light. However, after red light, leaf applications were much less effective than applications to the stem.

In normal seedlings, isotope dilution and GC–MS procedures showed that the GA_1 concentration was higher by a factor of 10 or more in the growing region of the stem than in the leaf. Red/far-red treatments caused little change in GA_{19} concentration. Compared to the red-treated plants, the far-red-treated plants had lower GA_{20} concentrations in the leaves and growing region while GA_1 rose in the stems and decreased in the leaves. Low P_{FR} concentrations (far-red treated) appear to facilitate the GA_{20} to GA_1 conversion process and, possibly, the movement of GA_1 and GA_{20} from the leaf to stem by unknown mechanisms. These results indicate that attempts to correlate whole-plant GA contents with growth control can miss important aspects of GA metabolism and movement.

Other GAs identified by full-scan mass spectrometry were GA_{53}, GA_{44}, and GA_8.

Acknowledgments. We thank Drs. M. Brenner, A. Crozier, P. Davies, J. MacMillan, R Pharis, B.O. Phinney, N. Takahashi, and J.A.D. Zeevaart for generously providing samples of gibberellins. R.H. Thompson gave valuable technical assistance and Earl Booth provided seeds of cowpea. Dr. D. Jones is acknowledged for technical assistance with mass spectrometry. The National Science Foundation is acknowledged for financial assistance.

References

1. Phinney BO. Gibberellin A_1, dwarfism and the control of shoot elongation in higher plants. In: Crozier A, Hillman JR, eds. The biosynthesis and metabolism of plant hormones. Cambridge: Cambridge University Press, 1984: pp. 17–41.
2. MacMillan J. Gibberellin deficient mutants of maize and pea and the molecular basis of gibberellin action. In: Hoad GV, Lenton JR, Jackson MB, et al., eds. Hormone action in plant development. London: Butterworths, 1987: pp. 73–87.
3. Ingram TJ, Reid JB, MacMillan J. The quantitative relationship between gibberellin A_1 and internode growth in *Pisum sativum* L. Planta. 1986; 168:414.
4. Downs, RJ. Photoreversibility of leaf and hypocotyl elongation of dark grown red kidney bean seedlings. Plant Physiol. 1955; 30:468–473.
5. García-Martínez JL, Keith B, Bonner B, et al. Phytochrome regulation of the response to endogenous gibberellins by epicotyls of *Vigna sinensis*. Plant Physiol. 1986; 85:212–216.
6. Campell BR, Bonner BA. Evidence for phytochrome regulation of gibberellin A_{20} 3β-hydroxlation in shoots of dwarf (*lele*) *Pisum sativum* L. Plant Physiol. 1986; 82:909–915.
7. Gilmour SJ, Zeevaart JAD, Schwenen L, et al. Gibberellin metabolism in cell-free extracts from spinach leaves in relation to photoperiod. Plant Physiol. 1986; 82:190–195.

8. Sponsel VM. Gibberellins in dark- and red light-grown shoot and tall cultivars of *Pisum sativum*. Planta. 1986; 168:119.
9. García-Martínez JL, Rappaport L. Contribution of the leaves to gibberellin-induced epicotyl elongation in cowpea. J Plant Growth Regul. 1982; 1:129–137.
10. Jones RL, Phillips IDJ. Organs of gibberellin synthesis in light-grown sunflower plants. Plant Physiol. 1966; 41:1381–1386.
11. Rappaport L, Bonner BA, García-Martínez JL. Sites of phytochrome and GA interaction in the control of first internode elongation in cowpea, *Vigna sinensis*, L. In: Abstracts for the 13th International Conference on Plant Growth Substances, Calgary, 1988; Abstract No. 363.
12. Hedden P. Gibberellins. In: Rivier L, Crozier A, eds. Principles and practice of plant hormone analysis, Vol.1. London: Academic Press, 1987: pp. 61–64.
13. García-Martínez JL, Rappaport L. Physiology of gibberellin-induced elongation of epicotyl explants from *Vigna sinensis*. J Plant Growth Regul. 1984; 2:197–208.

CHAPTER 28

Role of Gibberellins in Phytochrome-Mediated Lettuce Seed Germination

Y. Inoue

1 Introduction

It is well established that the germination of "Grand Rapids" lettuce seed is induced by red light (R) and inhibited by subsequent far-red light (FR).[1] This R/FR photoreversibility was the clue that led to the discovery of phytochrome, the ubiquitous photoreceptor chromoprotein of photomorphogenesis in green plants. While there have been many publications on this photoreaction,[2] the mechanism of phytochrome action in the control of lettuce seed germination is still unknown.[3]

We have recently reported that the removal of the fruit wall from the seed induced the dark germination of lettuce seed[4]; also, the lack of germination of the intact seed (fruit) was due to the presence of inhibitors in the fruit wall.[4] Two substances were isolated and identified as inhibitors.[5] One was abscisic acid (ABA), and the other was tetrahydrophthalimide (THPI). The content of ABA in the fruit wall of a lettuce seed was about 0.33 µg/g. Application of the ($\pm$) cis–trans isomer of ABA at a concentration of 3×10^{-6} M to the decoated lettuce seed in the dark supprerssed seed germination. It was concluded that ABA in the fruit wall is physiologically active in controlling lettuce seed germination. Tetrahydrophthalimide inhibited germination at a concentration of ca. 10^{-3} M. Concentrations of THPI in the fruit wall were found to be less than 10^{-6} M and varied with the seed source. The THPI present in the fruit wall is probably artificial in origin, that is, not native to the plant. Thus, THPI has no physiological role in the control of lettuce seed germination.

On the other hand, gibberellin (GA) is known to mimic the effect of red light on the germination of lettuce seed.[6] The present study reports on the role of GA in the phytochrome-mediated photocontrol of lettuce seed (*Lactuca sativa* L. cv. Grand Rapids) germination.

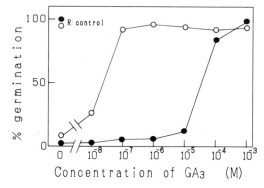

Fig. 1. Effect of GA_3 on the germination of intact (●) and punctured (○) lettuce seed in the dark. Gibberellin A_3 was given from the beginning of imbibition. R, red light (3.3 W/m², 5 min) control given 3 h after sowing. Germination was determined 2 days after sowing (Inoue and Yamane[9])

2 Results

2.1 Substitution of GA for Red Light Effects

The concentration of GA_3 required for maximal germination is greater than 10^{-4} M.[7,8] The dose–response curve of GA_3 for the intact seed used in the present study showed that the induction concentration of GA_3 was 10^{-4} M; 10^{-3} M GA_3 was required for full substitution of a brief red light irradiation[9] (Fig. 1). In contrast, a concentration of 10^{-7} M GA_3 induced maximal germination when seeds were punctured (one hole was punched in the middle portion using a fine tungsten needle) (Fig. 1)[9]. This difference in minimum GA_3 concentration for the induction of a saturation level of germination is probably due to the low permeability of GA in the structures that surround the embryo.

A brief red light irradiation also overcomes the inhibitory effects of ABA and THPI on the germination of decoated seed.[5] Germination of decoated seed in the dark was inhibited by 10^{-6} M ABA and 3×10^{-4} M THPI. The effects of these inhibitors were overcome by the application of GA_3 at a concentration of 10^{-5} M.[9] This result showed that GA could mimic the effect of red light in nullifying the effects of germination inhibitors contained in the fruit wall of lettuce seed.

2.2 Effects of GA Synthesis Inhibitors on the Germination of Decoated Lettuce Seed

If endogenous GA is essential for the photoinduction of lettuce seed germination, application of GA synthesis inhibitors should suppress light-induced germination. Application of 10^{-3} M AMO-1618 [2'-isopropyl-4'-(trimethylammonium chloride)-5'-methylphenyl piperidine-1-carboxylate] to decoated seed inhibited germination in the dark.[9] Red light did not result in reversal of this inhibition of germination.[9] A similar effect was observed[9] using the plant growth retardant S-3307 [(E)-1-(4-chlorophenyl)-4,4-dimethyl-2-(1,2,4-triazol-1-yl)-1-penten-3-ol], a compound which inhibits the oxidation of kaurene to kaurenoic acid.[10] The (S)-(E) form of the compound [(+)-S-3307] was about 7- to 70-fold more effective than the (R)-(E) form [(−)-S-3307] in the inhibition of GA biosynthesis.[10] In the case of lettuce seed germination, (+)-S-3307 is about 10-fold more effective than (−)-S-3307. In this case also, red light irradiation did not reverse the inhibition of germination.[9] Both GA synthesis inhibitors [AMO-1618, (+)-S-3307] effectively suppressed the germination of decoated lettuce seed irrespective of red light irradiation. By contrast, inductive red light irradiation overcame the inhibition of ABA and THPI in the same experimental system.[5]

The germination of decoated seed, inhibited by AMO-1618 (10^{-3} M) or (+)-S-3307 (5×10^{-5} M), was recovered by the addition of GA$_3$ at a concentration of 10^{-5} M.[9] Red light irradiation did not reverse this germination suppression. These results suggested that the effects of AMO-1618 and (+)-S-3307 were due to the inhibition of GA biosynthesis, which implies that endogenous GA is essential for the photoinduction of lettuce seed germination.

2.3 Comparison of the Time Course of Seed Germination Induced by Red Light and GA$_3$

The results described above suggest that red light increases GA levels in the seed, which in turn cancels the inhibitory effects of ABA in the fruit wall. Evidence against this scheme is that the germination of lettuce seed induced by GA$_3$ apparently occurs more slowly than does germination induced by red light.[11] In the latter work, intact seed were used, and the permeability of GA$_3$ was not considered. It is now known that exogenous GA$_3$ does not easily diffuse to the target site in an intact seed, as seen in Fig. 1. We have now determined the time course of seed germination using punctured seeds as follows. Six hours after sowing, punctured seeds were given red light (3.3 W/m^2, 5 min) or GA$_3$ (10^{-3} M). Germination became obvious at 11.5 h after sowing and maximal at about 19 h after sowing in the seeds to which 10^{-3} M GA$_3$ was added[9] (Fig. 2). The time course of seed germination induced by red light shifted to about 1 h later than that induced by GA$_3$ (10^{-3} M)[9] (Fig. 2). These results clearly show that the period required for lettuce seed germination induced by a high concentration of exogenous GA$_3$ was about 1 h shorter than that for germination induced by red light.

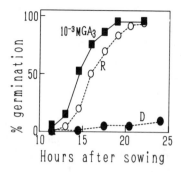

Fig. 2. Time course of punctured lettuce seed germination induced by red light (3.3 W/m², 5 min) (○) or 10^{-3} M GA₃ (■) given 6 h after sowing. Gibberellin A₃ was applied continuously until determination of the germination rate indicated in the abscissa (Inoue and Yamane[9]) (●); dark control

2.4 Determination of GA Levels in Lettuce Seed

The effect of red light irradiation on the concentration of GAs in lettuce seed was determined.[12] Since GA_{19}, GA_{20}, GA_1, and 3-*epi*-GA_1 were identified by combined gas chromatography–mass spectrometry (GC–MS) as endogenous components of lettuce seeds, we determined the levels of GA_1 and its precursors, GA_{20} and GA_{19}, using combined gas chromatography–selected ion monitoring (GC–SIM) with deuterium-labeled GA_1, GA_{20}, and GA_{19} as internal standards.[12]

The seed cultured for 3 h in the dark contained 46 ng GA_1/100 g dry seed equivalent (Table 1). This value did not change with an additional 3-h dark incubation. A brief red light irradiation increased the level of GA_1 to about 3 times that of a dark control (compare 6D with 3D/R/3D in Table 1). A brief far-red light irradiation, given after red light, completely suppressed

Table 1. Content of GAs in lettuce (cv. Grand Rapids) seed cultured under different light conditions determined by GC–SIM. Deuterium-labeled GAs were added as an internal standard at the end of the culture (Tsuji et al.[12])

| Light condition[a] | GA content (ng/100 g dry seed equivalent) | | | Germination[b] |
	GA_{19}	GA_{20}	GA_1	
3D	590	4300	46	–
6D	460	4200	41	–
3D/R/3D	570	5500	120	+
3D/R/FR/3D	640	3900	42	–

[a] 3D, 3 h in the dark; 6D, 6 h in the dark; R, red light (3.3 W/m², 5 min); FR, far-red light (5 W/m², 10 min)
[b] –, Germination was suppressed; +, germination was induced

the effect of red light, and the level of GA_1 remained at the dark control level (Table 1). The content of GA_{20} in the dark-incubated samples was about 100 times higher than that of GA_1. The content of GA_{20} was also increased by red light irradiation to a level 1.3 times that of the dark control (Table 1). Red light had no effect on the levels of GA_{19} (Table 1).

This result clearly showed that the levels of GA_1 increased within 3 h of a brief irradiation of red light. Only this red light irradiation could induce seed germination; other light treatments shown in Table 1 did not induce it.

3 Discussion

The present results showed that (1) exogenous GAs substitute for inductive red light in the germination of lettuce seed; (2) the synthesis of GAs is essential for the photoinduction of germination; and (3) inductive red light induces an increase in the levels of endogenous GA_1. These results suggest that the light signal perceived by phytochrome controls the level of endogenous GA_1. This endogenous GA_1 counteracts the inhibitory effect of ABA present in the fruit wall. As a result, seed germination occurs (Fig. 3).

While reports in the literature have suggested that the inhibitory effect of ABA on lettuce seed germination was not overcome by GA,[13,14] they used seed treated with a high concentration of GA_3 to induce dark germination. In addition, the concentration of exogenous ABA used was bout 10 times higher than the minimum concentration necessary for the complete inhibition of dark germination in decoated seed.

The other evidence against the scheme shown in Fig. 3 was the more

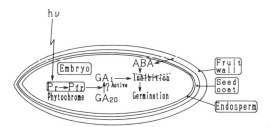

Fig. 3. Schematic representation of photoinduction mechanism of lettuce seed germination. Germination of lettuce seed is suppressed in the dark by ABA in the fruit wall. Inductive red light is absorbed by phytochrome and induces phototransformation from red-light-absorbing form (Pr) to far-red-absorbing form (Pfr) induces the increment of GA_1 level within 3 h after light irradiation, probably by activation of the step from GA_{20} to GA_1. Gibberellin A_1 cancels inhibitory effect of ABA and induces seed germination

rapid decrease in sensitivity to GA than that to red light during dark incubation.[15] We have not observed this phenomenon in experiments with punctured seed (unpublished data). In fact, punctured seed incubated for 4 days under intermittent far-red light irradiation (15-min FR/6-h dark interval) germinated not only under red light but also in the presence of GA. Thus, the effects reported in the literature were probably due to GA permeability changes during dark incubation.

Previous reports on the presence of GA-like substances[16] and their red-light-induced increases were based on bioassays.[17,18] In the report of Köhler,[17] total GA-like activity was reported to increase within 2 h of red light treatment to levels about 5 times those of seed grown in the dark. Bianco and Bulard[18] reported the "identification" of GA_4, GA_7, and GA_9, based on thin-layer chromatography behavior. These authors reported that the amount of free "GA_9" increased about 100 times within 30 min of light treatment. We did not detect any GA_9 in lettuce seed based on GC–MS studies, and the data in Table 1 show no such extreme change in levels of GAs. These previous reports should be reexamined using current isolation and identification methods. The red-light-induced increases in GA_1 levels (Table 1) are probably too small to be of significance. The data in Table 1 were obtained using whole lettuce seed. Unfortunately, we could not determine the primary target organ of the light effect. If we can determine the level of GA_1 in the primary target organ, difference of it between light-induced and dark incubated seeds must be significant.

Lettuce seed cultured in the dark contained large amounts of GA_{20}. The inductive effect of exogenous GA_{20} on punctured seed in the dark was about 1/1000 of that of GA_1.[12] These results suggested that the biosynthesis of GA_1 in the dark is blocked at the step $GA_{20} \rightarrow GA_1$, and red light irradiation releases the block. We are now studying the metabolism of radiolabeled GA_1, GA_{20}, and GA_{19} to clarify the phytochrome-controlled step(s) of GA_1 synthesis.

Acknowledgments. This work was conducted in collaboration with Dr. H. Yamane, Miss H. Tsuji, Mr. M. Nakayama, Dr. I. Yamaguchi, and Professor N. Takahashi of Department of Agricultural Chemistry, The University of Tokyo. I would like to thank them for their thorough and painstaking efforts and helpful advice during this study.

References

1. Borthwick HA, Hendricks SB, Parker MW, et al. A reversible photoreaction controlling seed germination. Proc Natl Acad Sci USA. 1952; 38:662–666.
2. Bewley JD, Black M. Physiology and biochemistry of seeds in relation to germination, Vol. 2. Berlin: Springer-Verlag, 1982.
3. Cone JW, Kendrick RE. Photocontrol of seed germination. In: Kendrick RE, Kronenberg GHM, eds. Photomorphogenesis in plants. Dordrecht: Martinus Nijhoff Publishers, 1986: pp. 443–465.

4. Huang YJ, Sakaguchi S, Inoue Y. Interaction of germination inhibitor contained in fruit wall and phytochrome in lettuce seed germination. Plant Cell Physiol. submitted.
5. Inoue Y, Yamane H, Huang YJ, et al. Characterization of phytochrome-dependent germination inhibitors extracted from fruit wall of lettuce seed. Plant Cell Physiol. submitted.
6. De Greef JA, Fredericq H. Photomorphogenesis and hormones. In: Shropshire W, Jr Mohr H, eds. Encyclopedia of plant physiology, new series, Vol. 16A. Berlin: Springer-Verlag, 1983; pp. 401–427.
7. Kahn A, Goss JA. Effect of gibberellin on germination of lettuce seed. Science. 1957; 125:645–646.
8. Ikuma H, Thimann KV. Action of gibberellic acid on lettuce seed germination. Plant Physiol. 1960; 35:557–566.
9. Inoue Y, Yamane H. Effects of gibberellin synthesis inhibitors on the germination of photoblastic lettuce seed. J. Plant Growth Regul. submitted.
10. Izumi K, Kamiya Y, Sakurai A, et al. Studies of sites of action of a new plant growth retardant (E)-1-(4-chlorophenyl)-4,4-dimethyl-2-(1,2,4-triazol-1-yl)-1-penten-3-ol (S-3307) and comparative effects of its stereoisomers in a cell-free system from *Cucurbita maxima*. Plant Cell Physiol. 1985; 26:821–827.
11. Lewak S, Khan AA. Mode of action of gibberellic acid and light on lettuce seed germination. Plant Physiol. 1977; 60:575–577.
12. Tsuji H, Yamane H, Inoue Y, et al. Red light induced increase of gibberellin levels in photoblastic lettuce seed. J. Plant Growth Regul. submitted.
13. Khan AA. Inhibition of gibberellic acid-induced germination by abscisic acid and reversal by cytokinins. Plant Physiol. 1968; 43:1463–1465.
14. Bewley JD, Fountain DW. A distinction between the actions of abscisic acid, gibberellic acid and cytokinins in light-sensitive lettuce seed. Planta. 1972; 102:368–371.
15. Vidaver W, Hsiao AI. Actions of gibberellic acid and phytochrome on the germination of Grand Rapids lettuce seeds. Plant Physiol. 1974; 53:266–268.
16. Blumenthal-Goldschmidt S, Lang A. Presence of gibberellin-like substances in lettuce seed. Nature. 1960; 183:815–816.
17. Köhler D. Veränderungen des Gibberellingehaltes von Salatsamen nach Belichtung. Planta. 1966; 70:42–45.
18. Bianco J, Bulard C. Influence of light treatment on gibberellin (GA_4, GA_7 and GA_9) content of *Lactuca sativa* L. cv. Grand Rapids achenes. Z Pflanzenphysiol. 1981; 101S:189–194.

CHAPTER 29

Inhibitors of Gibberellin Biosynthesis: Applications in Agriculture and Horticulture

W. Rademacher

1 Introduction

Plant growth regulators (PGRs) represent only a comparatively small portion (approximately 4%) of the pesticide market. In 1987, the sales of such compounds, including desiccants and defoliants, accounted for some U.S. $720 million.[1,2] Most of the plant growth regulators are compounds that reduce longitudinal shoot growth—plant growth retardants. In some cases, for instance, in barley, growth retardation can be induced by ethephon, an ethylene-releasing compound. Daminozide, which is active in apple, peanuts, and some ornamentals, is supposed to inhibit the translocation of gibberellins (GAs)[3,4] and lead to a more rapid degradation of these hormones.[5] However, the majority of compounds used to reduce shoot growth act primarily by blocking certain steps in the biosynthetic pathway leading to vegetative growth-active GAs. These include well-established compounds, such as chlormequat chloride (CCC) and mepiquat chloride, as well as several new inhibitors of GA biosynthesis that have been found during recent years. It is the intention of this paper to give an up-to-date survey of the most relevant compounds and also identify further possibilities for their practical application.

2 Inhibitors of GA Biosynthesis

At present, three different groups of inhibitors of GA biosynthesis in higher plants are known:
- "Onium" compounds
- Compounds with a nitrogen-containing heterocycle
- Cyclohexanetriones

Each of these groups inhibits GA biosynthesis at different stages (Fig. 1). More details of their biochemical modes of action are given elsewhere[6-8] and in other chapters of this volume.

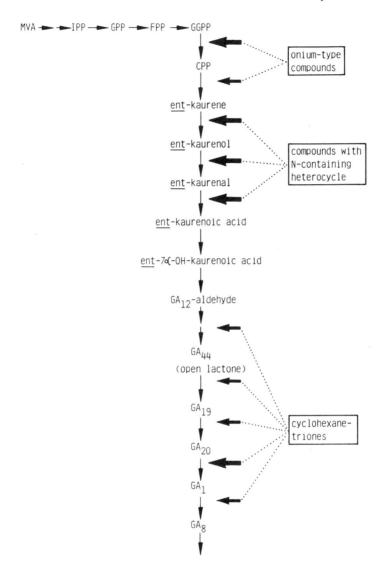

Fig. 1. Principal steps involved in the biosynthetic formation of GA_1 and points of inhibition by plant growth retardants (lower degrees of activity are indicated by smaller arrows) (MVA, mevalonic acid; IPP, isopentenylpyrophosphate; GPP, geranyl-PP; FPP, farnesyl-PP; GGPP, geranylgeranyl-PP; CPP, copalyl-PP)

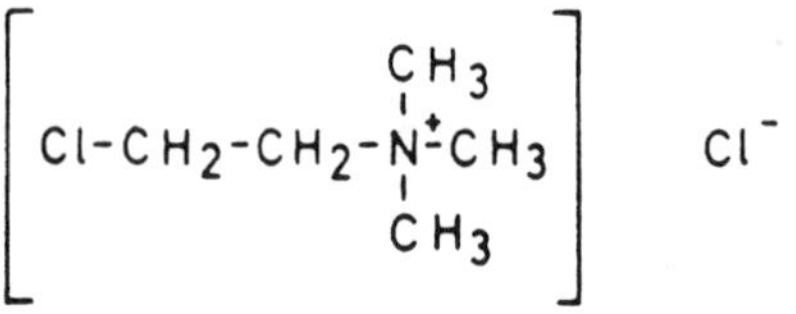

$$\left[Cl-CH_2-CH_2-\overset{\overset{CH_3}{|}}{\underset{\underset{CH_3}{|}}{N^+}}-CH_3 \right] \quad Cl^-$$

(1)

$$\left[\underset{H_3C \quad CH_3}{\overset{\bigcirc}{N^+}} \right] \quad Cl^-$$

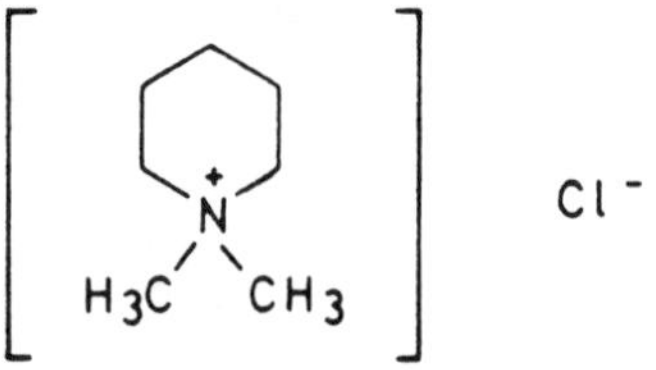

(2)

Fig. 2. "Onium"-type plant growth retardants: (1) chlormequat chloride (CCC) and (2) mepiquat chloride

2.1 "Onium" Compounds

Several compounds interfere with the biosynthetic steps directly before *ent*-kaurene. These substances can generally be classified as the group of "onium" compounds, possessing a positively charged ammonium, phosphonium, or sulfonium moiety. Chlormequat chloride and mepiquat chloride (Fig. 2) have found widespread application. AMO-1618, chlorphonium chloride, certain trimethylammonium iodides, BTS 44 584, and LAB 140 810 are further representatives of this group.[8]

Inhibition of the formation of sterols and other terpenoids has been reported as a side activity of, for instance, CCC, AMO-1618, and chlorphonium chloride.[9] "Onium" compounds retard shoot growth only in a restricted number of plant species. Therefore, CCC and mepiquat chloride have found major uses only in cereals, cotton, and some ornamental plant species.

2.2 Compounds with N-Containing Heterocycles

Several plant growth retardants inhibit the oxidative reactions leading from *ent*-kaurene to *ent*-kaurenoic acid. The structural feature common to all these compounds is a heterocyclic ring containing sp^2-hybridized nitrogen. It is assumed that the lone electron pair of this nitrogen binds to the protoheme iron of a cytochrome P-450, thereby blocking a monooxygenase that catalyzes *ent*-kaurene oxidation.[10]

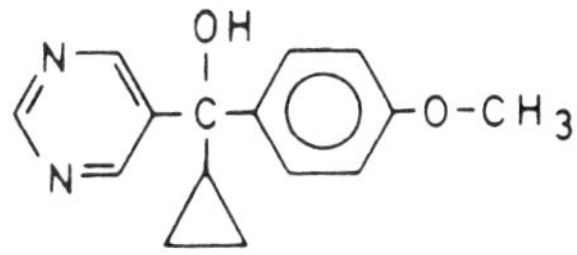

(1)

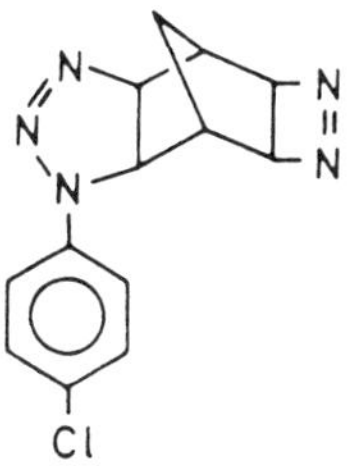

(2)

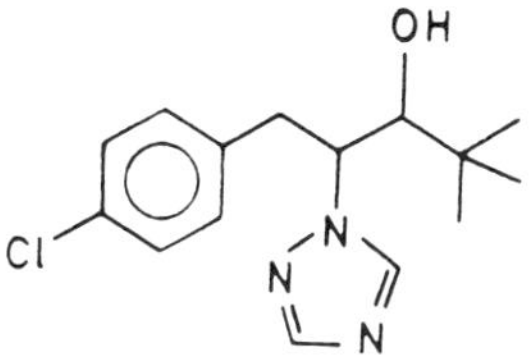

(3)

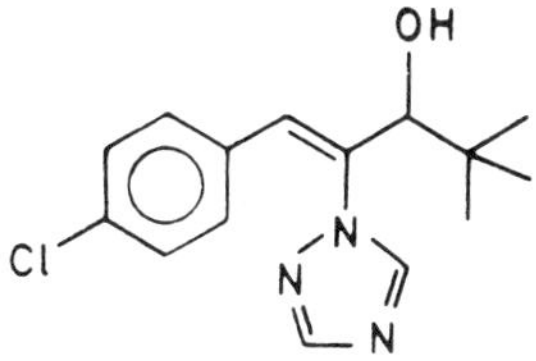

(4)

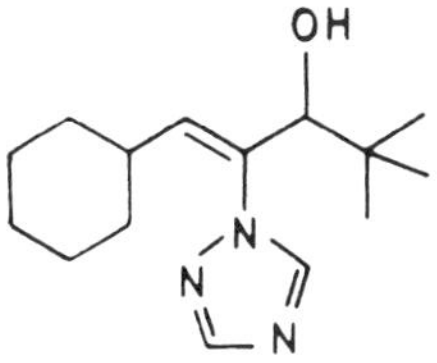

(5)

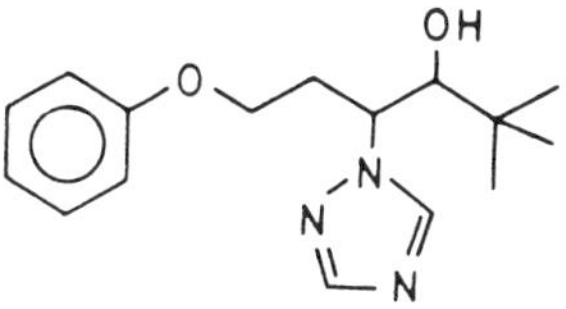

(6)

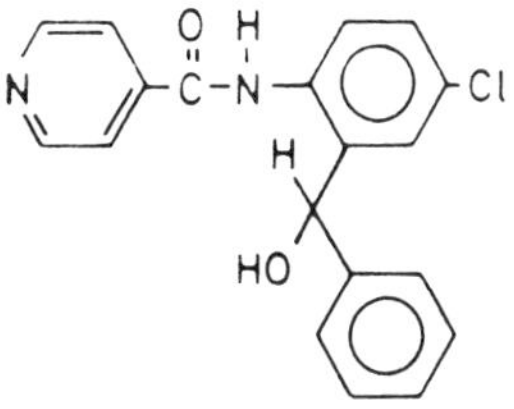

(7)

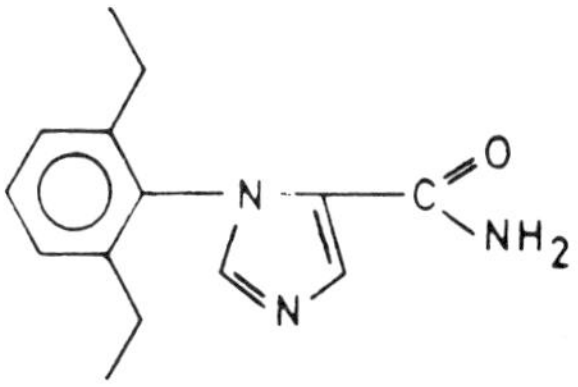

(8)

Fig. 3. Plant growth retardants of the pyrimidine, norbornanodiazetine, triazole, 4-pyridine, and imidazole types: (1) ancymidol, (2) tetcyclacis, (3) paclobutrazol, (4) uniconazole, (5) triapenthenol, (6) BAS 111..W, (7) inabenfide, (8) HOE 074784

Ancymidol, a pyrimidine derivative, has been known for several years and has found some commercial application.[11] Additionally, the activity of the related flurprimidol (EL 500) has been described.[12] The plant growth-retarding activity of certain norbornanodiazetines was first described in 1980.[13] Tetcyclacis (LAB 102 883, BAS 106..W) became the most important representative of this group.[14] Reduction of shoot growth by triazoles was first reported for the fungicides triadimefon and triadimenol.[15] After this, other triazoles were chemically optimized as plant growth retardants: compounds coded LAB 117 682, LAB 129 409, and LAB 130 827,[14,16] paclobutrazol (PP 333),[17] uniconazole (S-3307, XE-1019),[18,19] LAB 150 978,[20] triapenthenol (RSW 0411),[21] and BAS 110..W and BAS 111..W.[22] Inabenfide (CGR-811), a 4-substituted pyridine derivative,[23] also inhibits the oxidation of *ent*-kaurene.[10] Finally, imidazoles are known to inhibit the biosynthetic steps between *ent*-kaurene and *ent*-kaurenoic acid; examples are 1-*n*-decylimidazole and 1-geranylimidazole[24] and HOE 074784.[25] The structures of the most relevant compounds are given in Fig. 3.

Clear evidence is available that, besides inhibiting *ent*-kaurene oxidation, the above-mentioned compounds can also affect other cytochrome P-450-type oxygenases, although in most cases at a far lower degree of activity. In particular, the inhibition of sterol formation by the blocking of 14α-demethylation and the inhibition of the oxidative inactivation of ABA must be mentioned. It has also been reported that the oxidative metabolism of certain xenobiotics can be inhibited.

2.3 Cyclohexanetriones

Certain cyclohexanetriones represent a new class of plant growth retardants. The first two patent applications date back to 1983.[26,27] A typical representative of this group, which is covered by both patent applications, is LAB 198 999 (Fig. 4).

Independent work by Nakayama et al. (Chapter 30 in this volume), Hedden (Chapter 10), and Graebe et al. (Chapter 6) has shown that cyclohexanetriones block late steps in the biosynthetic pathway leading to active GAs. A primary target appears to be the 3β-hydroxylation step (e.g., from GA_{20} to GA_1), although other reactions are also affected. All reactions inhibited by cyclohexanetriones are catalyzed by dioxygenases that require 2-oxoglutarate as a cosubstrate. It is tempting to assume that structural similarities between these retardants and 2-oxoglutarate are responsible for the observed activity. Further evidence for such correlations can be derived from the fact that cyclohexanetriones also block the formation of anthocyanins (D.R. Carlson and W. Rademacher, unpublished), the biosynthesis of which also involves 2-oxoglutarate-dependent dioxygenases.

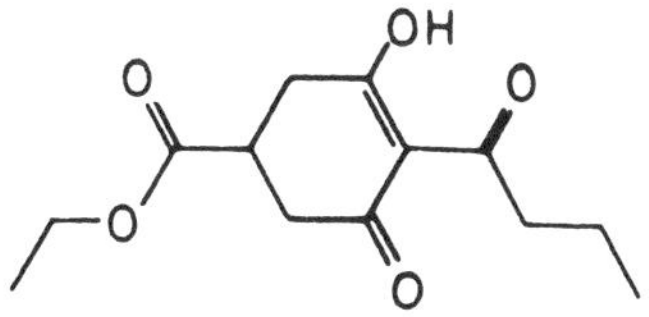

Fig. 4. Cyclohexanetrione-type plant growth retardant: LAB 198 999

3 Practical Applications

The plant growth retardants described above have already found, or are candidates for, practical application in agriculture or horticulture. In most cases, the growth-retarding effect is the only point of interest. This would primarily exploit the potential of such compounds to reduce the endogenous levels of vegetative growth-active GAs. However, side effects must not be overlooked since they may add to the benefit of an application. In a given crop, the usefulness of a given compound will therefore also be determined by the combination of morphoregulation with such side effects. In certain cases, side effects may even be of primary relevance.

In the following section, standard applications of GA-biosynthesis inhibitors will be described as well as their potential for further uses. With the availability of new compounds, some of these potentials will very likely be realized in the near future. In contrast, other potential indications appear to be more of academic interest at present.

3.1 Control of Lodging in Cereals

The control of lodging in cereals is by far the largest use for plant growth regulators at present.[2] In intensive wheat production, the use of chlormequat chloride has been a standard practice for many years. Other small grains such as barley, rye, or oats do not respond sufficiently to this compound. Therefore, combinations of chlormequat chloride or mepiquat chloride with ethephon or ethephon alone are used in these cases.[28]

Compounds such as the triazoles paclobutrazol,[29] uniconazole,[30] triapenthenol,[21] and BAS 110..W and BAS 111..W[22] have also shown promising results in different cereal species. The same is true for the new cyclohexanetriones (W. Rademacher and J. Jung, unpublished). However, one must not overlook the fact that new compounds have to offer advanced solutions at competitive prices in order to replace existing products. This may be especially difficult in this segment of the market, in which, for instance, the price of chlormequat chloride is relatively low.

Table 1. Concentrations of different growth retardants required for a 50% reduction of shoot growth in rice (cv. Girona) after root application (KI_{50})

Compound	Type	KI_{50} (M)
Ethephon		$\gg 10^{-2}$
Chlormequat chloride	Onium	1.5×10^{-2}
Mepiquat chloride	Onium	6.5×10^{-3}
Ancymidol	Pyrimidine	2.3×10^{-7}
Paclobutrazol	Triazole	1.8×10^{-7}
LAB 150 978	Triazole	1.2×10^{-7}
BAS 111..W	Triazole	1.0×10^{-5}
Tetcyclacis	Norbornanodiazetine	1.2×10^{-7}
LAB 198 999	Cyclohexanetrione	1.5×10^{-6}

3.2 Control of Lodging in Rice

Lodging is also a problem in intensive cultivation of rice. Chlormequat chloride, mepiquat chloride, and ethephon are not suitable in this case. However, good results have been obtained with inabenfide,[31] which is now being used commercially. Other inhibitors of *ent*-kaurene oxidation are also candidates for use in rice. Particularly paclobutrazol,[17] uniconazole,[30] triapenthenol,[21] and HOE 07478425 should be mentioned here. The cyclohexanetriones, which are very active in rice, are also worth attention.

Table 1 gives a comparison of the activities of a selection of compounds in reducing shoot growth of rice seedlings (experimental details are given elsewhere[16]). These in vitro data do not necessarily reflect the usefulness of a compound under practical conditions. Nonetheless, it is evident that chlormequat chloride or ethephon are surpassed in activity by more modern compounds.

3.3 Reduction of Vegetative Growth in Cotton

Mepiquat chloride has been used in cotton since about 1980. It controls excessive vegetative growth and allows for high-density plantations with better insect control in a crop canopy more accessible to the spray of pesticides. In addition to this, the incidence of boll rot is reduced and boll maturity is earlier and more uniform. All this leads to economic advantages which are most manifest when mechanical picking is used.[32,33]

3.4 Improvement of Canopy Structure and Yield in Oilseed Rape

Paclobutrazol, BAS 111..W and triapenthenol reduce the height of oilseed rape plants and lead to a canopy structure that allows a better penetration of light to the lower parts of plants. Yield increases of about 10% are commonly observed. Reduced lodging and fewer fungal infections (cf. Section 3.11) also occurred.[22,34,35]

3.5 Reduction of Vegetative Shoot Growth in Orchard Trees

Compounds such as paclobutrazol, uniconazole, triapenthenol, LAB 150 978, and flurprimidol offer possibilities for application in different pome and stone fruits, in citrus, and in nut trees.[36-41] The general target for growth retardants in orchard trees is to alter assimilate partitioning in favor of generative growth (flowers, fruits) at the expense of vegetative shoot growth. Less shoot growth would lower the costs required for pruning. Additionally, it would also be of relevance in fruit trees, such as sweet cherries or plums, for which no dwarfing rootstocks or compact scion cultivars are available. This would allow for efficient high-density plantations as, for instance, in apples. In many cases, reduction of vegetative growth will promote the process of flower initiation. This may lead to yield increases and may also allow crop scheduling and the rectification of biennial bearing.

A general difficulty especially encountered with the use of triazoles in orchard trees lies in the type of application. In most cases, foliar treatments are less efficient than soil treatments since these compounds are more readily transported in the xylem. However, soil applications make growth regulation relatively difficult since the availability of a compound to the tree will be strongly influenced by parameters such as the mobility and persistence of the compound in the soil. The persistence is relatively high for some compounds,[39,41,42] which may cause additional problems. Trunk painting offers an alternative type of application.[37]

3.6 Growth Reduction in Other Woody Plants

Trees that grow, for instance, under power or telephone lines have to be trimmed regularly. Growth retardants such as paclobutrazol or uniconazole may be used to save part of this expense. Application by stem injection has been proposed for administering the active ingredients.[42,43] This may help to avoid chemical contact with nontarget plants and reduce residues in the environment. Just as in the case of trees, some of the costs required for trimming of hedges and other shrubs may also be saved.

3.7 Regulation of Turf Growth

Considerable sums of money are spent on mowing turf grasses, for instance, in private lawns, golf courses, and public greens. Compounds such as paclobutrazol and flurprimidol have shown good reduction of shoot height in different turf species. However, these substances do not sufficiently control unwanted seedhead formation. Combinations with compounds that have more a herbicidal character, for example, mefluidide or amidochlor, may offer more advanced solutions.[44-46]

3.8 Height Regulation in Ornamentals

Chlormequat chloride, ancymidol, and, more recently, paclobutrazol and uniconazole are used professionally to control shoot height in poinsettias, chrysanthemums, and other ornamentals. Favorable results can also be achieved with virtually all other inhibitors of GA biosynthesis.[30,47,48] However, the use of cyclohexanetriones might be restricted in this application, because of their adverse effect on anthocyan formation.

3.9 Improving the Quality of Seedlings Used for Transplanting

Tetcyclacis, applied as a seedsoaking, considerably improves the quality of rice seedlings used for mechancial transplanting: the shortened seedlings can be kept at a size optimal for transplanting for a longer period of time; such seedlings are more resistant to transplantation stress; the performance of such seedlings under field conditions is superior to that of nontreated ones.[49] Similar effects have also been observed in seedlings of maize and other plant species used for mechanical transplantation (P.E. Schott and M. Luib, unpublished).

3.10 Improved Resistance to Climatic Stress

Low temperatures lead to chilling or frost damage in many crop plants.[50,51] In the USSR, CCC is used as a seed treatment for large areas of winter and spring wheat to prevent frost damage.[52] Improved resistance to low temperatures has also been found, for instance, in cucumber and zucchini squash treated with paclobutrazol,[53] in tomato treated with uniconazole,[54] in oats and rice under the influence of tetcyclacis,[55,56] and in winter oilseed rape treated with BAS 111 . . W.[35]

In many areas of the world, water is a limiting factor that severely restricts crop productivity. In addition to other measures, attempts are being made to use plant growth regulators to overcome drought stress.[57] Thus, several investigations have shown that plants performed better under conditions of drought after they had been treated with compounds such as CCC, triadimefon, flurprimidol, paclobutrazol, uniconazole, and tetcyclacis.[58-62]

The positive effects of plant growth retardants on stress resistance may be explained in several ways. First, a purely physical effect should be considered: plants with retarded shoot growth may have their growing points protected in the ground and expose less leaf surface to unfavorable conditions. Improved root growth may be capable of absorbing increased amounts of water. In addition, it must not be overlooked that increased levels of ABA can be induced by certain inhibitors of cytochrome P-450-dependent oxygenases, such as tetcyclacis[63] or BAS 111 . . W.[64] It is known that ABA improves the resistance of plants to low temperatures[65] and to drought.[66]

3.11 Improved Resistance to Fungal Infections

In oilseed rape, the rate of infection with different fungi was reduced by treatments with triapenthenol and BAS 111..W.[34,35,67] In melon seedlings, resistance to fusarium wilt was improved by paclobutrazol and ancymidol.[68] While some of the observed effects are thought to be due to direct fungicidal activity of the growth retardants,[69] evidence is also available that other phenomena, such as delayed senescence,[35] lead to an increased resistance of the plants.

3.12 Increase of Herbicidal Activity

Several pesticides and other xenobiotics are metabolized in plants by monooxygenases that are cytochrome P-450 dependent.[70] Blocking the metabolism of a herbicidal agent will accordingly increase its activity. Thus, it has been demonstrated recently that the oxidation of the herbicide chlortoluron is inhibited in cell suspensions of cotton and wheat by compounds such as tetcyclacis and paclobutrazol.[71,72] In intact seedlings of soybeans, the inactivation of the herbicide bentazon is also significantly delayed by tetcyclacis.[73] Further studies will, however, be required to find out whether such effects can be used under practical conditions.

4 Conclusions

Many inhibitors of GA biosynthesis are known to date. Several of these compounds have already found practical application in agriculture and horticulture. It can be forecast that, due to their different physiological properties, further indications will be covered by the newer compounds.

References

1. Geissbuehler H, Kerber E, Mueller U, et al. Prospects for chemical plant growth regulation: An industrial view point. In: Hawkins AF, Stead AD, Pinfield NJ, eds. Plant growth regulators for agricultural and amenity use. Thornton Heath: British Crop Protection Council, Monograph No. 36, 1987: pp. 11–17.
2. Baylis A. Economic aspects of plant growth regulators. In: Pharis RP, Rood SB, eds. Plant growth substances 1988. Heidelberg: Springer-Verlag, 1990 in press.
3. Menhenett R. Evidence that daminozide, but not two other growth retardants, modifies the fate of applied gibberellin A_9 in *Chrysanthemum morifolium* Ramat. J Exp Bot. 1980; 31:1631–1642.
4. Menhenett R. Interaction of the growth retardants daminozide and piproctanyl bromide and gibberellins A_1, A_3, A_{4+7}, A_5 and A_{13} in stem extension and inflorescence development in *Chrysanthemum morifolium* Ramat. Ann Bot. 1981; 47:359–370.

5. Takeno K, Legge RL, Pharis RP. Effect of the growth retardant B-9 (SADH) on endogenous GA level, and transport and conversion of exogenously applied [^{3}H]GA$_{20}$ in Alaska pea. Plant Physiol. 1981; 67(Suppl):581.
6. Graebe JE. Gibberellin biosynthesis and control. Ann Rev Plant Physiol. 1987; 38:419–465.
7. Hedden P. The action of plant growth retardants at the biochemical level. In: Pharis RP, Rood SB, eds. Plant growth substances 1988. Heidelberg: Springer-Verlag, 1990: in press.
8. Rademacher W. Gibberellins: Metabolic pathways and inhibitors of biosynthesis. In: Boeger P, Sandmann G, eds. Target sites for herbicide action. Boca Raton, Florida: CRC Press, 1989: pp. 127–145.
9. Douglas TJ, Paleg LG. Inhibition of sterol biosynthesis and stem elongation of tobacco seedlings induced by some hypocholesterolemic agents. J Exp Bot. 1981; 32:59–68.
10. Rademacher W, Fritsch H, Graebe JE, et al. Tetcyclacis and triazole-type plant growth retardants: Their influence on the biosynthesis of gibberellins and other metabolic processes. Pestic Sci. 1987; 21:241–252.
11. Thomas TH, ed. Plant growth regulator potential and practice. Croydon: British Crop Protection Council, 1982: p. 177.
12. Sterrett JP, Tworkoski TJ. Flurprimidol: Plant response, translocation and metabolism. J Amer Soc Hort Sci. 1987; 112:341–345.
13. Jung J, Koch H, Rieber N, et al. Zur wachstumsregulierenden Wirkung von Triazolin- und Aziridinderivaten des Norbornenodiazetins. J Agron Crop Sci. 1980; 149:128–136.
14. Rademacher W, Jung J, Graebe JE, et al. On the mode of action of tetcyclacis and triazole growth retardants. In: Menhenett R, Lawrence DK, eds. Biochemical aspects of synthetic and naturally occurring plant growth regulators. Wantage: British Plant Growth Regulator Group, Monograph No. 11, 1984: pp. 1–11.
15. Buchenauer H, Roehner E. Effect of triadimefon and triadimenol on growth of various plant species as well as on gibberellin content and sterol metabolism in shoots of barley seedlings. Pestic Biochem Physiol. 1981; 15:58–70.
16. Rademacher W, Jung J. Comparative potency of various synthetic plant growth retardants on the elongation of rice seedlings. J Agron Crop Sci. 1981; 150:363–371.
17. Lever BG, Shearing SJ, Batch JJ. PP 333—a new broad spectrum growth retardant. In: British Crop Protection Conference—Weeds 1982, Vol. 1. Croydon: British Crop Protection Council, 1982: pp. 3–10.
18. Izumi K, Yamaguchi I, Wada A, et al. Effects of a new plant growth retardant (E)-1-(4-chlorophenyl)-4,4-dimethyl-2-(1,2,4-triazol-1-yl)-1-penten-3-ol (S-3307) on the growth and gibberellin content of rice plants. Plant Cell Physiol. 1984; 25:611–617.
19. Izumi K, Kamiya Y, Sakurai A, et al. Studies of sites of action of a new plant growth retardant (E)-1-(4-chlorophenyl)-4,4-dimethyl-2-(1,2,4-triazol-1-yl)-1-penten-3-ol (S-3307) and comparative effects of its stereoisomers in a cell-free system from *Cucurbita maxima*. Plant Cell Physiol. 1985; 26:821–827.
20. Jung J, Rentzea C, Rademacher W. Plant growth regulation with triazoles of the dioxanyl type. J Plant Growth Regul. 1986; 4:181–188.

21. Luerssen K, Reiser W. Triapenthenol—a new plant growth regulator. Pestic Sci. 1987; 19:153–164.
22. Jung J, Luib M, Sauter H, et al. Growth regulation in crop plants with new types of triazole compounds. J Agron Crop Sci. 1987; 158:324–332.
23. Shirakawa N, Tomioka H, Takeuchi M, et al. Effect of a new plant growth retardant N-[4-chloro-2-(α-hydroxybenzyl)phenyl]isonicotinamide (CGR-811) on the growth of rice plants. In: Plant growth regulators in agriculture. Taipei: Food and Fertilizer Technology Center for the Asian and Pacific Region, 1986: pp. 1–17.
24. Wada K. New gibberellin biosynthesis inhibitors, 1-n-decyl- and 1-geranylimidazole: inhibitors of (−)-kaurene 19-oxidation. Agric Biol Chem. 1978; 42:2411–2413.
25. Buerstell HW, Hacker E, Schmierer R. HOE 074 784 and analogues: A new synthetic group of highly active plant growth retardants in cereals (esp. rice) and rape. In: Cooke AR, ed. Proceedings of the plant growth regulator society of America. Ithaca, New York: Plant Growth Regulator Society of America, 1988: p. 185.
26. Motojima K, Miyazawa T, Toyokawa Y, et al. New cyclohexane derivatives having plant-growth regulating activities, and uses of these derivatives. Japanese Patent Application JP 71264/83, 1983 (quoted in European Patent Application EP 0 123 001).
27. Brunner HG. Cyclohexandion-carbonsäurederivate mit herbizider und das Pflanzenwachstum regulierender Wirkung. Swiss Patent Application CH 2693183, 1983 (quoted in European Patent Application EP 0 126 713).
28. Jung J, Rademacher W. Plant growth regulating chemicals—cereal grains. In: Nickell LG, ed. Plant growth regulating chemicals, Vol. 1. Boca Raton, Florida: CRC Press, 1983: pp. 253–271.
29. Froggatt PJ, Thomas WD, Batch JJ. The value of lodging control in winter wheat as exemplified by the growth regulator PP 333. In: Hawkins AF, Jeffcoat B, eds. Opportunities for manipulation of cereal productivity. Wantage: British Plant Growth Regulator Group, Monograph No. 7, 1982: pp. 71–87.
30. Oshio H, Izumi K. S-3307, a new plant growth retardant—its biological activities, mechanism and mode of action. In: Plant growth regulators in agriculture. Taipei: Food and Fertilizer Technology Center for the Asian and Pacific Region, 1986: pp. 200–208.
31. Nakamura K. Newly developed PGR seritard (inabenfide) an anti-lodging agent for paddy rice. Jpn Pestic Inf. 1987; 51:23–26.
32. Kerby TA, Hake K, Keeley M. Cotton fruiting modification with mepiquat chloride. Agron J. 1986; 78:907–912.
33. Walter H, Gausman HW, Ritting FR, et al. Effect of mepiquat chloride on cotton plant leaf and canopy structure and dry weights of its components. In: Brown JM, ed. Proceedings of the Beltwide Cotton Production Research Conference. St. Louis: National Cotton Council, 1980: pp. 32–35.
34. Child RD, Butler DR, Sims IM, et al. Control of canopy structure in oilseed rape with growth retardants and consequences for yield. In: Hawkins AF, Stead AD, Pinfield NJ, eds. Plant growth regulators for agricultural and amenity use. Thornton Heath: British Crop Protection Council, Monograph No. 36, 1987: pp. 21–35.

35. Luib M, Koehle H, Hoeppner P, et al. Further results with BAS 111 04 W, a new growth regulator for use in oilseed rape. In: Hawkins AF, Stead AD, Pinfield NJ, eds. Plant growth regulators for agricultural and amenity use. Thornton Heath: British Crop Protection Council, Monograph No. 36, 1987: pp. 37–43.
36. Aron Y, Monselise SP, Goren R, et al. Chemical control of vegetative growth in citrus trees by paclobutrazol. HortScience. 1985; 20:96–98.
37. Curry EA, Williams MW, Reed AN. Triazole bioregulators to control growth of fruit trees: An update. Bull Plant Growth Regul Soc Am. 1987; 15:4–7.
38. Quinlan JD. Use of paclobutrazol in orchard management to improve efficiency of fruit production. In: Hawkins AF, Stead AD, Pinfield NJ, eds. Plant growth regulators for agricultural and amenity use. Thornton Heath: British Crop Protection Council, Monograph No. 36, 1987: pp. 149–153.
39. Davis, TD, Steffens GL, Sankhla N. Triazole plant growth regulators. In: Janick J, ed. Horticultural reviews, Vol. 10. Portland, Oregon: Timber Press, 1988: pp. 63–105.
40. Harty AR, van Staden J. The use of growth retardants in citriculture. Isr J Bot. 1988; 37:155–164.
41. Steffens GL. Gibberellin biosynthesis inhibitors: Comparing growth-retarding effectiveness on apple. J Plant Growth Regul. 1988; 7:27–36.
42. Sterrett JP. XE-1019: Plant response, translocation, and metabolism. J Plant Growth Regul. 1988; 7:19–26.
43. Sterrett JP. Paclobutrazol: A promising growth inhibitor for injection into woody plants. J Amer Soc Hort Sci. 1985; 110:4–8.
44. Elkins DM. Growth regulating chemicals for turf and other grasses. In: Nickell LG, ed. Plant growth regulating chemicals, Vol. II. Boca Raton, Florida: CRC Press, 1983: pp. 113–130.
45. Kaufmann JE. Biological responses of amenity grasses to growth regulators. In: Hawkins AF, Stead AD, Pinfield NJ, eds. Plant growth regulators for agricultural and amenity use. Thornton Heath: British Crop Protection Council, Monograph No. 36, 1987: pp. 99–118.
46. Kavanagh T. Ringing the changes on three growth retardant chemicals for amenity grass management. In: Hawkins AF, Stead AD, Pinfield NJ, eds. Plant growth regulators for agricultural and amenity use. Thornton Heath: British Crop Protection Council, Monograph No. 36, 1987: pp. 135–145.
47. Barrett JE. Chrysanthemum height control by ancymidol, paclobutrazol and EL 500 dependent on medium composition. HortScience. 1982; 17:896–897.
48. Wulster GJ, Gianfagna TJ, Clarke BB. Comparative effects of ancymidol, propiconazol, triadimefon, and Mobay RSW 0411 on lily height. HortScience. 1987; 22:601–602.
49. Schott PE, Knittel H, Klapproth H. Tetcyclacis: A new bioregulator for improving the development of young rice plants. In: Ory RL, Ritting FR, eds. Bioregulators—chemistry and uses. Washington, DC: American Chemical Society, 1984: pp. 45–63.
50. Wilson JM. The economic importance of chilling injury. Outlook Agric. 1984; 14:197–203.
51. Andrews CJ. Low-temperature stress in field and forage crop production—an overview. Can J Plant Sci. 1987; 67:1121–1133.

52. Zadoncev AI, Pikus GR, Grincenko AL. CCC in der Pflanzenproduktion. Berlin, GDR: VEB Deutscher Landwirtschaftsverlag, 1977.
53. Wang CY. Modification of chilling susceptibility in seedlings of cucumber and zucchini squash by the bioregulator paclobutrazol (PP333). Sci Hort. 1985; 26:293–298.
54. Senaratna T, Mackay CE, McKersie BD, et al. Uniconazole-induced chilling tolerance in tomato and its relationship to antioxidant content. J Plant Physiol. 1988; 133:56–61.
55. Anderson HM, Huband NDS. Improvement of winter hardiness and seedling growth of oats with seed dressings of tetcyclacis. In: Hawkins AF, Stead AD, Pinfield NJ, eds. Plant growth regulators for agricultural and amenity use. Thornton Heath: British Crop Protection Council, Monograph No. 36, 1987: pp. 45–50.
56. Chu C, Hwang SJ, Lee TM. Hormonal regulation of cold-resistance in rice seedlings. In: Proceedings of the 5th Seminar on Science and Technology—Phytohormones. Nara, Japan: Interchange Association, Japan, 1987: pp. 125–154.
57. Rademacher W, Maisch R, Liessegang J, et al. Water consumption and yield formation in crop plants under the influence of synthetic analogues of abscisic acid. In: Hawkins AF, Stead AD, Pinfield NJ, eds. Plant growth regulators for agricultural and amenity use. Thornton Heath: British Crop Protection Council, Monograph No. 36, 1987: pp. 53–66.
58. Plaut Z, Halevy AH. Regeneration after wilting, growth and yield of wheat plants, as affected by two growth retarding compounds. Physiol Plant. 1966; 19:1064–1071.
59. Fletcher RA, Nath V. Triadimefon reduces transpiration and increases yield in water stressed plants. Physiol Plant. 1984; 62:422–426.
60. Atkinson D. Effects of some plant growth regulators on water use and uptake of mineral nutrients by tree crops. Acta Hort. 1986; 179:395–404.
61. Vaigro-Wolff AL, Warmund MR. Suppression of growth and plant moisture stress of forsythia with flurprimidol and XE-1019. HortScience. 1987; 22:884–885.
62. Shanahan JF, Nielsen CD. Influence of growth retardants (anti-gibberellins) on corn vegetative growth, water use, and grain yield under different levels of water stress. Agron J. 1987; 79:103–109.
63. Zeevaart JAD. Biosynthesis and catabolism of abscisic acid. In: Pharis RP, Rood SB, eds. Plant growth substances 1988. Heidelberg: Springer-Verlag, 1990: in press.
64. Carlson DR, Jung J, Rademacher W. Direct effects of growth retardants on plant water consumption. In: Pharis RP, Rood SB, eds. Abstracts for the 13th International Conference on Plant Growth Substances, Calgary, 1988; Abstract No. 67.
65. Rikin A, Richmond AE. Amelioration of chilling injuries in cucumber seedlings by abscisic acid. Physiol Plant. 1976; 38:95–97.
66. Davies WJ, Mansfield TA. The role of abscisic acid in drought avoidance. In: Addicott FT, ed. Abscisic acid. New York: Praeger, 1983: pp. 237–268.
67. Stinchcombe GR, Hutcheon JA, Jordan VWL. Effects of growth regulators on light leaf spot and yield components in oilseed rape. In: 1986 British Crop

Protection Conference—Pests and Diseases, Vol 3. Croydon: British Crop Protection Council Publications, 1986: pp. 1009–1015.
68. Cohen R, Yarden O, Katan J, et al. Paclobutrazol and other plant growth-retarding chemicals increase resistance of melon seedlings to fusarium wilt. Plant Pathol. 1987; 36:558–564.
69. Fletcher RA, Hofstra G, Gao J. Comparative fungitoxic and plant growth regulating properties of triazole derivatives. Plant Cell Physiol. 1986; 27:367–371.
70. Lamoureux GL, Frear DS. Current problems, trends, and developments in pesticide metabolism in plants. In: Greenhalgh R, Roberts TR, eds. Pesticide science and biotechnology. Oxford: Blackwell, 1987: pp. 455–473.
71. Cole DJ, Owen WJ. Influence of monooxygenase inhibitors on the metabolism of the herbicides chlortoluron and metolachlor in cell suspension cultures. Plant Sci. 1987; 50:13–20.
72. Canivenc MC, Cagnac B, Cabanne F, et al. Induced changes of chlorotoluron metabolism in wheat cell suspension cultures. Plant Physiol Biochem. 1989; 27:193–201.
73. Fritsch H, Rademacher W, Retzlaff G. Inhibition of plant growth, gibberellin biosynthesis and cinnamate 4-monooxygenase by selected growth regulators. In: Greuter W, Zimmer B, Behnke HD, eds. Book of Abstracts, XIV International Congress of Botany, Berlin, FRG, 1987; Abstract No.2-113b-5.

CHAPTER 30

Studies on the Action of the Plant Growth Regulators BX-112, DOCHC, and DOCHC-Et

I. Nakayama, T. Miyazawa, M. Kobayashi, Y. Kamiya, H. Abe, and A. Sakurai

1 Introduction

A number of plant growth retardants are known to inhibit gibberellin (GA) biosynthesis in plants.[1] These growth retardants inhibits specific steps between mevalonic acid and GA_{12}-aldehyde.[2] The growth inhibition caused by these retardants is alleviated by exogenously applied GAs, indicating that the inhibition of GA biosynthesis is the primary action of these retardants.[2-4]

Recently, a new class of plant growth regulators, calcium 3,5-dioxo-4-propionylcyclohexanecarboxylate (BX-112) and its free acid and ethyl ester derivatives, DOCHC and DOCHC-Et (Fig. 1), have been synthesized. These compounds retard the growth of a wide variety of plants (Table 1).[5] In terms of agricultural usage, these compounds have the promising characteristics of rapid effects following foliar treatment; also, they lack residual activity in terms of subsequent rotational crops. These retardants can be used as lodging-preventing agents for cereals and as growth retardants for flowering plants.[5] Preliminary experiments showed that the retardation by these compounds was reversed by exogenously applied GA_3, suggesting that the compounds are inhibitors of GA biosynthesis. In this paper, we report on the mode of action of BX-112 and its derivatives using rice seedlings and cell-free systems from immature seeds.

2 Materials and Methods

2.1 Rice Seedling Bioassay

Rice (*Oryza sativa*) seeds (dwarf cultivar, Tan-ginbozu) were soaked in water for 2 days in the dark at 30 °C. Germinated seeds were used for assays by the dipping method[3] or the microdrop method.[6] For the former method, water containing the retardants and steviol was used as the culture

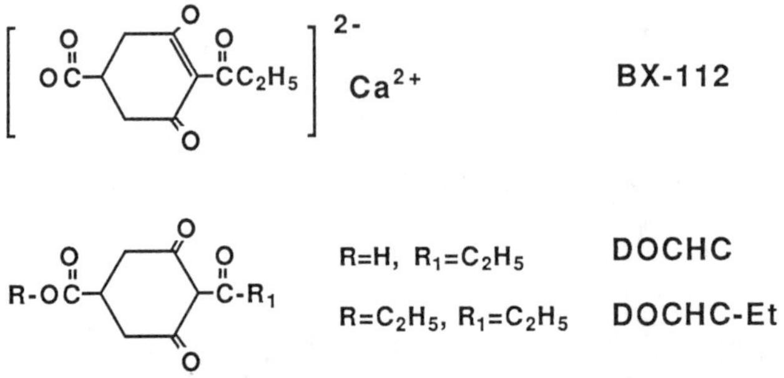

Fig. 1. Structures of calcium 3,5-dioxo-4-propionylcyclohexanecarboxylate (BX-112) and its free acid (DOCHC) and ethyl ester (DOCHC-Et) derivatives

solution. For the microdrop method, 1 µl of acetone–water (1:1) solution containing the retardant and GAs was applied to each seedling.

2.2 Analysis of Endogenous GAs in Rice

Rice seedlings (normal cultivar, Harebare) were grown in a nursery box (60 × 30 × 3 cm) filled with greenhouse soil. At the four-leaf stage, DOCHC was sprayed at doses of 0.3, 3, and 30 mg/m^2. Leaves and shoots were harvested 6 days after treatment and analyzed for GAs. Deuterated GA_1, GA_{19}, and GA_{20} were added as internal standards before extraction.[7] Contents of GA_1, GA_{19}, and GA_{20} were determined by combined gas chromatography–selected ion monitoring (GC–SIM)[8] using a DB-1 capillary column (0.25 mm × 30 m).[9] The conditions for analyzing and converting the samples were as described in previous papers.[8,9]

2.3 Labeled Substrates

[^{14}C]Mevalonic acid was prepared from [2-^{14}C]mevalonic acid lactone (1.96 GBq/mmol, Amersham Japan Ltd.) by hydrolysis with 0.01 N KOH for 15 min at 40 °C. [^{14}C]Gibberellin A_{12} (7.29 GBq/mmol) was prepared from [2-^{14}C]mevalonic acid using a cell-free preparation from *Cucurbita maxima*.[10] [^{14}C]Gibberellin A_9 (6.66 GBq/mmol) was prepared from [^{14}C]GA_{12} by incubation with a *Pisum sativum* cell-free preparation.[11] [17-^{13}C,^{3}H$_2$]Gibberellin A_{20} (1.79 GBq/mmol)[12] was a gift from Professor B.O. Phinney of the University of California, Los Angeles.

2.4 Preparation of Cell-Free Extracts and Incubation

Cell-free enzyme preparations were prepared from immature seeds of *C. maxima*,[10] *Phaseolus vulgaris*,[7] and *P. sativum*.[11] Incubation of

Table 1. Growth retardation effects of BX-112 on various plants. Plants were treated with BX-112 at a concentration of 3000 ppm by foliar application, and the plant heights were measured at 2 to 4 weeks after treatment

Group	Common name	Scientific name	Retardation effects[a]
Crops	Sugar beet	*Beta vulgaris*	++
	Cucumber	*Cucumis sativus*	+++
	Cotton	*Gossypium* sp.	++
	Soybean	*Glycine max*	++
	Barley	*Hordeum vulgare*	+++
	Rice	*Oryza sativa*	+++
	Wheat	*Triticum aestivum*	+++
	Maize	*Zea mays*	++
Turfs	Velvet bent grass	*Agrostis* sp.	+++
	Bermuda grass	*Cynodon* sp.	+++
Flowers and Trees	Cockscomb	*Celosia cristata*	+++
	Chrysanthemum	*Chrysanthemum morifolium*	++
	Carnation	*Dianthus caryophyllus*	+++
	Azalea	*Rhododendron* sp.	+
	Oak	*Quercus* sp.	+

[a] Retardation effects on plant heights: −, no effect; +, less than 20%; ++ 20 to 50%; +++ more than 50%

[^{14}C]mevalonic acid with a cell-free preparation from *C. maxima* was carried out according to the method described by Izumi et al.[3] [^{14}C]Gibberellin A$_{12}$ and [17-^{13}C,^{3}H$_2$]GA$_{20}$ were incubated with the soluble enzyme preparation from *P. vulgaris*.[13] The enzyme preparation from *P. sativum*[11] was used for 2β-hydroxylation of [17-^{13}C,^{3}H$_2$]GA$_{20}$ and [^{14}C]GA$_9$ and 13-hydroxylation of [^{14}C]GA$_{12}$. A solution of DOCHC in methanol (2 μl) was added to each enzyme incubation mixture (198 μl) containing cofactors.[7,10,11] After incubation at 30 °C for 2 h, the mixture was acidified with 5 N HCl (pH 2.5) and extracted with ethyl acetate. The ethyl acetate extracts were analyzed by reverse-phase high-performance liquid chromatography (HPLC) using a Nucleosil 5C$_{18}$ column (4.6 × 100 mm) or a Nucleosil 5N(CH$_3$)$_2$ column (4.6 × 50 mm) with on-line radio-counting.[14] Each radioactive peak was collected and the radioactivity was accurately counted using a liquid scintillation counter. The conversion ratios were determined by dividing the radioactivity of the products by the total radioactivity. Identification of the products was by full-scan GC–mass spectrometry (GC–MS) using published conditions.[7]

3 Results and Discussion

To study the effects of DOCHC on rice, a comparison was made with a well-known growth retardant, ancymidol.[2] The growth retardation by

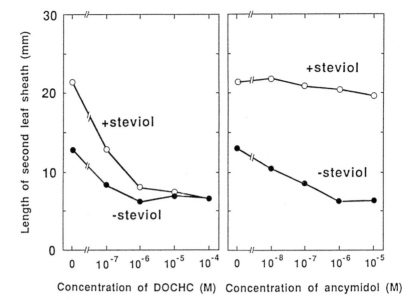

Fig. 2. Effects of DOCHC or ancymidol on the shoot elongation of rice caused by steviol. Steviol (6.2×10^{-4} M) and DOCHC or ancymidol were dissolved in 2 ml of water, and 8 uniformly germinated seeds (cv. Tan-ginbozu) were placed in a vial. After incubation under continuous light for 9 days at 30 °C, length of second leaf sheath was measured

DOCHC was not reversed by the addition of steviol (steviol, *ent*-13-hydroxykaurenoic acid, is not endogenous to rice; however, it is known to be metabolized to a bioactive GA-like substance in rice[15]). However, when ancymidol, an inhibitor of *ent*-kaurene oxidation,[2] was applied to rice seedlings, the inhibition was reversed by the addition of steviol (Fig. 2). These results suggest that DOCHC inhibits a step(s) subsequent to *ent*-kaurenoic acid in the GA biosynthetic pathway.

Gibberellins A_{19}, A_{20}, and A_1 were exogenously applied with various doses of BX-112 to the rice seedlings. BX-112 inhibited the shoot elongation caused by GA_{19} and GA_{20}. However, BX-112 did not inhibit the shoot elongation caused by GA_1 (Fig. 3). Gibberellins A_{19} and A_{20} are precursors to GA_1; all three are native to rice, and GA_1 is thought to be the active GA in the control of shoot elongation.[15-18] These results suggest that BX-112 inhibits the 3β-hydroxylation of GA_{20} to GA_1.[19] However, it is unknown whether or not BX-112 also inhibits the conversion of GA_{19} to GA_{20}. At high doses of BX-112 (250 to 2500 ng/plant), the shoot elongation caused by GA_1 was promoted (Fig. 3). This promotion of shoot elongation may be due to the inhibition of inactivation by 2β-hydroxylation of GA_1.

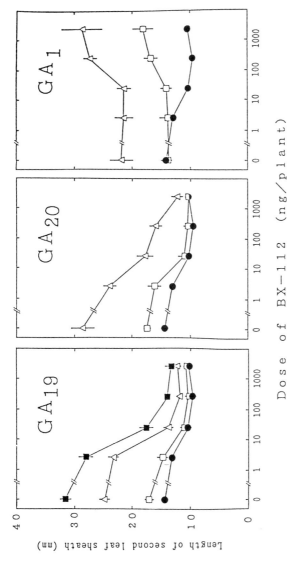

Fig. 3. Effects of BX-112 on shoot elongation of rice caused by exogenously applied GA_1, GA_{19}, and GA_{20}. BX-112 and GAs were applied to 2-day-old seedlings (cv. Tanginbozu) using the microdrop method.[6] The length of second leaf sheath was measured 3 days after application. Each point represents the mean ± SE ($n = 5$). Symbols: ●, control without GA; □, with 1 ng/plant GA; △, with 10 ng/plant GA; ■, with 100 ng/plant GA

Table 2. Concentrations of GAs in whole shoots of rice. DOCHC was sprayed at four leaf stage, and shoots were harvested 6 days after treatment. Concentrations of GAs were analyzed by GC–SIM with deuterated internal standards

DOCHC (mg/m^2)	Plant height (mm)	Length of 5th leaf sheath (mm)	Concentration (ng/g fr wt)[a]		
			GA_{19}	GA_{20}	GA_1
Control	385	134	14.3	0.67	0.26
0.3	302	99	16.2	1.09	0.25
3	271	77	20.9	2.09	0.23
30	248	57	21.4	2.12	0.10

[a] For GC–SIM conditions and kinds of ions used for quantification, see refs. 7 and 8

The effect of DOCHC on the GA levels in rice was analyzed by GC–SIM (Table 2). The compound was applied at the four-leaf stage and the plants were harvested 6 days after treatment. When DOCHC was sprayed at a dose of 30 mg/m^2, the length of the fifth leaf sheath was reduced to 43% of that of the control. The content of GA_1 was decreased to 38% of that of the control. However, the levels of GA_{19} and GA_{20} were increased to 150% and 316% of the control values, respectively. These results suggest that DOCHC inhibits the 3β-hydroxylation of GA_{20} in rice seedlings and that shoot growth is related to the content of GA_1 rather than GA_{19} and GA_{20}. The results provide further support that GA_1 is active per se in promoting shoot growth in rice.[18]

Since the 3β-hydroxylase appears to be the enzyme most affected by DOCHC, enzyme inhibition was studied using a cell-free preparation from immature seeds of *P. vulgaris*[7]. In this system, [17-^{13}C,^{3}H$_2$]GA$_{20}$ is converted to [17-^{13}C,^{3}H$_2$]GA$_1$ and [17-^{13}C,^{3}H$_2$]GA$_6$ via [17-^{13}C,^{3}H$_2$]GA$_5$. The 3β-hydroxylation of GA_{20} to GA_1 is catalyzed by a 2-oxoglutarate-dependent dioxygenase.[20,21] DOCHC inhibition was observed at a concentration of 1×10^{-5} M (Fig. 4). Since oxidation of C-20 and 2β-hydroxylation of GAs are catalyzed by the same type of enzymes, that is, 2-oxoglutarate-dependent dioxygenases,[2] the effects of DOCHC on these enzymes were also studied. To study the effects of DOCHC on the oxidation of C-20 of GAs, [^{14}C]GA$_{12}$ was incubated with DOCHC and the soluble enzyme preparation from *P. vulgaris*.[7] In a control incubation, [^{14}C]GA$_{12}$ was converted to [^{14}C]gibberellins A$_{15}$, A$_{24}$, A$_9$, A$_4$, and A$_{37}$. The production of GA_4 and GA_{37} was reduced by DOCHC. Gibberellin A$_{15}$, one of the intermediates, was accumulated (Fig. 5). To study the effect of DOCHC on 2β-hydroxylase, [^{14}C]GA$_9$ and [17-^{13}C,^{3}H$_2$]GA$_{20}$ were incubated with DOCHC and a soluble enzyme preparation from immature seeds of *P. sativum*.[11] The 2β-hydroxylase was also inhibited approximately 60% by DOCHC at a concentration of 1×10^{-4} M.

13-Hydroxylation of [^{14}C]GA$_{12}$ to GA_{53}, which is catalyzed by a micro-

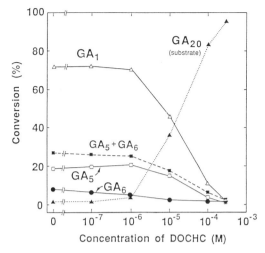

Fig. 4. The effects of DOCHC on the conversion of [17-^{13}C,^{3}H$_2$]GA$_{20}$ (833 Bq) to [17-^{13}C,^{3}H$_2$]GA$_1$, [17-^{13}C,^{3}H$_2$]GA$_5$, and [17-^{13}C,^{3}H$_2$]GA$_6$ using a soluble enzyme preparation from *P. vulgaris* (200 μl) containing 5 mM 2-oxoglutarate, 5 mM ascorbate, and 0.5 mM FeSO$_4$ at the final concentration. The products were separated by reverse-phase HPLC[16] and were identified by GC-MS. Radioactivity was counted by a liquid scintillation counter

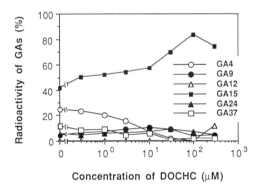

Fig. 5. The effects of DOCHC on the conversion of [^{14}C]GA$_{12}$ to gibberellins A$_4$, A$_9$, A$_{15}$, A$_{24}$, and A$_{37}$. [^{14}C] Gibberellin A$_{12}$ (1000 Bq) was incubated with a soluble enzyme preparation of *P. vulgaris* (200 μl) containing DOCHC and the same cofactors as described in the legend to Fig. 4. The products were separated by reverse-phase (Nucleosil 5C$_{18}$) and ion exchange [Nucleosil N(CH$_3$)$_2$] HPLC and identified by full-scan GC–MS

somal enzyme,[11] was not inhibited by DOCHC at a concentration of 3×10^{-4} M in the cell-free system from *P. sativum*. When the cell-free system from the endosperm of *C. maxima*[3] was treated with DOCHC, the conversion of [^{14}C]mevalonic acid to GA_{12}-aldehyde was also not inhibited by DOCHC at a concentration of 1×10^{-4} M.

The primary action of DOCHC is the inhibition of 2-oxoglutarate-dependent dioxygenases of GA biosynthesis. This might be speculated from the structural similarity of this type of retardants to 2-oxoglutarate. For further study on the mechanism of enzyme inhibition, it is necessary to use purified enzymes.

4 Conclusion

BX-112 and its derivatives are specific inhibitors of 2-oxoglutarate-dependent dioxygenases in GA biosynthesis. The primary action of BX-112 was the inhibition of the 3β-hydroxylation step in GA biosynthesis, as shown by in vivo and in vitro assays. Other oxidation steps catalyzed by 2-oxoglutarate-dependent dioxygenases in GA biosynthesis were also inhibited by the compound, but 3β-hydroxylation appears to be most sensitive to the inhibitor. Our study gives further support for the hypothesis that GA_1 is an active GA in controlling shoot growth[18] and that GA_{19} and GA_{20} are activated through their metabolism to GA_1 in rice. The compounds are promising chemical probes of the biosynthesis and mode of action of GAs.

Acknowledgment. The authors wish to thank Professor B.O. Phinney (UCLA) for providing [17-^{13}C,^{3}H$_2$]-labeled GAs.

References

1. Jung J, Rademacher W. Plant growth regulating chemicals—cereal grains. In: Nickell LG, ed. Plant growth regulating chemicals, Vol. 1. Boca Raton, Florida: CRC Press, 1983: pp. 253–271.
2. Graebe JE. Gibberellin biosynthesis and control. Ann Rev Plant Physiol. 1987; 38:419–465.
3. Izumi K, Yamaguchi I, Wada A, et al. Effects of a new growth retardant (*E*)-1-(4-chlorophenyl)-4,4-dimethyl-2-(1,2,4-triazol-1-yl)-1-penten-3-ol (S-3307) on the growth and gibberellin content of rice plants. Plant Cell Physiol. 1984; 25:611–617.
4. Rademacher W, Fritsch H, Graebe JE, et al. Tetcyclasis and triazol-type plant growth retardants: Their influence on the biosynthesis of gibberellins and other metabolic process. Pestic Sci. 1987; 21:241–252.
5. Motojima K, Miyazawa T, Toyokawa Y, et al. Cyclohexane derivatives having plant-growth regulation activities, and uses of these derivatives. United States Patent US 4,560,403, 1985.

6. Murakami Y. A new rice seedling bioassay for gibberellins, "microdrop method", and its use for testing extracts of rice and morning glory. Bot Mag Tokyo. 1968; 81:33–43.
7. Takahashi M, Kamiya Y, Takahashi N, et al. Metabolism of gibberellins in a cell-free system from immature seeds of *Phaseolus vulgaris* L. Planta. 1986; 168:190–199.
8. Kobayashi M, Sakurai A, Saka H, et al. Quantitative analysis of endogenous gibberellins in normal and dwarf cultivars of rice. Plant Cell Physiol. 1989; 30:963–969.
9. Kobayashi M, Kamiya Y, Sakurai A, et al. Metabolism of gibberellins in cell-free extracts of anthers from normal and dwarf rice. Plant Cell Physiol. 1990; 31:289–293.
10. Graebe JE, Hedden P, Gaskin P, et al. Biosynthesis of gibberellins A_{12}, A_{15}, A_{24}, A_{36} and A_{37} by a cell-free system from *Cucurbita maxima*. Phytochemistry. 1974; 13:1433–1440.
11. Kamiya Y, Graebe JE. The biosynthesis of all major pea gibberellins in a cell-free system from *Pisum sativum*. Phytochemistry. 1983; 22:681–689.
12. Fujioka S, Yamane H, Spray CR, et al. Qualitative and quantitative analyses of gibberellins in vegetative shoots of normal, *dwarf*-1, *dwarf*-2, *dwarf*-3, and *dwarf*-5 seedling of *Zea mays* L. Plant Physiol. 1988; 88:1367–1372.
13. Kwak SS, Kamiya Y, Takahashi M, et al. Metabolism of [^{14}C]GA$_{20}$ in a cell-free system from developing seeds of *Phaseolus vulgaris* L. Plant Cell Physiol. 1988; 29:707–711.
14. Kwak SS, Kamiya Y, Sakurai A, et al. Partial purification of gibberellin 3β-hydroxylase from immature seeds of *Phaseolus vulgaris* L. Plant Cell Physiol. 1988; 29:935–943.
15. Murakami Y. Dwarfing genes in rice and their relation to gibberellin biosynthesis. In: Carr DJ, ed. Plant growth substances 1970. Berlin: Springer-Verlag, 1972: pp. 166–174.
16. Suzuki Y, Kurogouchi S, Murofushi N, et al. Seasonal changes of GA_1, GA_{19} and abscisic acid in three rice cultivars. Plant Cell Physiol. 1981; 22:1085–1093.
17. Kobayashi M, Yamaguchi I, Murofushi N, et al. Endogenous gibberellins in immature seeds and flowering ears of rice. Agric Biol Chem. 1984; 48:2725–2729.
18. Phinney BO. Gibberellin A_1, dwarfism and the control of shoot elongation in higher plants. In: Crozier A, Hillman JR, eds. The biosynthesis and metabolism of plant hormones. Cambridge: Cambridge University Press, 1984: pp. 17–41.
19. Nakayama I, Miyazawa T, Kobayashi M, et al. Effects of a new plant growth regulator prohexadione calcium (BX-112) on shoot elongation caused by exogenously applied gibberellins in rice (*Oryza sativa* L.) seedlings. Plant Cell Physiol. 1990; 31:195–200.
20. Kamiya Y, Takahashi M, Takahashi N, et al. Conversion of gibberellin A_{20} to gibberellins A_1 and A_5 in a cell-free system from *Phaseolus vulgaris*. Planta. 1984; 162:154–158.
21. Albone KS, Gaskin P, MacMillan J, et al. Enzymes from seeds of *Phaseolus vulgaris* L.: Hydroxylation of gibberellin A_{20} and A_1 and 2,3-dehydrogenation of gibberellin A_{20}. Planta. 1989; 177:108–115.

CHAPTER 31

Studies on Sites of Action of the Plant Growth Retardant 4'-Chloro-2'-(α-Hydroxybenzyl)isonicotinanilide (Inabenfide) in Gibberellin Biosynthesis

T. Miki, T. Ichikawa, Y. Kamiya, M. Kobayashi, and A. Sakurai

1 Introduction

4'-Chloro-2'-(α-hydroxybenzyl)isonicotinanilide (inabenfide; Fig. 1) is a growth retardant for rice.[1] The retardation is reversed by the application of gibberellin A_3 (GA_3), and inabenfide was found not to interfere with gibberellin (GA)-induced elongation,[2] suggesting that inabenfide inhibits the biosynthesis of GA.

Kobayashi et al.[3] have shown that GA_{19} is a major GA and GA_1, GA_{20}, GA_{29}, and GA_{53} are minor GAs in the vegetative stems of rice. Among these GAs, GA_1 was considered to be the physiologically active GA in regulating the vegetative growth. As a first step toward investigating the effect of inabenfide on GA biosynthesis, the kinds and amounts of endogenous GAs in rice were determined after treatment with inabenfide.

Immature seeds of some higher plants have been used to prepare active cell-free systems.[4] We have investigated the effects of inabenfide on GA biosynthesis using cell-free extracts from *Cucurbita maxima*, *Phaseolus vulgaris*, and *Pisum sativum*. Inabenfide has two isomers, the (S)-form and the (R)-form (Fig. 1). They are easily oxidized in rice to 2'-benzoyl-4'-chloroisonicotinanilide (compound [1] in Fig. 1).[5] The effects of these isomers and compound [1] on GA biosynthesis have also been studied, and the difference in inhibitory activity on GA biosynthesis and growth retardant activity on pumpkin seedlings has been compared. Based on the results obtained, we discuss the growth retarding activity of inabenfide on intact plants.

2 Materials and Methods

2.1 Analysis of Endogenous GAs in Rice

Rice (*Oryza sativa* cv. Koshihikari) was grown for 3 weeks in a nursery box (60 × 30 × 3 cm) filled with artificial soil uniformly treated with inaben-

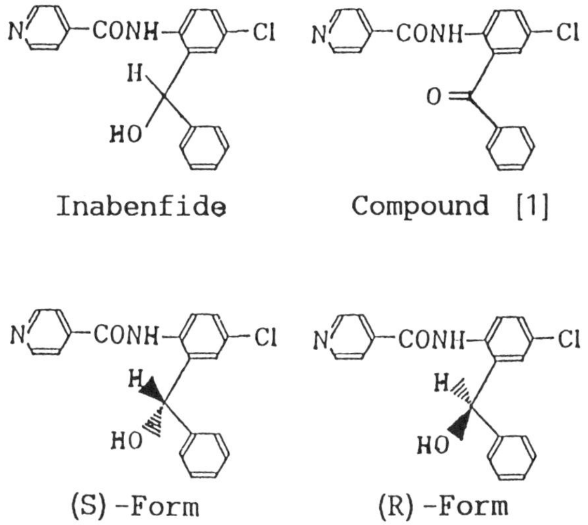

Fig. 1. Structures of inabenfide and its isomers and 2'-benzoyl-4'-chloroisonicotinanilide (compound [1])

fide. The plants heights were measured and the contents of GA_1, GA_{19}, and GA_{20} in the leaves and shoots were analyzed by combined gas chromatography–selected ion monitoring (GC–SIM) using an Incos 50 mass spectrometer connected to a Hewlett-Packard HP-5890 gas chromatograph equipped with a DB-1 capillary column (0.25 mm × 30 m). Deuterated GA_1, GA_{19}, and GA_{20}, prepared by the method of MacMillan and Willis[6] and Duri and Hanson,[7] were used as internal standards for the identification and quantification of each GA.

2.2 Cell-Free Systems

Cell-free systems from *C. maxima*, *P. vulgaris*, and *P. sativum* were prepared essentially as described by Graebe et al.,[8] Takahashi et al.,[9] and Kamiya and Graebe.[10]

2.3 Preparation of Substrates

R-[2-^{14}C]Mevalonic acid was prepared from *R*-[2-^{14}C]mevalonic acid lactone (53 mCi/mmol, Amersham Japan Ltd.) by hydrolysis with 0.01 *N* KOH for 15 min at 40 °C. [^{14}C]*ent*-Kaurene, [^{14}C]*ent*-kaurenol, [^{14}C]*ent*-kaurenal, [^{14}C]*ent*-kaurenoic acid, and [^{14}C]GA_{12} were prepared biosynthetically from *R*-[2-^{14}C]mevalonic acid by incubation with the *C. maxima* cell-free system. The specific radioactivity of the *ent*-kaurene, its oxidation products, and GA_{12}, determined by combined GC–mass spectrometry

(GC–MS), was approximately 150 mCi/mmol and 120 mCi/mmol, respectively.

2.4 Incubation and Assay

R-[2-^{14}C]Mevalonic acid was incubated with enzyme extracts from *C. maxima* according to the method described by Izumi et al.[11] [^{14}C]Gibberellin A_{12} was incubated with cell-free preparations from *P. vulgaris* and *P. sativum* using methods modified from those described by Takahashi et al.[9] and Ropers et al.[12] The reaction mixtures were extracted and analyzed by high-performance liquid chromatography (HPLC) with on-line radiocounting.[13,14] Each radioactive peak was collected and identified conclusively by GC–MS as described above. The percent of ^{14}C incorporated into the product was determined by dividing the radioactivity of the product by the total radioactivity recovered. The inhibition of oxidation of [^{14}C]*ent*-kaurene, [^{14}C]*ent*-kaurenol, [^{14}C]*ent*-kaurenal, and [^{14}C]*ent*-kaurenoic acid were also investigated by similar procedures using the *C. maxima* cell-free system, and the percent inhibition was calculated by dividing the radioactivity of the oxidized products by the total radioactivity recovered.

2.5 Pumpkin Seedling Assay

Serial dilutions of test compounds in methanol (1 ml) were uniformly applied to 100 ml of vermiculite in plastic pots (90 mm top diameter, 60 mm bottom diameter, 80 mm high). Then three seeds of *C. maxima* were placed in the pots. Plant heights were measured following growth under continuous illumination from fluorescent lamps (12,000 lux) for 14 days at 26 °C. Growth retardant activity was determined by dividing the mean of the heights of treated plants by that of the control plants.

3 Results and Discussion

Rice seedlings treated with inabenfide decreased in height with increasing amounts of added inabenfide. Seedlings were about half the height of the control at 0.3 g of inabenfide per nursery box (Fig. 2). The contents of GA_{19}, GA_{20}, and GA_1 in leaves and shoots of these rice seedlings are shown in Fig. 3. Untreated control leaves and shoots contained 3.7 ng of GA_{19}, 0.42 ng of GA_{20}, and 0.09 ng of GA_1 per g fr wt. The GA_{19} and GA_{20} contents of the leaves and shoots were drastically decreased with inabenfide treatment. This was especially true for the GA_{19} content, which decreased to about one-tenth of that of the control (0.37 ng/g fr wt) when the seedlings were treated with 0.3 g of inabenfide. The level of GA_1 was also decreased by inabenfide.

The step(s) in GA biosynthesis inhibited by inabenfide was investigated

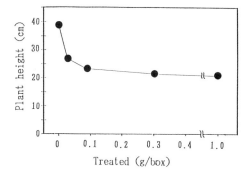

Fig. 2. Dose–response curve for inabenfide in terms of plant height of rice

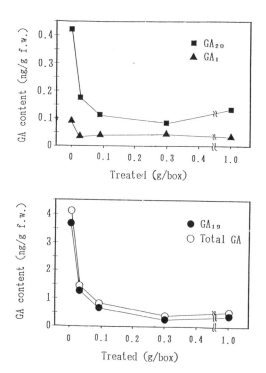

Fig. 3. Content of gibberellins A_{19}, A_{20}, and A_1 in rice treated with inabenfide

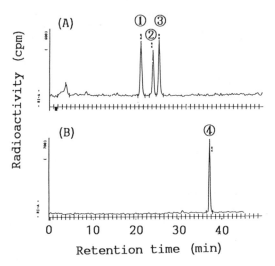

Fig. 4. HPLC on-line radiochromatograms of the incubation products of R-[2-^{14}C]mevalonic acid (10^5 dpm, 53 mCi/mmol) with *C. maxima* enzyme extracts and cofactors: (A) Control; (B) In presence of inabenfide (10^{-5} M). Peaks: 1, *ent*-6α,7α-dihydroxykaurenoic acid; 2, GA_{12}; 3, GA_{12}-aldehyde; 4, *ent*-kaurene

using cell-free systems described in Section 2.2. When R-[2-^{14}C]mevalonic acid was incubated in the cell-free system from *C. maxima* endosperm together with appropriate cofactors, three peaks corresponding to *ent*-6α,7α-dihydroxykaurenoic acid, GA_{12}-aldehyde, and GA_{12} were detected in the radiochromatogram (Fig. 4A). In the presence of inabenfide (10^{-5} M), the radiochromatogram showed only one peak, which corresponded to *ent*-kaurene (Fig. 4B). The effects of inabenfide concentration on the conversion of [^{14}C] mevalonic acid to *ent*-kaurene, GA_{12}-aldehyde, and GA_{12} were determined by measuring the percentage of the ^{14}C incorporated into these compounds (Fig. 5). Clearly, inabenfide inhibited the conversion of mevalonic acid to GA_{12}-aldehyde and GA_{12} at concentrations over 3×10^{-7} M, resulting in the accumulation of *ent*-kaurene. The effects of inabenfide on the oxidation of [^{14}C]*ent*-kaurene, [^{14}C]*ent*-kaurenol, [^{14}C]*ent*-kaurenal, and [^{14}C]*ent*-kaurenoic acid were also examined and are shown in Fig. 6. The oxidations of *ent*-kaurene, *ent*-kaurenol, and *ent*-kaurenal were blocked by inabenfide at concentrations over 3×10^{-7} M. In contrast, the oxidation of *ent*-kaurenoic acid was not affected even at a concentration of 3×10^{-5} M. The effect of inabenfide on GA biosynthesis subsequent to GA_{12} was investigated using cell-free systems from the embryos of *P. vulgaris* and *P. sativum*. In control experiments, the *P. vulgaris* cell-free system converted GA_{12} to GA_{15}, GA_{37}, and GA_4, and the *P. sativum* cell-free system converted GA_{12} to GA_{53} (Fig. 7). In the presence of in-

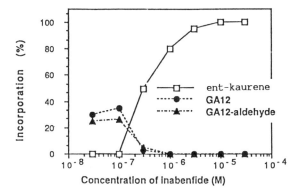

Fig. 5. Incorporation of R-[2-^{14}C]mevalonic acid (10^{-5} dpm, 53 mCi/mmol) into *ent*-kaurene, GA_{12}-aldehyde, and GA_{12}, at varying concentrations of inabenfide

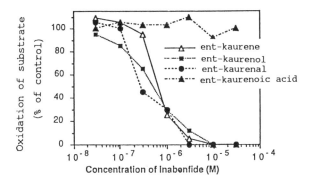

Fig. 6. Effects of inabenfide on the oxidation of [^{14}C]*ent*-kaurene, [^{14}C]*ent*-kaurenol, [^{14}C]*ent*-kaurenal, and [^{14}C]*ent*-kaurenoic acid

abenfide, even at a concentration of 3×10^{-5} M, metabolism of GA_{12} was not inhibited. These results show that GA biosynthesis subsequent to GA_{12} is not affected by inabenfide. The conversion of *ent*-kaurene to GA_{12} is catalyzed by microsomal enzymes that require NADPH and O_2.[15,16] Microsomal enzyme preparations of pea embryo also catalyze C-13 hydroxylation of GA_{12} to GA_{53}.[10] Successive oxidations of GA_{12} to GA_4 and GA_{53} to GA_1 are catalyzed by soluble enzymes that require α-ketoglutarate, Fe^{2+}, O_2, and ascorbate.[10] Our experiments showed that inabenfide specifically inhibits the microsomal enzyme(s) catalyzing the oxidation of *ent*-kaurene, *ent*-kaurenol, and *ent*-kaurenal. Other growth retardants, namely, uniconazole (S-3307), paclobutrazol (PP-333), tetcyclasis, and ancymidol, also are known to inhibit these three oxidation steps

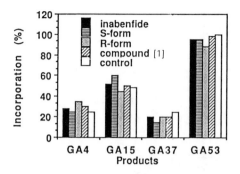

Fig. 7. Incorporation of [^{14}C]GA$_{12}$ into GA$_4$, GA$_{15}$, GA$_{37}$, and GA$_{53}$ in the presence of inabenfide (racemic form), (S)- and (R)-forms, and compound [1], at a concentration of $3 \times 10^{-5}M$. Gibberellins A$_4$, A$_{15}$, A$_{37}$ were produced in the incubation of [^{14}C]GA$_{12}$ with a cell-free system from *P. vulgaris*. Gibberellin A$_{53}$ was produced in the incubation of [^{14}C]GA$_{12}$ with a cell-free system from *P. sativum*

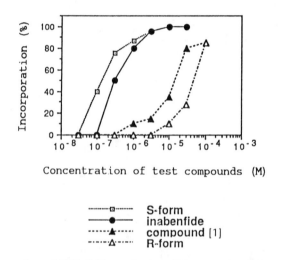

Fig. 8. Incorporation of R-[2-^{14}C]mevalonic acid into *ent*-kaurene at varying concentrations of inabenfide (racemic form) (●), (S)-form (□), (R)-form (△), and compound [1] (▲)

subsequent to *ent*-kaurene,[11,17-19] although they have different chemical structures.

The effects of (S)- and (R)-forms of inabenfide and compound [1] on GA biosynthesis were also investigated with the three cell-free systems described above. In the *C. maxima* cell-free system, these compounds were applied at concentrations ranging from 3×10^{-8} M to 10^{-4} M (Fig. 8). The concentrations for 50% incorporation of [^{14}C]mevalonic acid into *ent*-kaurene were approximately as follows: (S)-form, 10^{-7} M; (R)-form, 3 ×

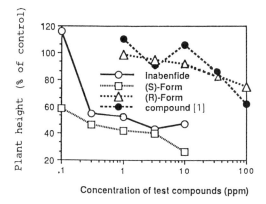

Fig. 9. Dose–response curves of inabenfide (racemic form), (S)- and (R)-forms, and compound [1] in the pumpkin seedling assay

10^{-5} M; racemic form, 3×10^{-7} M; compound [1], 10^{-5} M. The (S)-form was the most active for inhibiting the oxidation of *ent*-kaurene, the racemic form being the second most active, while the effectiveness of compound [1] and the (R)-form was low. The effect of these compounds on GA biosynthesis after GA_{12} was examined using cell-free systems from *P. vulgaris* and *P. sativum* (Fig. 7). The conversion of GA_{12} to GA_{15}, GA_{37}, GA_4, and GA_{53} was not inhibited at a concentration of 3×10^{-5} M.

In the pumpkin seedling assay, treatment with 0.3 ppm (μg/g of vermiculite) of inabenfide (racemic form) and with 0.1 ppm of the (S)-form reduced the plant height to 60% of that of the control. At 100 ppm, the (R)-form and compound [1] reduced the plant height to 70% of that of the control (Fig. 9). The (S)-form was the most effective retardant, the racemic form being the second most effective; compound [1] and the (R)-form had the least retardation. Because this retardant activity in the pumpkin seedling assay correlated with the inhibitory activity on *ent*-kaurene oxidation in the *C. maxima* cell-free system, it was concluded that inabenfide reduces the plant height by inhibiting the three oxidation steps subsequent to *ent*-kaurene.

The effects of (S)- and (R)-forms of inabenfide are similar to those of uniconazole and paclobutrazol.[11,17] The (S)-form of inabenfide was more active than the (R)-form in both the enzyme assay and the seedling assay (Figs. 8 and 9). Interestingly, although the inhibitory activity of uniconazole, paclobutrazol, and inabenfide on the oxidation of *ent*-kaurene was approximately the same using cell-free systems (Fig. 10), the growth-retarding effects of these compounds on intact plants are different.[20] The effects of inabenfide are slow, while those of uniconazole and paclobutrazol are rapid. Further study on the penetration, translocation, and catabolism of these compounds in intact plants is necessary before their different activities can be understood.

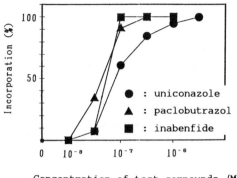

Fig. 10. Inhibitory activity of paclobutrazol, uniconazole, and inabenfide on the incorporation of R-[2-^{14}C]mevalonic acid into *ent*-kaurene

4 Conclusions

The plant height of rice was decreased with increasing concentrations of inabenfide. The content of GA_{19}, GA_{20}, and GA_1 in the leaf and vegetative shoots of these rice seedlings was also reduced. These results suggest that inabenfide inhibits GA biosynthesis. The inhibitory sites of inabenfide in GA biosynthesis were investigated using cell-free systems from *Cucurbita maxima*, *Phaseolus vulgaris*, and *Pisum sativum*. Inabenfide specifically inhibited the oxidation of *ent*-kaurene, *ent*-kaurenol, and *ent*-kaurenal, but did not affect later steps in the GA biosynthetic pathway.

The (S)- and (R)-forms of inabenfide and 2'-benzoyl-4'-chloroisonicotinanilide also inhibited the same oxidation steps. The inhibitory activity on GA biosynthesis and the growth retardant activity of these compounds were examined using cell-free systems and seedlings of *C. maxima*. The (S)-form was the most active, the racemic form the second most active and the (R)-form and 2'-benzoyl-4'-chloroisonicotinanilide the least active.

It is suggested that the growth retardant activity of inabenfide is due to the inhibition of the three oxidation steps from *ent*-kaurene to *ent*-kaurenoic acid.

References

1. Shirakawa N, Tomioka H, Fukazawa M, et al. Growth regulating activity on rice seedlings and antilodging effect of 4'-chloro-2'-(α-hydroxybenzyl)isoniconanilide (inabenfide). J Pestic Sci. 1987; 12:599–607 (in Japanese).
2. Shirakawa N, Tomioka H, Fukazawa M, et al. Effects of a new plant growth regulator "inabenfide" (CGR-811) on the growth of rice plants. Chem Regul Plants. 1987; 22:142–152 (in Japanese).
3. Kobayashi M, Yamaguchi I, Murofushi N, et al. Fluctuation and localization of endogenous gibberellins in rice. Agric Biol Chem. 1988; 52:1189–1194.

4. Graebe JE. Gibberellin biosynthesis and control. Ann Rev Plant Physiol. 1987; 38:419–465.
5. Ichikawa T, Sugiyama H, Iwane Y. Metabolism of inabenfide in rice seedlings. J Pestic Sci. 1987; 12:509–512 (in Japanese).
6. MacMillan J, Willis CL. N.m.r.assignments of ring A hydrogens in gibberellin A_4 methyl esters and some derivatives. J Chem Soc Perkin Trans 1. 1984; 351–355.
7. Duri ZJ, Hanson JR. Stereochemistry of reduction of a gibberellin delta-1 unsaturated ketone. J Chem Soc Perkin Trans 1. 1984; 363–366.
8. Graebe JE, Hedden P, Gaskin P, et al. Biosynthesis of gibberellins A_{12}, A_{15}, A_{24}, A_{36} and A_{37} by a cell-free system from *Cucurbita maxima*. Phytochemistry. 1974; 13:1433–1440.
9. Takahashi M, Kamiya Y, Takahashi N, et al. Metabolism of gibberellins in a cell-free system from immature seeds of *Phaseolus vulgaris* L. Planta. 1986; 168:190–199.
10. Kamiya Y, Graebe JE. The biosynthesis of all major pea gibberellins in a cell-free system from *Pisum sativum*. Phytochemistry. 1983; 22:681–689.
11. Izumi K, Kamiya Y, Sakurai A, et al. Studies of sites of action of a new plant growth retardant (*E*)-1-(4-chlorophenyl)-4,4-dimethyl-2-(1,2,4-triazol-1-yl)-1-penten-3-ol (S-3307) and comparative effects of its stereoisomers in a cell-free system from *Cucurbita maxima*. Plant Cell Physiol. 1985; 26:821–827.
12. Ropers HJ, Graebe JE, Gaskin P, et al. Gibberellin biosynthesis in a cell-free system from immature seed from *Pisum sativum*. Biochem Biophys Res Commun. 1978; 80:690–697.
13. Kamiya Y, Takahashi N, Graebe JE. The loss of carbon-20 in C_{19}-gibberellin biosynthesis in a cell-free system from *Pisum sativum* L. Planta. 1986; 169:524–528.
14. Kwak SS, Kamiya Y, Sakurai A, et al. Partial purification and characterization of gibberellin 3β-hydroxylase from immature seeds of *Phaseolus vulgaris* L. Plant Cell Physiol. 1988; 29:935–943.
15. Dennis DT, West CA. Biosynthesis of gibberellins. III The conversion of (−)-kaurene to (−)-kaurene-19-oic acid in endosperm of *Echinocystis macrocarpa* Greene. J Biol Chem. 1967; 242:3293–3300.
16. Hedden P. In vitro metabolism of gibberellins. In: Crozier A, ed. The biochemistry and physiology of gibberellins, Vol. 1. New York: Praeger, 1983: pp. 99–150.
17. Hedden P, Graebe JE. Inhibition of gibberellin biosynthesis by paclobutrazol in cell-free homogenates of *Cucurbita maxima* endosperm and *Malus pumila* embryos. J Plant Growth Regul. 1985; 4:111–122.
18. Rademacher W, Fritsch H, Graebe JE, et al. Tetcyclacis and triazole-type plant growth retardants: Their influence on the biosynthesis of gibberellins and other metabolic processes. Pestic Sci. 1987; 21:241–252.
19. Coolbaugh RC, Hirano SS, West CA. Studies on the specificity and site of action of α-cyclopropyl-α-(*p*-methoxyphenyl)-5-pyrimidine methyl alcohol (ancymidol), a plant growth regulator. Plant Physiol. 1978; 62:571–576.
20. Takeuchi Y, Konnai M, Takematu T. Bioassay methods of growth retardants. Chem Regul Plants. 1987; 22:130–141 (in Japanese).

CHAPTER 32

Effects of the Growth Retardant Uniconazole-P on Endogenous Levels of Hormones in Rice Plants

K. Izumi and H. Oshio

1 Introduction

Recently, several triazole-type plant growth retardants have been introduced for use in agriculture and horticulture. Uniconazole-P (an active optical isomer of uniconazole), also known as S-3307D and XE-1019D, is one such growth retardant. The characteristics of triazole-type growth retardants, in particular, uniconazole-P, are their potent activities in a wide spectrum of plants. The physiological nature of the triazoles as well as their practical applications have been studied intensively,[1-5] and it is now accepted that their primary mode of action is inhibition of gibberellin (GA) biosynthesis. In fact, all the published studies on the effects of triazoles on the GA content in plants indicate reductions of GA levels.[1,6-8]

The effect of triazoles on plants is not restricted to retardation of growth. They also produce interesting physiological effects:

1. Promotion of flower bud formation in woody plant species.[9,10]
2. Enhancement of femaleness in cucumber.[11]
3. Increase in the chlorophyll content of leaves and the delay of senescence.[2,12]

These effects of triazoles are similar to those reported for other types of growth retardants.[13-15] How are these plant responses brought about by triazoles? Are such hormonelike effects brought about solely by reduction of the GA content in plants? To answer these questions, we studied the effect of our triazole-type growth retardant, uniconazole-P, on the endogenous hormones in rice.[16] We here summarize the results of our studies of the effects of uniconazole and uniconazole-P on plant morphology and on the content of endogenous plant hormones. On the basis of our results, we discuss the interaction of endogenous hormones and the physiological phenomena induced by the triazoles.

Fig. 1. Effect of uniconazole on rhododendron. *Left*, untreated control; *right*, plant treated with 50 ppm uniconazole. (Photograph courtesy of Dr. M. Kunishige)

2 Morphological Effects on Plants

2.1 Promotion of Flower Bud Formation

The effect of uniconazole on rhododendron is shown in Fig. 1. Two-year-old plants of *Rhododendron catawbiense* cv. America were sprayed with a 50-ppm solution of uniconazole. The photograph shown in this figure was taken during the subsequent flowering season. In nature, it takes three years or more for rhododendron to come into bloom; the test rhododendron treated with uniconazole bloomed earlier than usual and had many more flowers as well as a compact shape. The enhancement of flower bud formation by uniconazole is not restricted to rhododendron. Woody species such as the azalea, camellia, pear, apple, and mango form multiple flower buds when treated with uniconazole or uniconazole-P.[9] Moreover, *Epiphyllum*, a species of cactus, initiates a greater number of flower buds when uniconazole is applied (Kobayashi Y, personal communication). The increase in the number of flower buds brought about by uniconazole treatment may be common not only to woody plants but to perennial plant species as well. The most favorable time for the application of this compound for the promotion of flower buds of common woody species is from the end of flowering to the beginning of the second flush of growth (May to August). The number of flowers produced is increased in the next flowering season, and the plant has a compact shape.

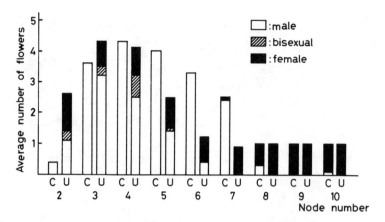

Fig. 2. Effect of uniconazole on sex expression in cucumber. Cucumber plants (cv. Sagami-hangiro) that were raised in a greenhouse were sprayed with a 50-ppm solution of uniconazole at the first leaf stage. Averages for 10 plants. C, Untreated control; U, uniconazole (50 ppm) treatment

For annuals, the increase in the number of flowers is not usually accompanied by growth retardation. Flowering in some plants is, however, sometimes delayed at high doses or prevented, as in bulbous plants such as tulips and lilies.

2.2 Modification of Sex Expression in Cucumber

In cucumber, the number of female flowers produced increases when plants are treated with uniconazole or uniconazole-P (Fig. 2). In monoecious cucumber varieties, male flowers differentiate at the lower nodes, after which female flowers differentiate at higher ones. When uniconazole is applied, female flowers are borne on the lower nodes, the percentage of female flowers being increased. Ethylene is known to enhance femaleness and GA to inhibit it.[17] As uniconazole and uniconazole-P are inhibitors of GA biosynthesis, they would exert on cucumber an effect that is the reverse of that of GA.

2.3 Increase in Leaf Chlorophyll Content and Delay of Senescence

In general, applications of growth retardants increase the chlorophyll content of leaves and delay senescence.[18] Uniconazole and uniconazole-P also have these effects. The chlorophyll content of the second true leaves of cucumbers (cv. Sagami-hangiro) that had been sprayed with uniconazole solution during the first-leaf stage is shown in Table 1. An application of uniconazole retarded plant growth, and the chlorophyll content per unit

Table 1. Effects of uniconazole on the chlorophyll content of the second true leaf of cucumber

Concentration of uniconazole (ppm)	Plant height (% of control)	Chlorophyll content (% of control)
0	100 (72.4 cm)	100 (37.2 µg/cm)
10	75	138
50	32	149

Table 2. Endogenous content of IAA in rice shoots treated with uniconazole-P

Dose of uniconazole-P (mg/m² nursery box)	Plant height (% of control)	Fresh weight (% of control)	Content of IAA (ng/g fr wt)
0	100 (19.9 cm)	100 (101 mg)	15.6
1	82	89	13.1
3	53	62	11.3
10	34	45	15.0

leaf area increased. Consequently, the color of the leaves became darker green when the leaves were treated with uniconazole or uniconazole-P.

3 Effects on the Endogenous Phytohormone Content in Rice Seedlings

Rice seeds were sown in nursery boxes, and a definite amount of uniconazole-P was applied to the surface of the soil. Seedlings were raised for about 3 weeks in a greenhouse, after which the plant hormone contents of their shoots were investigated.

3.1 IAA Content

The endogenous content of indole-3-acetic acid (IAA) in rice shoots was measured fluorometrically after purification by high-performance liquid chromatography (HPLC). The IAA content ranged from 11.3 to 15.6 ng/g fr wt; there were no significant differences in the IAA content between shoots treated with uniconazole-P and untreated shoots (Table 2).

3.2 Cytokinin Content

Trans-zeatin (t-Z), *cis*-zeatin (c-Z), *trans*-ribosyl zeatin (t-RZ), *cis*-ribosyl zeatin (c-RZ), and N^6-(Δ^2-isopentenyl)adenosine (i^6Ado) are the main cytokinins present in rice shoots.[19] We measured their levels using com-

Table 3. Endogenous content of cytokinins in rice shoots treated with uniconazole-P

Dose of uniconazole-P (mg/m² nursery box)	Plant height (% of control)	Content of cytokinins (ng/g fr wt)				
		t-Z	c-Z	t-RZ	c-RZ	i⁶Ado
0	100 (14.6 cm)	0.08	0.77	0.04	2.9	0.09
1	59	0.12	0.89	0.05	1.4	0.09
10	33	0.27	0.82	0.12	1.6	0.14

Table 4. Endogenous content of cis-ABA and trans-ABA in rice shoots treated with uniconazole-P

Dose of uniconazole-P (mg/m² nursery box)	Plant height (% of control)	Fresh weight (% of control)	Content of ABA (ng/g fr wt)	
			cis-ABA	trans-ABA
0	100 (19.6 cm)	100 (116 mg)	11.7	1.9
1	80	95	8.1	1.4
3	53	66	10.0	2.1
10	36	52	14.6	1.6

bined gas chromatography–selected ion monitoring (GC–SIM) according to the methods of Takagi et al.[19] and Nakagawa et al.[20]

In rice shoots, the c-Z and c-RZ levels are higher than those of t-Z and t-RZ.[19] Our analysis confirmed this (Table 3). In addition, there were clear changes in the cytokinin content; the amounts of t-Z and t-RZ (thought to be biologically important cytokinins) increased with the severity of growth retardation. In the untreated control shoots, the content of endogenous t-Z was 0.08 ng/g fr wt, and that of t-RZ, 0.04 ng/g fr wt. In contrast, shoots treated with uniconazole-P (10 mg/m²) contained 0.27 ng of t-Z and 0.12 ng of t-RZ per g fr wt. The amounts of t-Z and t-RZ in uniconazole-P-treated shoots were 3.4 and 3 times greater than in the control.

3.3 ABA Content

Gas chromatography–selectron capture detection (GC–ECD) was used to determine the amounts of endogenous abscisic acid (ABA) after purification with HPLC. In addition to cis-ABA (biologically active ABA), our analysis showed a small quantity of trans-ABA. Table 4 shows the amounts of cis- and trans-ABA in rice shoots treated with uniconazole-P. The cis- and trans-ABA levels were similar for both the control and the uniconazole-P treatments, evidence that uniconazole-P does not affect the endogenous contents of ABA in rice shoots.

Table 5. Amounts of ethylene evolved from rice seedlings incubated after uniconazole-P treatment

Concentration of uniconazole-P (ppm)	Plant height (% of control)	Fresh weight (% of control)	Evolution of ethylene (10^{-5} mol)	
			per g fresh wt	per plant
0	100 (8.7 cm)	100 (25 mg)	1.33	0.033
0.01	79	88	1.46	0.032
0.03	66	75	1.72	0.032
0.1	46	62	2.33	0.036
0.3	36	56	2.39	0.033

3.4 Evolution of Ethylene

The effect of uniconazole-P on ethylene synthesis was investigated. Twenty rice seeds that had germinated and been maintained in water for 2 days were transplanted to a vial (195 ml) that contained 30 ml of 1% agar medium with a known concentration of uniconazole-P. Vials were incubated for 6 days at 25 °C. The ethylene that evolved in the vial during incubation was analyzed by gas chromatography. The total volumes of ethylene produced were nearly the same for the control and uniconazole-P-treated seedlings on a per plant basis (Table 5). The fresh weight of the rice, however, decreased with increasing doses of uniconazole-P. The amount of ethylene produced (calculated on a fresh weight basis) increased as the concentration of uniconazole-P increased. The evolution of ethylene at 0.3-ppm uniconazole-P was 1.8 times that of the control.

4 Discussion

Our analysis of plant hormone levels in rice plants showed that uniconazole-P increased ethylene evolution and the amounts of active forms of cytokinins (*t*-Z and *t*-RZ). The increases in *t*-Z and *t*-RZ were particularly marked on both the fresh weight and the per plant basis.

All the reports of the effects of triazole-type growth retardants on plant hormones known to us indicate that triazole-type growth retardants have no, or only very weak, effects on the content of IAA in a variety of plant materials,[8,21] and these effects on endogenous content of IAA seem to be secondary. For ABA and ethylene, the reports are inconsistent. Rajasekaran et al.[21] reported that paclobutrazol, a triazole-type growth retardant of ICI, did not alter the endogenous content of ABA in young leaves of *Pennisetum purpureum*, whereas Wang et al.[22] reported that paclobutrazol reduced the endogenous ABA content of apple seedling leaves by about one-third. Grossmann et al.[8] also reported that immunoassays showed that LAB 150 978 (a triazole) and tetcyclacis (a norbornenodiazetine type of growth retardant from BASF) lowered the levels of ABA-like materials in the primary leaf, epicotyl, and root of soybean. The effects of triazoles on endogenous ABA content appear to differ with the plant species and experimental conditions.

Ethylene production is, in general, reduced by growth retardants. Sauerbrey et al.[23,24] reported that LAB 150 978 and tetcyclacis reduced ethylene production in sunflower seedlings and in suspensions of sunflower cells. Moreover, Abbas et al.[25] reported that triadimefon, a triazole-type fungicide with growth-retarding activity, reduced ethylene production and the endogenous content of 1-aminocyclopropane-1-carboxylic acid (ACC), an intermediate of ethylene biosynthesis, in cucumber seedlings. Our results do not agree with theirs. Further study is needed to ascertain the reasons why.

The most obvious effect of triazoles on hormone content other than GA levels is an increase in the amounts of cytokinins. A bioassay by Fletcher and Arnold[26] showed that triadimefon increased cytokinin content in the cotyledons and roots of treated cucumber seedlings. The study by Grossmann et al.[8] on the effects of LAB 150 978 and tetcyclacis on the hormone content of soybean seedlings indicates that the levels of t-RZ and dihydrozeatin-riboside type cytokinin were elevated markedly by treatment with these retardants. Their results agree well with ours for uniconazole-P. Seemingly, an increase in cytokinin is a common result of treatment with triazole-type growth retardants.

Some of the physiological effects of the triazoles are very similar to those of the cytokinins. Cytokinins have been reported to act positively in the formation of flower buds by woody plants[27]; moreover, they delay senescence and increase the chlorophyll content of leaves. These similarities offer indirect evidence that the physiological effects of the triazoles are brought about, at least in part, by the increase in the endogenous cytokinin content.

The physiological effects of the triazoles cannot, however, be accounted for solely by the increase in cytokinins. Gibberellins are known to inhibit the flowering of woody angiosperms and to promote maleness in cucurbits.[28] It is not surprising that triazoles act in the opposite way to GAs because they reduce the endogenous GA content of plants. Thus, the question remains as to whether the decrease in GA content or the increase in cytokinin content is the primary cause of the physiological phenomena produced by the triazoles.

Chen[29] examined the correlation of plant hormone levels to the flowering of mango and suggested that reduced shoot growth (low GA activity) and increased cytokinin activity are associated with flower bud formation. Banno et al.[30] reported that an application of the growth retardant succinic acid-2,2-dimethyl hydrazide (SADH) to Japanese pear increased flowering. Their analyses of the nutrient elements and endogenous growth regulators present indicated a decrease in GA content in shoots and auxiliary buds. They also reported that a supply of cytokinin to auxiliary buds initiated flower bud formation. They concluded that endogenous growth regulators, in particular, GAs and cytokinins, are deeply involved in flower bud formation. Taking into account these studies on the correlation of plant morphology and hormones, the decrease in the content of GAs and the increase in that of cytokinins might be responsible for the physiological phenomena produced by the triazoles.

How do triazole-type growth retardants increase the cytokinin content in plants? Do they affect the biosynthetic pathways of cytokinins directly, or is it the reduction of the GA content that stimulates cytokinin synthesis? If we are to understand fully the mechanisms of action of triazole-type growth retardants in plants, these questions must be answered.

References

1. Izumi K, Yamaguchi I, Wada A, et al. Effects of a new plant growth retardant (*E*)-1-(4-chlorophenyl)-4,4-dimethyl-2-(1,2,4-triazol-1-yl)-1-penten-3-ol (S-3307) on the growth and gibberellin content of rice plants. Plant Cell Physiol. 1984; 25:611–617.
2. Dalziel J, Lawrence DK. Biochemical and biological effects of kaurene oxidase inhibitors, such as paclobutrazol. In: Monograph No. 11, British Plant Growth Regulator Group. 1984: pp. 43–57.
3. Izumi K, Kamiya Y, Sakurai A, et al. Studies of sites of action of a new plant growth retardant (*E*)-1-(4-chlorophenyl)-4,4-dimethyl-2-(1,2,4-triazol-1-yl)-1-penten-3-ol (S-3307) and comparative effects of its stereoisomers in a cell-free system from *Cucurbita maxima*. Plant Cell Physiol. 1985; 26:821–827.
4. Hedden P, Graebe JE. Inhibition of gibberellin biosynthesis by paclobutrazol in cell-free homogenates of *Cucurbita maxima* endosperm and *Malus pumila* embryos. J Plant Growth Regul. 1985; 4:111–122.
5. Rademacher W, Fritsch H, Graebe JE, et al. Tetcyclacis and triazole-type growth retardants: Their influence on the biosynthesis of gibberellins and other metabolic processes. Pestic Sci. 1987; 21:241–252.
6. Rademacher W, Yung J, Graebe JE, et al. On the mode of action of tetcyclacis and triazole growth retardants. In: Monograph No. 11, British Plant Growth Regulator Group. 1984: pp. 1–11.
7. Jung J, Luib M, Sauter H, et al. Growth regulation in crop plants with new types of triazole compounds. J Agron Crop Sci. 1987; 158:324–332.
8. Grossmann K, Kwiatkowski J, Siebecker H, et al. Regulation of plant morphology by growth retardants. Plant Physiol. 1987; 84:1018–1021.
9. Aoki K. Methods for the use of a growth retardant S-07 in horticulture. Agriculture and Horticulture. 1983; 58:699–703 (in Japanese).
10. Raese JT, Burts EC. Increased yield and suppression of shoot growth and mite population of 'd'Anjou' pear trees with nitrogen and paclobutrazol. Hort Sci. 1983; 18:212–214.
11. Oshio H, Izumi K. S-3307, a new plant growth retardant, its biological activities, mechanism and mode of action. In: Plant growth regulators in agriculture. Food and Fertilizer Technology Center Book Series No. 34. 1986:p. 198–208. Taipei.

12. Gao J, Hofstra G, Fletcher RA. Anatomical changes induced by triazoles in wheat seedlings. Can J Bot. 1988; 66:1178–1185.
13. Stuart NM. Initiation of flower buds in rhododendron after application of growth retardants. Science. 1961; 134:50–52.
14. Mishra RS, Pradhan B. The effect of (2-chloroethyl) trimethyl ammonium chloride on sex expression in cucumber. J Hort Sci. 1970; 45:29–31.
15. Koranski DS, Struckmeyer BE, Beck GE. The role of ancymidol in *Clerodendrum* flower initiation and development. J Amer Soc Hort Sci. 1978; 103:813–815.
16. Izumi K, Nakagawa S, Kobayashi M, et al. Levels of IAA, cytokinins, ABA and ethylene in rice plants as affected by a gibberellin biosynthesis inhibitor, uniconazole-P. Plant Cell Physiol. 1988; 29:97–104.
17. Atsmon D, Tabbak C. Comparative effects of gibberellin, silver nitrate and aminoethoxyvinyl glycine on sexual tendency and ethylene evolution in the cucumber plant (*Cucumis sativus* L). Plant Cell Physiol. 1979; 20:1547–1555.
18. Cathey HM. Physiology of growth retarding chemicals. Ann Rev Plant Physiol. 1964; 15:271–302.
19. Takagi M, Yokota T, Murofushi N, et al. Fluctuation of endogenous cytokinin contents in rice during its life cycle—quantification of cytokinins by selected ion monitoring using deuterium-labeled internal standards. Agric Biol Chem. 1985; 49:3271–3277.
20. Nakagawa S, Tjokrokusumo DS, Sakurai A, et al. Endogenous levels of gibberellin, IAA and cytokinins in tobacco crown gall tissues of different morphologies. Plant Cell Physiol. 1987; 28:485–493.
21. Rajasekaran K, Hein MB, Vasil IK. Endogenous abscisic acid and indole-3-acetic acid and somatic embryogenesis in cultured leaf explants of *Pennisetum purpureum* Schum. Plant Physiol. 1987; 84:47–51.
22. Wang SY, Sun T, Ji ZL, et al. Effect of paclobutrazol on water stress-induced abscisic acid in apple seedling leaves. Plant Physiol. 1987; 84:1051–1054.
23. Sauerbrey E, Grossmann K, Jung J. Influence of growth retardants on the internode elongation and ethylene production of sunflower plants. Physiol Plant. 1987; 70:8–12.
24. Sauerbrey E, Grossmann K, Jung J. Ethylene production by sunflower cell suspensions. Plant Physiol. 1988; 87:510–513.
25. Abbas S, Fletcher RA, Murr DP. Alteration of ethylene synthesis in cucumber seedlings by triadimefon. Can J Bot. 1989; 67:278–280.
26. Fletcher RA, Arnold V. Stimulation of cytokinins and chlorophyll synthesis in cucumber cotyledons by triadimefon. Physiol Plant. 1986; 66:197–201.
27. Zeevaart JAD. Physiology of flower formation. Ann Rev Plant Physiol. 1976; 27:321–348.
28. Pharis RP, King RW. Gibberellins and reproductive development in seed plants. Ann Rev Plant Physiol. 1985; 36:517–568.
29. Chen WS. Endogenous growth substances in relation to shoot growth and flower bud development of mango. J Amer Soc Hort Sci. 1987; 112:360–363.
30. Banno K, Hayashi S, Tanabe K. Effects of SADH and shoot-bending on flower bud formation, nutrient components and endogenous growth regulators in Japanese pear (*Pyrus secotina* Rehd.). J Jpn Soc Hort Sci. 1985; 53:365–372 (in Japanese with English summary).

CHAPTER 33

Inconsistency Between Growth and Endogenous Levels of Gibberellins, Brassinosteroids, and Sterols in *Pisum Sativum* Treated with Uniconazole Antipodes

T. Yokota, Y. Nakamura, N. Takahashi, M. Nonaka, H. Sekimoto, H. Oshio, and S. Takatsuto

1 Introduction

Stereochemistry is important in determining the biological activity of triazole agrochemicals.[1-3] Uniconazole is a triazole growth retardant that has one chiral carbon carrying a hydroxyl. In rice seedlings, it has been demonstrated that the *S*-isomer of uniconazole (equivalent to uniconazole-P) exhibits plant growth-retarding activity by inhibiting the biosynthesis of gibberellins (GAs).[4,5] Recently, it was found that *S*-uniconazole increases the content of *trans*-zeatin and *trans*-ribosylzeatin and enhances the evolution of ethylene.[6] Application of racemic uniconazole to mung bean seedlings reduces ethylene production and increases spermine levels.[7] Thus, the mechanism of action of *S*-uniconazole seems more complicated than that postulated so far for triazoles.[8] The *R*-isomer exhibits fungicidal activity by inhibiting the biosynthesis of fungal sterols (cf. ref. 9). Triazole stereoisomers are also known to affect the biosynthesis of plant sterols in some arable crops.* It was therefore expected that *R*-uniconazole would also modify sterol biosynthesis. Furthermore, it was also of interest to examine whether or not uniconazoles modify the biosynthesis of the steroidal plant hormones, the brassinosteroids (BRs). In order to understand the growth retardation mechanism of *S*-uniconazole, we analyzed the shoots of *Pisum sativum* L. treated with *S*- and *R*-uniconazoles in terms of the levels of the endogenous GAs. BRs, and sterols.

---

*References 1, 2, 10, and 11.

2 Materials and Methods

2.1 Treatment with Uniconazoles

Seeds of *P. sativum* L. cv. Holland were imbibed for 2 days, sown on 3 December, and grown under greenhouse conditions. A 50-μl solution of *S*-uniconazole (0.5 μg) in 80% ethanol (EtOH) was put on the second leaf above the cotyledonary node 6, 8, 12, 15, and 19 days after the sowing. A 50-μl solution of *R*-uniconazole (25 μg) was likewise applied 6, 8, 12, 13, 15, and 19 days after the sowing. Control plants were treated with 80% EtOH. In each treatment, 230 shoots were harvested by cutting just above the cotyledonary nodes 22 days after the sowing. The yields of the shoots were 952 g in the control, 637 g in the treatment with *S*-uniconazole, and 753 g in the treatment with *R*-uniconazole.

2.2 Extraction and Solvent Partitionings

Plant materials were extracted with methanol (MeOH). The extract was reduced to an aqueous phase in vacuo, and the pH was adjusted to 8 by adding an equal volume of 0.5 M K_2HPO_4 prior to partitioning against $CHCl_3$ ($\times 3$). Then the pH of the aqueous phase was adjusted with aqueous HCl to 3 prior to partitioning against ethyl acetate (EtOAc) ($\times 4$). The $CHCl_3$ fraction was evaporated, dissolved in hexane, and partitioned against 80% MeOH ($\times 2$).

2.3 Purification of GAs

Each EtOAc fraction (200 g fr wt equiv) was dissolved in 0.1 M phosphate buffer, pH 8, mixed with [$^2H_5$]GA_1 (20 ng), [$^2H_2$]GA_{19} (500 ng), and [$^2H_5$]GA_{20} (500 ng),[12] and passed through a column of insoluble polyvinylpyrrolidone (bed volume, 29 ml) equilibrated with the same buffer. The first 170-ml eluate was collected, acidified to pH 3, and extracted with EtOAc. The EtOAc fraction was purified with a short column of Sepralyte (DEA)[12] prior to high-performance liquid chromatography (HPLC) on a 5-μm Develosil ODS column (6 $\times$ 200 mm) and eluted at a flow rate of 1.5 ml with aqueous MeOH containing 0.1% acetic acid (AcOH). The MeOH concentration was 45% for the first 15 min, then elevated to 70% over 5 min, and thereafter was maintained at 70%. Fractions were collected every minute. Fractions 5–6 and 14–23 were separately pooled for combined gas chromatography–selected ion monitoring (GC–SIM) analysis of GA_1, GA_{19} and GA_{20}, respectively (Fig. 1). For the identification of GAs in the control plants, the EtOAc fraction equivalent to 543 g fr wt was processed as described above, without addition of deuterated standards.

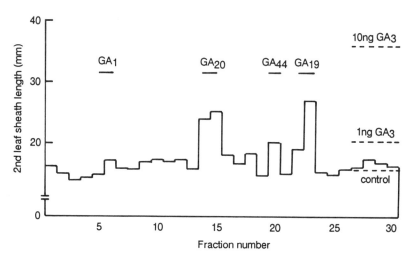

Fig. 1. Separation of GAs after reverse-phase HPLC of the extract of *P. sativum* shoots (control) as examined by Tan-ginbozu microdrop test. Deuterated standards of GA_1, GA_{19}, and GA_{20} were added to the EtOAC fraction prior to purification

2.4 Purification of BRs

Each 80% MeOH fraction (200 g fr wt equiv) was dissolved in $CHCl_3$, mixed with [$^2H_6$]castasterone (50 ng) and [$^2H_6$]brassinolide (5 ng),[13] charged on a column of silica gel (8.4 g), and eluted with $CHCl_3$ (80 ml), 3% MeOH in EtOAc (80 ml), and 7% MeOH in $CHCl_3$ (160 ml). The last fraction was chromatographed on Sephadex LH-20.[14] Fractions 34–38 were collected, dissolved in 80% MeOH, and passed through Sep-Pak ODS, which was further eluted with 80% MeOH (8 ml). The eluate was chromatographed on a column of 5-μm Develosil ODS (8 × 250 mm) at a flow rate of 3 ml/min at 40 °C. Fractions were collected every minute. For the identification of endogenous BRs in the control plants, the 80% MeOH fraction (548 g fr wt equiv) was processed as described above, without addition of deuterated standards.

2.5 Purification of Sterols

Each hexane fraction (400 g fr wt equiv) was successively subjected to saponification, hexane extraction, silica gel chromatography, and acetylation to give a fraction of sterol acetates,[14] which was subjected to combined GC–mass spectrometry (GC–MS) and GC–SIM. In SIM, base peaks of the various sterol acetates were monitored. For quantitation by GC–SIM, an internal standard, 5α-cholestane (100 μg), was mixed with one-third the amount of the acetates dissolved in 100 ml of hexane.

2.6 GC–MS and GC–SIM

a Gibberellins

Gibberellins were converted to methyl ester trimethylsilyl ethers (MeTM-Si) prior to analysis. Selected ion monitoring was carried out using a Hewlett-Packard quadrupole spectrometer.[12] The column temperature was 120 °C for the first 2 min, then was raised to 216 °C at 16 °C/min and held at 216 °C for 5 min, and then was raised to 280 °C at 8 °C/min. Full mass spectra of GA methyl esters were taken with a JEOL instrument, DX-303.[15] The column temperature was 140 °C for the first 2 min and then was elevated to 200 °C at a rate of 32 °C/min and then to 280 °C at a rate of 4 °C/min.

b Brassinosteroids

Brassinosteroids were analyzed using the DX-303 instrument after conversion to bismethaneboronates, as already reported.[15] The column temperature was 175 °C for the first 2 min and then was raised to 275 °C at 32 °C/min and further to 290 °C at 2 °C/min.

c Sterols

Acetates of sterols were analyzed with the Hewlett-Packard instrument used for GA analysis. The column temperature was 175 °C for the first minute and then was elevated to 295 °C at 32 °C/min.

3 Results

The biological activity of S-uniconazole for rice seedlings is 70-fold higher than that of R-uniconazole.[5] In order to see distinct growth retardation due to R-uniconazole, it was applied on *P. sativum* at a dosage 50 times higher than that of S-uniconazole. S-Uniconazole suppressed the growth of *P. sativum* more than R-uniconazole (Table 1). The effects of S- and

Table 1. Effects of S- and R-uniconazoles on the growth of *P. sativum* L. ($n = 10$)

		Treated with:	
	Control	S-Uniconazole	R-Uniconazole
Plant height (cm)	23.4	8.8 (38)	13.2 (56)
Leaf number	6.1	5.4 (89)	5.8 (95)
Shoot weight (g/plant)	4.0	2.6 (65)	3.4 (85)
Leaf area (cm²)	115.5	81.8 (71)	94.9 (82)

Figures in parentheses represent percent of the control value

Table 2. Effects of S- and R-uniconazoles on the content of GAs (ng/g fr wt), BRs (ng/g fr wt), and sterols (μ/g fr wt) in shoots of *P. sativum*

	Control	Treated with:	
		S-Uniconazole	R-Uniconazole
GA_1	0.12	0.044 (37)	0.028 (23)
GA_{19}	1.53	0.070 (5)	0.093 (6)
GA_{20}	4.35	1.64 (38)	1.28 (29)
Castasterone	0.90	0.48 (54)	0.30 (34)
Campesterol (plus 24β isomer)	15.5	14.9 (96)	14.7 (95)
Stigmasterol	46.0	43.8 (95)	43.4 (94)
Sitosterol	77.2	74.3 (96)	65.2 (84)
Isofucosterol	28.4	23.0 (80)	21.0 (74)

Figures in parentheses represent percent of control value

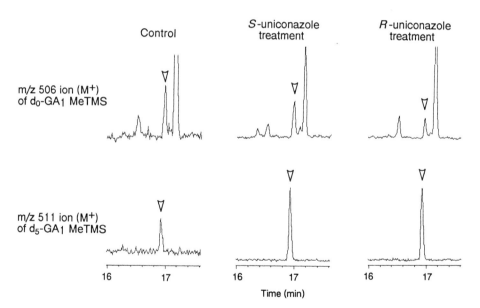

Fig. 2. GC–SIM quantitation of GA_1 MeTMSi derived from HPLC fractions 5 and 6 of *P. sativum* shoots treated with uniconazoles

R-uniconazoles on the content of GAs, BRs, and sterols are shown in Table 2.

The shoots of *P. sativum* are known to contain GA_1, GA_8, GA_{19}, GA_{20}, GA_{29}, and GA_{29} catabolite.[16,17] In the present study, the presence of GA_{19}, GA_{20}, and GA_{44} in the shoots of the cultivar Holland was demonstrated by full mass spectra (data not shown). The amounts of GA_1, GA_{19}, and GA_{20} were measured by GC–SIM using deuterated standards, as shown

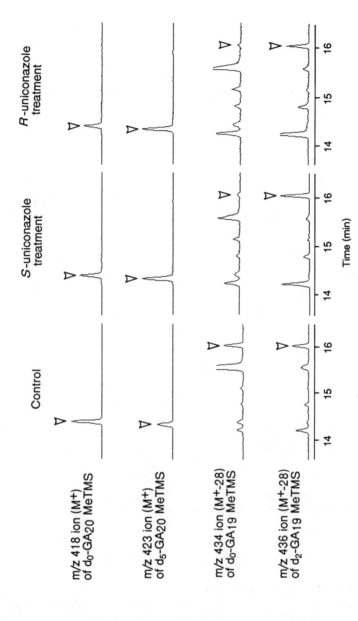

Fig. 3. GC–SIM quantitation of GA_{19} MeTMSi and GA_{20} MeTMSi from HPLC fractions 14–23 of *P. sativum* shoots treated with uniconazoles

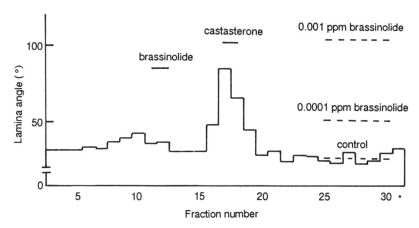

Fig. 4. Distribution of BR activity after reverse-phase HPLC of the extract of control shoots of *P. sativum* as examined by the rice lamina inclination test. *Washing with 80% acetonitrile

in Figs. 2 and 3. In control shoots, the amount of GA_1 is one order of magnitude lower than that of GA_{19} and GA_{20}. *R*-Uniconazole reduced the content of GA_1 and GA_{20} more than *S*-uniconazole. Interestingly, both antipodes of uniconazole reduced the content of GA_{19} more selectively than that of GA_1 and GA_{20}. An analogous effect has been observed in rice seedlings treated with a racemic uniconazole.[4]

The 80% methanol fraction, in which BRs are known to be partitioned, was analyzed for endogenous BRs. After the reverse-phase HPLC, a single peak of biological activity was observed (Fig. 4). Analysis of the active fractions by GC–MS led to the identification of castasterone, a putative precursor of brassinolide[18]: m/z (rel. int.) 512 (M^+, 54%), 155 (100%). Quantitation by GC–SIM using an internal standard (Fig. 5) revealed that the content of castasterone was 0.9 ng/g fr wt in the control plants and, after treatment with *S*- and *R*-uniconazoles, was reduced to 54% and 34% of the content in the controls, respectively (Table 2). Brassinolide could not be found by GC–MS (SIM) or bioassay.

Analysis by GC–MS revealed that all sterol acetate fractions obtained from control and treated shoots contain campesterol acetate, which presumably includes the 24β-isomer [m/z 382 (M^+-60, 100%), 255 (15%)], stigmasterol acetate [m/z 394 (M^+-60, 100%), 255 (78%)], sitosterol acetate [m/z 396 (M^+-60, 100%), 255 (41%), 213 (26%)] and isofucosterol acetate [m/z 394 (M^+-60, 30%), 296 ((100%), 253 (15%)] as major sterols. Figure 6A shows a mass chromatogram obtained from control shoots. Figure 6B shows a SIM profile for quantitation of sterol acetates of control shoots. The total amounts of sterols in shoots treated with *S*- and *R*-uniconazoles were reduced to 93% and 86%, respectively. The apparent de-

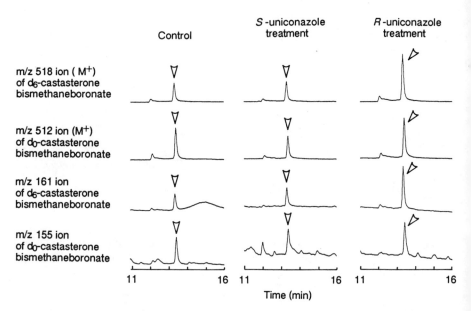

Fig. 5. GC–SIM quantitation of castasterone bismethaneboronate from HPLC fractions 17 and 18 of *P. sativum* shoots treated with uniconazoles

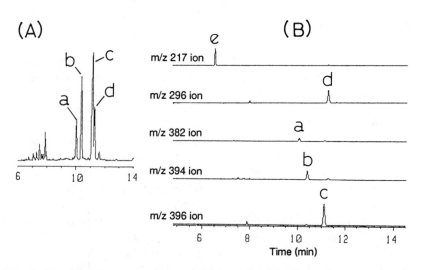

Fig. 6. Mass chromatogram (A) and SIM profile (B) of sterol acetates fraction derived from control shoots of *P. sativum*. a, 24-Methylcholesterol acetate; b, stigmasterol acetate; c, sitosterol acetate; d, isofucosterol acetate; e, 5α-cholestane (internal standard)

crease in the content of isofucosterol was observed in shoots treated with S- and R-uniconazoles (to a greater extent in the latter). Drastic elimination of isofucosterol has been observed in cultured cells of celery treated with a triazole, paclobutrazol.[19,20] The content of sitosterol was also decreased in shoots treated with R-uniconazole, but that of stigmasterol and campesterol was not. This is rather different from the data obtained for the celery cells mentioned above, in which the stigmasterol:sitosterol ratio was significantly lowered. Our finding that C_{29} sterols (sitosterol and isofucosterol) were decreased in the treatment with uniconazoles indicates that the second methylation at C-28 in the side chain is partially inhibited. This effect has been observed in plant cells cultured with the non-triazole-type fungicides, triarimol and triparanol.[21] Inhibition of the first methylation at C-24 in the side chain has been reported in oat shoots fed with tetcyclacis.[10]

4 Discussion

Gibberellin A_1 is believed to be active per se and to be responsible for shoot elongation.[22] Since S-uniconazole exhibited much stronger retardation activity than R-uniconazole, it was expected that the level of GA_1 would be more affected by S-uniconazole. However, the levels of GA_1 and also of GA_{20} were less reduced by S-uniconazole. Reassessment of the data published by Izumi et al.[4] revealed a result analogous to ours; namely, in rice seedlings treated with racemic uniconazole,[4] the content of endogenous GA_1 was not decreased as much as expected from the retardation rate. In this connection, we should refer to the finding of Köller[23] that growth reduction elicited by four stereoisomers of triadimenols (triazole) was not correlated with their inhibitory activity toward *Cucurbita* kaurene oxidase.[3,24]

On the other hand, considerable reduction of castasterone content was observed in *Pisum* shoots treated with both antipodes, suggesting that the altered metabolism of BRs is likely to be involved in the mechanism of action of S-uniconazole. The content of castasterone, however, was again less reduced in the S-uniconazole treatment. Furthermore, the content of bulk sterols was also less reduced in the S-uniconazole treatment. However, the reduction of the sterol contents in S- and R-uniconazole treatment was considered to be rather marginal to affect the growth.

The inconsistency between the growth retardation and endogenous level of GAs, castasterone, and sterols suggests that the effect elicited by S-uniconazole cannot be ascribed only to the limited synthesis of GAs, BRs, and sterols. Possible involvement of cytokinins, ethylene, and polyamines has also been suggested in other plants.[6,7] However, side effects derived from the modified metabolism of these growth substances do not seem to account for the growth retardation effect of S-uniconazole. Other unknown actions of S-uniconazole could be hypothesized, such as a suppression of unknown growth substances, an antagonism to plant hormones, or an inhibition of their movement or localization.

References

1. Lürssen K. The use of inhibitors of gibberellin and sterol biosynthesis to probe hormone action. In: Hoad GV, Lenton JR, Jackson MB, et al. eds. Hormone action in plant development—a critical appraisal. London: Butterworths, 1987: pp. 133–144.
2. Köller W. Isomers of sterol synthesis inhibitors: Fungicidal effects and plant growth regulator activities. Pestic Sci. 1987; 18:129–147.
3. Lenton JR. Mode of action of triazole growth retardants and fungicides—a progress report. News Bull British Plant Growth Regul Group. 1987; 9:1–12.
4. Izumi K, Yamaguchi I, Wada A, et al. Effects of a new plant growth retardant (E)-1-(4-chlorophenyl)-4,4-dimethyl-2-(1,2,4-triazol-1-yl)-1-penten-3-ol (S-3307) on the growth and gibberellin content of rice plants. Plant Cell Physiol. 1984; 25:611–617.
5. Izumi K, Kamiya Y, Sakurai A, et al. Studies of sites of action of a new plant growth retardant (E)-1-(4-chlorophenyl)-4,4-dimethyl-2-(1,2,4-triazol-1-yl)-1-penten-3-ol (S-3307) and comparative effects of its stereoisomers in a cell-free system from *Cucurbita maxima*. Plant Cell Physiol. 1985; 26:821–827.
6. Izumi K, Nakagawa S, Kobayashi M, et al. Levels of IAA, cytokinins, ABA and ethylene in rice plants as affected by a gibberellin biosynthesis inhibitor, uniconazole-P. Plant Cell Physiol. 1988; 29:97–104.
7. Hofstra G, Krieg LC, Fletcher RA. Uniconazole reduces ethylene and 1-aminocyclopropane-1-carboxylic acid and increases spermine levels in mung bean seedlings. J Plant Growth Regul. 1987; 8:45–51.
8. Rademacher W, Jung J, Graebe JE, et al. On the mode of action of tetcyclacis and triazole growth retardants. In: Menhenett R, Lawrence DK, eds. Biochemical aspects of synthetic regulators. Monograph No. 11, British Plant Growth Regulator Group. 1984: pp. 1–11.
9. Takano H, Oguri Y, Kato T. Antifungal and plant growth regulatory activities of enantiomers of (E)-1-(2,4-dichlorophenyl)-4,4-dimethyl-2-(1,2,4-triazol-1-yl)-1-penten-3-ol (S3308L). J Pestic Sci. 1986; 11:373–378.
10. Burden RS, Cooke DT, White PJ, et al. Effects of the growth retardant tetcyclacis on the sterol composition of oat (*Avena sativa*). Plant Growth Regul. 1987; 5:207–217.
11. Burden RS, Clark T, Holloway PJ. Effects of sterol biosynthesis-inhibiting fungicides and plant growth regulators on the sterol composition of barley plants. Pestic Biochem Physiol. 1987; 27:289–300.
12. Endo K, Yamane H, Nakayama M, et al. Endogenous gibberellin in the respective shoots of tall and dwarf cultivars of *Phaseolus vulgaris* L. Plant Cell Physiol. 1989; 30:137–142.
13. Takatsuto S, Ikekawa N. Synthesis of deuterio-labelled brassinosteroids, [26,28-^{2}H$_6$]brassinolide, [26,28-^{2}H$_6$]castasterone, [26,28-^{2}H$_6$]typhasterol and [26,28-^{2}H$_6$]teasterone. Chem Pharm Bull. 1986; 34:4045–4049.
14. Yokota T, Kim SK, Fukui, et al. Brassinosteroids and sterols from green

alga, *Hydrodictyon reticulatum*: Configuration at C-24. Phytochemistry. 1987; 26:503–506.
15. Park KH, Yokota T, Sakuma A, et al. Occurrence of castasterone, brassinolide and methyl 4-chloro-3-acetate in immature *Vicia faba* seeds. Agric Biol Chem. 1987; 51:3081–3086.
16. Davies PJ, Emschwiller E, Gianfagna TJ, et al. The endogenous gibberellins of vegetative and reproductive tissue of G2 peas. Planta. 1982; 154:266–272.
17. Ingram TJ, Reid JB, Potts WC, et al. Internode length in *Pisum*. IV. The effect of the Le gene on gibberellin metabolism. Physiol Plant. 1983; 59:607–616.
18. Yokota T, Takahashi N. Chemistry, physiology and agricultural application of brassinolide and related steroids. In: Bopp M, ed. Plant growth substances 1985. Heidelberg: Springer-Verlag, 1986: pp. 129–138.
19. Haughan PA, Lenton JR, Goad LJ. Sterol requirements and paclobutrazol inhibition of a celery cell culture. Phytochemistry. 1988; 27:2491–2500.
20. Haughan PA, Burden RS, Lenton JR, et al. Inhibition of celery cell growth and sterol biosynthesis by the enantiomers of paclobutrazol. Phytochemistry. 1989; 28:781–789.
21. Hosokawa G, Patterson GW, Lusby WR. Effects of triarimol, tridemorph and triparanol on sterol biosynthesis in carrot, tobacco and soybean suspension cultures. Lipids. 19; 1984:449–456.
22. Phinney BO, Spray C. Chemical genetics and the gibberellin pathway in *Zea mays* L. In: Wareing PF, ed. Plant growth substances 1982. London: Academic Press, 1982: pp. 101–110.
23. Köller W. Plant growth regulator activities of stereochemical isomers of triadimenol. Physiol Plant. 1987; 71:309–315.
24. Burden RS, Carter GA, Clark T, et al. Comparative activity of the enatiomers of triadimenol and paclobutrazol as inhibitors of fungal growth and plant sterol and gibberellin biosynthesis. Pestic Sci. 1987; 21:253–267.

CHAPTER 34

Gibberellin Increases Cropping Efficiency in Sour Cherry (*Prunus cerasus* L.)

M.J. Bukovac* and E. Yuda

1 Introduction

There are three important physiologically based factors, namely, inadequate fruit set, excessive early cropping of young trees, and stress (virus) alteration of the flowering and growth pattern of mature trees, that contribute to low cropping efficiency in the sour cherry.[1] The gibberellins (GAs) enhance fruit set[2,3] and modify the flowering behavior of a wide range of plant species.[4,5] Either effective enhancement of fruit set or modification of flower expression, or both, under commercial conditions, could potentially increase production efficiency.

Thus, in this study we focused on the effects of GA_3 in overcoming these physiological limitations in the sour cherry (*Prunus cerasus* L. cv. Montmorency) and on developing cultural practices for increasing cropping efficiency.

2 Effect on Fruit Set

The effects of foliar sprays of GA_3(72 to 576 μM) on fruit set were evaluated on fruiting branches of mature, cropping sour cherry trees over several growing seasons. Treatments were made as single or multiple sprays from anthesis to about 1 week before "June drop."

There was no consistent, significant increase in fruit set although occasionally the number of fruits developing to maturity was 5 to 10% greater than for the controls. A striking secondary effect of the GA_3 treatment was the inhibition of flower formation observed on the treated branches the following spring, particularly at concentrations of 216 μM and greater.[1]

---

*I dedicate this contribution to Professor Eiji Yuda, who passed away on April 2, 1989. I have lost not only a valued colleague but also a close friend, and the scientific community an outstanding scientist and leader.

Table 1. Effect of GA_3 on induction of parthenocarpy in sour cherry[a]

Treatment	Fruit set		Fruit size		Soluble solids (%)
	Initial (%)	Final (%)	Diameter (mm)	Fresh wt (g)	
Open pollinated	36	30	1.87 ± 0.08	3.69 ± 0.41	13.1 ± 1.7
Emasculated	0	0	—	—	—
GA_3	61	9	1.47 ± 0.20	1.79 ± 0.40	11.9 ± 1.4
GA_3 + NAA	60	30	1.22 ± 0.34	1.65 ± 0.86	10.8 ± 1.8
GA_3 + NAA + BA	46	39	1.80 ± 0.94	3.09 ± 0.42	11.8 ± 0.7

[a] GA_3 (576 μM) alone and in combination with NAA (158 μM) and BA (222 μM) was applied (total of about 200 μl/flower) 6 times at weekly intervals to emasculated flowers. Initial fruit set was determined 2 weeks after treatment and final fruit set after "June drop." Size and quality measurements were made at maturity on a random sample of 20 fruits per treatment. Data represent means ± standard errors

No commercial application of GAs for increasing fruit set in sour cherry could be visualized since the fruit set response was minimal and inconsistent, and the concomitant strong inhibition of flowering resulted in a significant decrease in yield the following year.

Further studies on fruit set were focused on the effect of GA_3 on induction of parthenocarpic fruit growth in sour cherry, a crop in which several other growth regulators failed to induce parthenocarpy [e.g., Indole-3-acetic acid, (IAA), 2,4-dichlorophenoxyacetic acid, (2,4-D), 2(1-naphthyl)-acetic acid (NAA), 2-chlorophenoxyacetic acid, β-naphthoxyacetic acid; M.J. Bukovac, unpublished data]. The results of numerous experiments on the activity of GA_3 alone or in combination with auxins and cytokinin are typified by the data in Table 1. Six multiple applications of GA_3 induced strong initial set, but only 9% of the treated flowers developed into seedless fruit that persisted to maturity compared to 30% for the open pollinated control. The activity of GA_3 was increased by addition of NAA and NAA plus BA (Table 1). Neither NAA nor BA was effective alone (data not presented). The parthenocarpic fruit produced by GA_3 or GA_3 + NAA was significantly smaller than the open pollinated fruits, but when BA was added to the GA_3 + NAA treatment, parthenocarpic fruits approached the size (96% by diameter, 84% by weight) of seeded fruit. Soluble solids of parthenocarpic fruit were less than for the seeded controls (Table 1).

Gibberellin A_3 is unique among the plant hormones in being effective in inducing parthenocarpy in the sour cherry, although fruit set was low (9%) compared to open pollinated populations (20–30%). These observations led to an investigation of seed-produced GAs in which GA_{32} was isolated from immature sour cherry seed.[6] Recently, Dr. H. Matsui (personal communication), working in our laboratory, has shown that a partially purified

Fig. 1. Photographs illustrating excessive flowering (A) and GA_3 inhibition of flowering (B) in young "Montmorency" sour cherry trees. See text for description.

extract of immature sour cherry seed containing GA_{32} was significantly more effective than GA_3 in inducing parthenocarpy. Initial fruit set (3 weeks after treatment) was 33, 66, and 50% and, at maturity, 28, 3, and 39% for open pollinated control, GA_3, and GA_{32}, respectively. It should be pointed out that there may be other GAs present in the partially purified seed extract. We now have evidence (M.J. Bukovac, E. Yuda, T. Yokota and N. Takahashi, unpublished data) for the presence of a biologically active polar GA, in addition to GA_{32}, in the butanol fraction.

Interestingly, GA_{32} has only been isolated from two other higher plant species, *P. persica* (peach)[7] and *P. armeniaca* (apricot).[8] In a preliminary evaluation by Coombe,[8] a GA_{32} extract from apricot seed was more active than GA_3 in inducing parthenocarpy in apricot. However, no fruits persisted to maturity following a single application of the extract.

3 Inhibition of Flower Formation in Newly Planted Trees

In general, excessive flower and fruit formation early in the development of fruit trees leads to inhibition of vegetative growth and to a delay in development of the trees as economic cropping units.[9-11] Newly planted sour cherry trees, particularly if subjected to stress (e.g., light, droughty soils, weed competition, low nutrient levels), often initiate flower buds in the first growing season. The trees flower (Fig. 1A) and set fruit in the

Table 2. Effect of time of application and concentration of GA_3 on flower inhibition (%) in young "Montmorency" sour cherry trees[a]

Time of application (weeks after anthesis)	Concentration (μM)		
	72	144	288
2	20	60	80
3	60	80	95
4	60	85	90
6	30	70	85

[a] Data represent ratings compared to nontreated trees and are average values from three studies on 2- and 3-year-old trees

second season, and this fruit, if not removed, may significantly inhibit vegetative growth. Early cropping may further lower vigor, and this, in turn, enhances flower formation and excessive cropping, further aggravating this condition. Thus, inhibition of flower formation in young trees prevents early cropping and promotes vegetative growth, leading to more rapid development of an orchard as a productive unit.

Based on our initial observation that foliar application of GA_3 for enhancement of fruit set inhibited flower formation and, thus, fruiting the following season, we conducted a series of experiments over several years evaluating concentration (72 to 288 μM) and time of application (1 to 5 weeks after anthesis) for inhibition of flowering in young (1 to 3 years old) trees growing under conditions favoring early cropping. Trees were sprayed with appropriate GA_3 solutions [containing 0.1% Ortho X-77 (Chevron Chemical Co., Richmond, California) as a surfactant] until fully wetted at the designated times. The flowering response was evaluated the following spring by either determining the number of flowers produced per tree or by rating flower density in relation to the controls.

The data summarized in Table 2 are representative of our results. A single foliar spray at 72 to 288 μM concentration resulted in 20% to almost complete (95%) inhibition of flowering. Marked inhibition was achieved at concentrations of 144 and 288 μM when applied 3 to 4 weeks after anthesis, that is, prior to initiation of flower buds (estimated to commence 4 to 6 weeks after full bloom).[12] A newly planted sour cherry tree in its second growing season showing excessive flower formation and one in its third growing season treated with GA_3 during the previous 2 seasons (single annual spray, 288 μM 3 weeks after anthesis) showing complete inhibition of flowering are illustrated in Fig. 1.

In similar studies, Stang and Weidman[13] assessed the effect of GA_3 inhibition of flowering on the vegetative development of young sour cherry trees. They found an 18% increase in the trunk cross-sectional area after 3 seasons of GA_3 treatment at 144 μM concentration 2 and 4 weeks after anthesis.

Although not well documented, an additional benefit might be derived from prevention of flowering in young sour cherry trees with GA_3. Growth and fruiting of cropping sour cherry trees are frequently reduced by a serious, widespread pollen-transmitted virus complex known as cherry yellows (complex of necrotic ring spot and prune dwarf viruses).[14-16] By prevention of flowering, one may delay the initial virus infection. It is commonly accepted that well-established trees may tolerate the infection and exhibit less "shock" than young trees just becoming establishing. Further, GA_3 may also have a direct effect on suppressing virus expression. Hull and Klos[17] demonstrated that multiple sprays of GA_3 (6 applications of 2.9 mM) reversed the virus-induced inhibition of vegetative growth in sour cherry. Similarly, virus-caused stunting of sweet corn, China aster, and crimson clover was overcome by weekly applications of GA_3 at 288 μM concentration.[18]

4 Promotion of Spur Formation and Cropping Efficiency in Mature Trees

The cherry yellows virus complex is not only the most serious virus disease of sour cherry,[16] but is also present in essentially all mature, cropping trees, since it is pollen transmitted.[14,15] This virus complex expresses itself in the shock phase by causing chlorosis and premature leaf abscission, neither being lethal. After the shock phase, the virus causes suppression of vigor and excessive lateral flower bud formation on current season's shoots (Fig. 2). Since flower buds in sour cherry are "pure" (entire meristem utilized in development of the flower), the node subtending the flower bud becomes "blind" once flowering and fruiting are completed; that is, no vegetative or reproductive tissue develops from this node. Consequently, extensive regions on the branches become nonproductive, and new flower buds are produced primarily on current season's shoots (Fig. 2A), further contributing to the "blind wood syndrome."[1] Thus, infected trees not only decrease in vigor, but also in productivity and cropping efficiency.

With GA_3, one can inhibit flower formation in lateral buds on current season's shoots (Figs. 2B and 3) to varying degrees. Such vegetative buds develop into spurs bearing 1 to 5 flower buds (potentially 3 flowers per bud) in the season after treatment, and flower and fruit the following year.[1] Most important, spurs can remain productive for 3 to 5 years depending on cultural practices employed. Thus, theoretically one can cause functional spurs to form with GA_3, reducing the development of "blind" wood and markedly altering the flowering pattern induced by the virus. It should be pointed out that other factors, notably various stresses, that lead to low vigor can also increase formation of "blind" wood and often appear to magnify the expression of the cherry yellows virus complex.

Preliminary studies[1,19] established the potential of attenuating the effect

34. GA Increases Sour Cherry Cropping Efficiency 355

Fig. 2. Photographs depicting extensive flowering from lateral buds (A) and partial inhibition of flower initiation leading to spur development (B) on 1-year-old shoots of virus-infected "Montmorency" sour cherry.

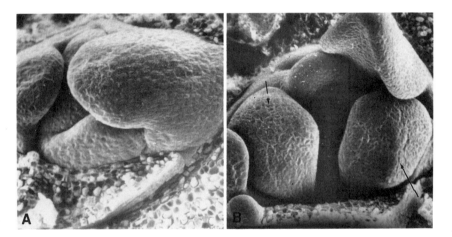

Fig. 3. Scanning electron micrographs of dissected lateral buds from current season's shoots of virus-infected "Montmorency" sour cherry, showing GA_3 inhibition of flowering (A) and flower primordia developing in a bud from a nontreated shoot (B, arrows).

of the virus with GA_3, but no widespread cultural practice emerged. Studies described below were undertaken to provide a more complete data base for development of a cultural practice to minize the effect of the virus, promote spur formation, and increase cropping efficiency of sour cherry.

The following bioassay was developed to establish the optimum concentration range, time of application, and biological activity of selected GAs. Mature, cropping sour cherry trees that were infected with the cherry yellows virus and showed "blind" wood symptoms were used. Branches were selected at the periphery of the tree with about 15 developing lateral and/or terminal shoots. About 3 weeks after anthesis, 10 shoots of comparable development (4 or 5 expanded leaves) were selected and marked, and the marked and adjacent branches were sprayed with the designated GA solution, fully wetting both surfaces of the leaves. Data were collected the following spring (just prior to anthesis) on the treated shoots that had developed the previous season (1 year old). Shoot extension, total number of nodes, and number of nodes with flower buds, vegetative buds, and lateral shoots were recorded. Usually, data were collected on 10 shoots per treatment and each treatment was replicated 10 times.

4.1 Effect of Concentration and Time of Application

Gibberellin A_3 was applied at 0, 43, 86, 129, and 172 μM at 3 weeks after anthesis. Flower initiation in lateral buds on current season's shoots was inversely and linearly related ($\hat{Y} = 94.6 - 0.47X$, $r^2 = 0.956$) to concentration (Fig. 4). The high flowering potential of the test trees is shown by the fact that 96% of the nodes on 1-year-old wood flowered.

The effect of time of application was established by treating with 72 μM GA_3 at 2, 3, 4, and 6 weeks after anthesis. This time interval was selected to bridge the period of flower initiation in "Montmorency" sour cherry.[12] Maximum inhibition of flower initiation occurred when GA was applied at

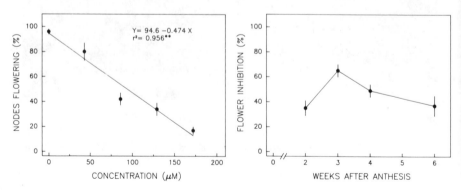

Fig. 4. Effect of concentration (left) and time of application (right) of GA_3 on flowering on 1-year-old shoots of virus-infected "Montmorency" sour cherry.

Table 3. Comparative biological activity of GA_1, GA_3, GA_4, GA_7, and GA_9 on inhibition of flower initiation of "Montmorency" sour cherry[a]

	% Nodes flowering					
Concentration (μM)	Control	GA_1	GA_3	GA_4	GA_7	GA_9
75	98a[b]	88abc	43d	87bc	80c	90abc
150	98a	72cd	23f	57de	54e	79bc

[a] Terminal shoots on virus-infected trees prone to excessive flower formation were sprayed 26 days after anthesis (4 or 5 expanded leaves present) and flowering nodes determined at anthesis the following spring
[b] Means within a row followed by similar letters are not significantly different at $P = 0.05$, Ducan multiple range test

3 weeks after anthesis (Fig. 4), a time when 4 to 5 leaves were unfolded and the 6th leaf was unfolding on current season's terminal and lateral shoots.

4.2 Comparative Biological Activity

The relative biological activity of GA_1, GA_3, GA_4, GA_7, and GA_9 was established at 75 and 150 μM concentrations (Table 3). Gibberellin A_3 had significantly greater activity in inhibiting flower initiation than GA_1, GA_4, GA_7, or GA_9. A similar order of activity was observed at both concentrations. Interestingly, there was no significant difference between GA_4 and GA_7 in inhibiting flower initiation in sour cherry, while GA_7 has been reported to be more active than GA_4 in the apple (cv. Cox's Orange Pippen).[20]

4.3 Orchard Studies—Effect on Shoot and Spur Development and Yield

Two long-term field studies were initiated to evaluate the feasibility of promoting spur development and thus increasing cropping efficiency of virus-infected trees with GA_3. Studies were established in two commercial orchards of mature sour cherry trees exhibiting a high incidence of virus symptoms ("blind" wood and excessive flowering of 1-year-old shoots) and testing positive for necrotic ring spot and/or prune dwarf viruses (ELISA test). The two studies differed in location, soil type, age of trees, and rates of GA_3 applied, but both addressed the objective of overcoming the virus limitations in sour cherry production. Since the results were similar, only one study will be described in detail.

Ten trees were assigned to each treatment, and each treatment was replicated 6 times. Gibberellin A_3 was applied about 3 weeks after anthesis as a high-volume foliar spray at 43 and 72 μM, concentrations projected from

Table 4. Effect of GA_3 on shoot and spur development of mature sour cherry ("Montmorency") trees infected with the cherry yellows virus complex[a]

Parameter	Concentration (μM)			L.S.D.[b] ($P = 0.05$)
	0	43	72	
Shoot length (cm)	19	20	20	NS
Total nodes (no.)	10	10	10	NS
Vegetative nodes (%)				
Lateral shoots	8	7	5	NS
Spurs	8	16	28	8
Flowering nodes (%)	84	77	67	6

[a] Trees were sprayed with GA_3 3 weeks after anthesis and data were collected the following season on 1-year-old shoots
[b] L.S.D., least significant difference

Table 5. Effect of GA_3 on sour cherry yield in the year after initial treatment and the mean annual yield for the following four consecutive years of treatment[a]

Time after treatment	Concentration (μM)		
	0	43	72
Year 1 (kg/tree)	27 ± 7[b]	31 ± 3	28 ± 5
Years 2–5 (mean annual yield)			
kg/tree	45 ± 8	56 ± 10	54 ± 10
% increase	0	24	20

[a] Mature virus-infected trees were sprayed annually with GA_3 at 3 weeks after anthesis. Year 1 represents the yield in the year following initial treatment, and years 2–5 the following 4 years during which cropping also occurred on spurs induced with GA_3
[b] Mean of 6 replications ± SE

the bioassays to induce 20 to 40% flower bud inhibition in current season's shoots. Gibberellin A_3 was applied annually for 5 growing seasons. Each year the following data were collected on 10 randomly selected 1-year-old shoots on each tree: length and number of nodes, number of vegetative and flowering nodes, and lateral shoots produced per shoot. Yield was determined by machine harvesting each treatment (10 trees) and measuring the total fruit produced.

Data presented in Table 4 represent the response obtained after the first year of treatment. Response in subsequent seasons was similar. Gibberellin A_3 had no effect on shoot extension or number of lateral shoots developing on 1-year-old wood (Table 4). The number of flowering nodes was significantly reduced, with a corresponding increase in the number of nodes producing vegetative buds developing into spurs.

The yield in the year following treatment was not significantly different from that of the control (Table 5). However, when GA$_3$ was applied annually, the mean annual yield over the remaining 4 years of the study was 45, 56, and 54 kg/tree for the control, 43 μM GA$_3$, and 72 μM GA$_3$, respectively, representing a 20 to 24% increase resulting from GA$_3$ treatment. One can visualize no treatment effect or even a decrease in cropping in the year following initial treatment if excessive flower inhibition is induced, but then yield and cropping efficiency should increase on continued treatment as new spurs are formed and commence fruiting (Table 5).

5 Conclusions

Gibberellin A$_3$ was not effective in increasing fruit set following foliar application, although direct repeated application to emasculated flowers induced low levels of parthenocarpy (9%).

Gibberellin A$_3$ applied at 144 to 288 μM 3 to 4 weeks after anthesis markedly inhibited flower formation in young sour cherry trees, leading to more rapid development of the tree.

Gibberellin A$_3$ overcame excessive lateral flower formation caused by the cherry yellows virus complex that leads to development of nonproductive branches and low cropping efficiency. Inhibition of flower formation in a portion of the lateral buds resulted in an increase in spurs bearing flower buds. Repeated annual applications of GA$_3$ (43 to 72 μM) increased spur density and cropping efficiency (about 20%) and suppressed the expression of the cherry yellows virus complex.

Acknowledgments. We thank Abbott Laboratories, North Chicago, Illinois, USA, for gifts of the GAs and financial support. These studies were supported inpart by the Michigan Agricultural Experiment Station, East Lansing, Michigan 48824, USA.

References

1. Bukovac MJ, Hull J, Kesner CD, et al. Prevention of flowering and promotion of spur formation with gibberellin increases cropping efficiency in 'Montmorency' sour cherry. Proc Mich State Hort Ann Rep. 1986; 116:122–131.
2. Crane JC, Primer PE, Campbell RC. Gibberellin induced parthenocarpy in *Prunus*. Proc Amer Soc Hort Sci. 1960; 75:129–137.
3. Webster AD, Goldwin GK. The effect of setting mixtures on the cropping, flowering and vegetative growth of sweet cherry. Acta Hort. 1984; 149:217–223.
4. Bradley MV, Crane JC. Gibberellin-induced inhibition of bud development in some species of *Prunus*. Science. 1960; 131:825–826.
5. Pharis RP, King RW. Gibberellins and reproductive development in seed plants. Ann Rev Plant Physiol, 1985; 36:517–568.
6. Bukovac MJ, Yuda E, Murofushi N, et al. Endogenous plant growth sub-

stances in developing fruit of *Prunus cerasus* L. VII. Isolation of gibberellin A_{32}. Plant Physiol. 1979; 63:129–132.
7. Yamaguchi I, Yokota T, Murofushi N, et al. Isolation and structure of a new gibberellin from immature seeds of *Prunus persica*. Agric Biol Chem. 1970; 34:1439–1441.
8. Coombe BG. GA_{32}: A polar gibberellin with high biological activity. Science. 1971; 172:856–857.
9. Bukovac MJ, Carpenter WS, Earl AR. Effects of excessive fruiting of young apple trees and a proposed chemical means of defruiting. Mich Agric Exp Sta Quart Bull. 1965; 47:364–372.
10. Proebsting EL Jr. A quantitative evaluation of the effect of fruiting on growth of Elberta peach trees. Proc Amer Soc Hort Sci. 1958; 71:103–109.
11. Wilcox JC. Field studies of apple tree growth and fruiting. II. Correlations between growth and fruiting. Sci Agric. 1937; 17:573–586.
12. Diaz DH, Rasmussen NP, Dennis FG Jr. Scanning electron microscope examination of flower bud differentiation in sour cherry. J Amer Soc Hort Sci. 1981; 106:513–515.
13. Stang EJ, Weidman RW. Economic benefit of GA_3 application on young tart cherry trees. HortScience. 1986; 21:78–79.
14. George JA, Davidson TR. Pollen transmission of necrotic ring spot and sour cherry yellows viruses from tree to tree. Can J Plant Sci 1963; 43:276–288.
15. Gilmer RM, Way RD. Pollen transmission of necrotic ring spot and prune dwarf viruses in sour cherry. Phytopathology. 1960; 50:624–625.
16. Keitt GW, Clayton CN. A destructive virus disease of sour cherry. Phytopathology. 1943; 33:449–468.
17. Hull J Jr, Klos EJ. Effect of gibberellin on virus-infected 'Montmorency' cherry trees. Phytopathol Mediterr. 1963; 2:205–208.
18. Maramorsch K. Reversal of virus-caused stunting in plants by gibberellic acid. Science. 1957; 126:551–552.
19. Basak W, Parker KG, Goodno RW, et al. Influence of gibberellin on bud differentiation of sour cherry and on damage by sour cherry yellows. Plant Dis Rep. 1962; 46:404–408.
20. Tromp J. Flower-bud formation in apple as affected by various gibberellins. J Hort Sci. 1982; 57:277–282.

CHAPTER 35

Prospects for Gibberellin Control of Vegetable Production in the Tropics

C.G. Kuo

1 Introduction

Vegetable production in the tropics is usually small-scale but labor-intensive. Many of the physiological processes and management regimes involved in vegetable production can be effected by gibberellins (GAs) in order to reduce production costs and to increase yield and its quality. The potential uses for GAs in vegetable production are summarized in Table 1. The principal purposes of employing GAs are to optimize vegetable production by modifying growth and development, to overcome abiotic stresses, and to enhance the quantitative and qualitative yield of vegetables.

Despite the diversity and importance of GAs within particular vegetable production systems, the extent to which virtually all of the uses described in Table 1 are commercially exploited is small in the temperate zones and essentially negligible in the tropics. Slow progress in introducing the use of GAs into actual practice is to be expected. The reasons for this are many, such as diversity in vegetable species, production systems, and consumption patterns, economic considerations, unreliability of effectiveness over seasons and physiological stages, and differences in varietal sensitivity. Furthermore, many of the GA-effected physiological processes of cool, dry season vegetables are readily overwhelmed by various abiotic stresses in the tropics.

In theory, GAs may be valuable either as alternatives or adjuncts to varietal improvement of vegetables. Gibberellins can achieve what breeding cannot achieve (e.g., overcoming a lack of flowering due to environmental limitations) or cannot achieve economically (e.g., hand-pollination of inbred lines). Furthermore, research on GAs may provide elucidation of certain problem areas, efficient selection criteria for breeding programs, and a practical guide for production systems. When coordinated with plant breeding in this way, GA physiology can support vegetable production with more realistic possibilities for the tropics. This report describes examples to illustrate these potential uses of GAs.

Table 1. Potential uses of GAs in vegetable production[38-42]

Physiological processes	Vegetables
1. Breaking dormancy & promoting germination	Celery, eggplant, lettuce, onion, potato, taro, tomato
2. Changing quality	Carrot, parsley, sweet potato
3. Controlling flowering	Carrot, crucifers, eggplant, lettuce
4. Enhancing fruit set and size	Cowpea, cucumber, eggplant, tomato
5. Increasing earliness	Artichoke
6. Improving seed production	Cabbage, cauliflower, carrot, celery, lettuce, onion
7. Promoting growth	Amaranth, azuki bean, celery, cucumber, edible chrysanthemum, lettuce, peppers, snapbean, soybean, soybean sprout, spinach, taro
8. Regulating sex expression	Cucurbits

2 Flowering and Seed Production of Chinese Cabbage

Most *Brassica* vegetables are quantitatively long-day plants and can be induced to flower by low temperature at 5 to 8 °C.[1-8] The effects of low temperature (vernalization) can be observed in the species of *B. campestris*, to which Chinese cabbage belongs, when moistened seeds or any ensuing growth states are chilled.[9,10] There are only slight photoperiodic fluctuations around 12 h and a small temperature flux in the tropics. Thus, the nullification or abating of vernalization requirement by Chinese cabbage, or alternatives to vernalization, should be an important factor in the control of flowering and seed production of Chinese cabbage in the tropics.

Gibberellin application replaces the requirement for long day or vernalization in some *Brassica* species.[11] It is effective in the induction and promotion of flowering of Chinese cabbage,* especially easy-to-flower varieties, when GAs are applied during seed imbibition or vegetative growth without vernalization.[10,13] There are cases in which GAs cause stem elongation without flowering in some difficult-to-flower varieties.[14] The endogenous GA content of vernalized Chinese cabbage seedlings also increases within the first few days under long-day conditions.[5] Gibberellins seem to play a role in the bolting of Chinese cabbage plants, but probably are not directly functional in initiating flowering.

Gibberellin application was proposed as an alternative to incomplete vernalization in order to bring about flowering in some difficult-to-flower *Brassica* species.[15,16] It is possible that breeding programs for *Brassica* species and their seed production may be enhanced by a combination of verna-

---

*References 8, 10, 12, and 13.

Table 2. Effect of GA_3 on days to flowering, percentage of plants flowering, and stem height of Chinese cabbage

Variety	GA_3			Control		
	Days to flowering[a]	Flowering[b] (%)	Stem ht[c] (cm)	Days to flowering[a]	Flowering[b] (%)	Stem ht[c] (cm)
B-6	—[d]	0	7 ± 2	—	0	3 ± 1
B-12	37 ± 1	87	28 ± 2	—	11	4 ± 0
B-14	—	0	6 ± 1	—	0	3 ± 0
B-31	34 ± 3	83	26 ± 3	—	28	4 ± 1
B-32	31 ± 1	100	23 ± 1	61 ± 4	78	5 ± 0
B-33	—	0	13 ± 2	—	0	3 ± 0
B-162	33 ± 2	100	25 ± 1	61 ± 5	50	4 ± 1
B-189	30 ± 0	100	26 ± 1	56 ± 1	50	5 ± 2
B-320	—	0	6 ± 2	—	0	2 ± 2

[a] Days counted from sowing
[b] Percentage of plants flowering at 70 days after sowing
[c] Measured at 55 days after sowing
[d] —, Did not reach 50% of plants flowering at 70 days after sowing

lization and GAs, thereby saving substantial time and energy. The combination of vernalization and GA application is appealing for the practical purpose of seed production; however, the effectiveness of the method depends on the crop variety. Most heat-tolerant (in terms of head formation at high temperature) Chinese cabbage varieties tend to flower easily[10]; therefore, the above method is applicable.[10,12]

For hybrid seed production of Chinese cabbage, on the other hand, the parent inbred must flower at or about the same time to maximize the pollen exchange. Applications of GAs or their inhibitors could aid the synchronization of flowering, which may otherwise occur at different times. The method of GA_3 spraying to hasten flowering of the late-flowering stock and Alar spraying to delay flowering is therefore employed.[10]

Furthermore, heat-tolerant Chinese cabbage varieties are known to be more sensitive to vernalization[10] and GA application (Table 2). This principle has been exploited as the screening technique for heat tolerance in segregating populations from breeding programs.[10,17] Here, GA application is superior to vernalization because of the time saved and because it is applicable in situations in which there is no access to cooling facilities.

3 Flowering of Sweet Potato

Most varieties of sweet potato, a short-day plant, are known to produce few or no flowers under natural photoperiodic conditions of the tropics. However, flowering can be induced by grafting sweet potato onto free-flowering varieties or onto other *Ipomoea* species.[18] Chemical induction of flowering, on the other hand, may provide an economic advantage over conventional grafting in breeding programs. There is evidence that GAs increase flowering in low-flowering sweet potato varieties.[19,20] However, $GA_{4/7}$, GA_5, or GA_7 was much more effective than GA_3.[20,21]

4 Fruit Set of Tomato at High Temperature

Gibberellins occur in reproductive organs of tomato.[22] Their levels fluctuate widely during development of the reproductive organ, but usually are reduced by high temperature.[23,24] They are required for the stamen development[25] and are able to reverse the detrimental effect of high temperature on petals, stamens, carpels, and locules.[26]

High temperature is clearly one of the factors in the tropics limiting fruit setting in tomato,[17] but can be partially overcome by GA applications.[22,27,28] Although GAs can bring about fruit set at high temperature, quality of fruits and predictability of response unfortunately cannot be guaranteed. Gibberellin A_3 has been shown to decrease the size of the developing fruit,[24,29,30] which can be attributed to reduced cell division.[30] In contrast, the largest increase in fruit size occurred with GA_5 and GA_7.[31] Gibberellin A_3 is also involved in the ripening process; it inhibits lycopene synthesis and delays chlorophyll degradation.[32] Differences in the location and timing of response and the hormonal stability are obviously major problems in the practical uses of GAs.

High temperatures limit fruit set because of a simultaneously and/or se-

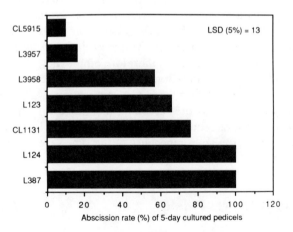

Fig. 1. Abscission rate (%) of excised pedicels without reproductive organs of heat-sensitive and tolerant tomatoes. Pedicels were obtained at anthesis and cultured in vitro for 5 days without growth regulators

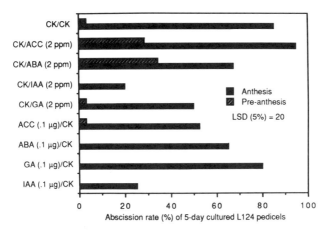

Fig. 2. Abscission rate (%) of excised pedicels without reproductive organs of a heat-sensitive tomato (L 124). Pedicels were obtained at 3 days before anthesis (pre-anthesis) and at anthesis. Abscission rates were measured 5 days after in vitro culturing with growth regulators applied basipetally/acropetally

quentially impaired series of reproductive processes such as pollen production and viability, ovule development, pollination, and pollen germination, and fertilization, resulting in floral or fruit abscission.[22] When reproductive organs were removed at anthesis and the excised pedicels were cultured in vitro, it was found that detached pedicels of heat-tolerant (normal fruit set at high temperature) tomato varieties hardly abscised after 5 days of culture (Fig. 1). Thus, the detached pedicel method provides an efficient screening method for heat tolerance in tomato. On the other hand, one of the functions of GAs in fruit set is to prevent pedicel abscission[33] by altering assimilate partitioning in favor of reproductive organs.[29] For a heat-sensitive tomato, GA_3 incorporated in the culture medium (acropetally) was able to prevent the abscission of the pedicels, isolated at anthesis, after 5 days of in vitro culturing (Fig. 2). Since there are varietal differences in the abscission of pedicels at high temperatures, which also reduce endogenous GAs in the reproductive organs, it may be that efforts of breeding programs for improving tomato fruit set at high temperatures might best be addressed by genetic manipulation of the GA responsiveness of specific tissue, such as the pedicel.

5 Accumulation of Assimilates by Storage Root of Sweet Potato

The total dry matter yield of the sweet potato storage root depends on photosynthetic activity of the leaf canopy source, the activity of the plant in translocating assimilates from the source to the storage-root sink, and the

Table 3. Effect of growth regulators on growth and storage-root yield of leaf cuttings of CN 1028-15

Growth regulator	Storage-root dry wt (g/cutting)	Leaf dry wt (g/cutting)	Total dry wt (g/cutting)	HI[a] (%)
CCC (3 mM)	4.01 bc[b]	0.65 a	5.32 ab	75 ab
GA$_3$ (3 mM)	3.03 cd	0.73 a	4.49 bc	67 b
BAS 106 (20 mg/liter)	4.38 ab	0.50 b	5.46 ab	80 a
Control	2.85 d	0.69 a	4.21 c	68 b

[a] HI: harvest index
[b] Mean separation within columns by Duncan multiple range test at 5% level

capacity of the storage root to accommodate or capture assimilates. With the use of rooted leaf cuttings, the relative importance and relation of these source–sink characteristics can be assessed. When 35 sweet potato clones that varied in dry matter production and storage-root yield were studied, a significant positive correlation was obtained between intact plants grown in the field and rooted leaf cuttings grown in the greenhouse.[34] The leaf cutting grafted with the standard source or sink has been exploited as the screening method for assessing the source potential and the sink strength.[35] The sink strength of the storage root has been shown to be a limiting factor under humid, tropical conditions.

Regarding the sink strength, there is some evidence that GAs may inhibit phloem unloading and sucrose accumulation in sink tissues.[36] In the sweet potato, soil drenching with GA$_3$ did not alter the ratio of shoot to storage root growth (Table 3). Instead, both CCC and BAS 106, inhibitors of GAs, increased total dry matter production and storage root yield. This suggests that endogenous GA affects the shoot/storage root ratio by reduction of the sink strength in the sweet potato. Gibberellin A$_3$ is known to decrease starch synthesis but increase exocellular polysaccharide synthesis of cultured sweet potato cells.[37] Thus, it would appear that the establishment of a strong sink in the storage root may be correlated with a reduced level of endogenous GAs. It remains to be seen whether there is a varietal difference in endogenous GAs in different genetic clones with various sink strengths as described earlier.

6 Conclusions

The range of practical applications of GAs available in vegetable production is clearly a broad one. Some of these techniques are already having a direct and significant impact, and their use is widespread in the temperate region. However, GA applications in vegetable production are still at the developmental stage in the tropics; the various problems related to their practical use remain to be solved. Time will be needed before the possible benefits of these types of research can be transferred to farmers' fields in the tropics.

In spite of the limitations, the examples illustrated here point out that the use of GAs can bring new opportunities to breeders and horticulturalists in removing or circumventing some of the barriers imposed by genetics and the abiotic stresses encountered in the tropics. The objectives of improving vegetable performance for the benefit of tropical farmers can be achieved sooner if both GA research and breeding programs are employed in combination or in a supplementary manner, rather than as alternatives.

References

1. Elers B, Wiebe HJ. Flower formation of Chinese cabbage. I. Response to vernalization and photoperiods. Sci Hort (Amst). 1984; 22:219–231.
2. Elers B, Wiebe HJ. Flower formation of Chinese cabbage. II. Antivernalization and short-day treatment. Sci Hort (Amst). 1984; 22:327–332.
3. Guttormsen G, Moe R. Effect of plant age and temperature on bolting in Chinese cabbage. Sci Hort (Amst). 1985; 25:217–224.
4. Guttormsen G, Moe R. Effect of day and night temperature at different stages of growth on bolting in Chinese cabbage. Sci Hort (Amst). 1985; 25:225–233.
5. Suge H, Takahashi H. The role of gibberellins in the stem elongation and flowering of Chinese cabbage, *Brassica campestris* var. *pekinensis* in their relation to vernalization and photoperiod. Rep Inst Agric Res Tohoku Univ. 1982; 22:15–34.
6. Thomas TH. Effects of decapitation, defoliation and stem girdling on axillary bud development in Brussels sprouts. Sci Hort (Amst). 1983; 20:45–51.
7. Wurr DCE, Akehurst JM, Thomas TH. A hypothesis to explain the relationship between low-temperature treatment, gibberellin activity, curd initiation and maturity of cauliflower. Sci Hort (Amst). 1981; 15:321–330.
8. Zee SY. The effect of gibberellic acid (GA_3) on plant growth before and after transplanting. Acta Hortic (The Hague). 1978; 72:185–190.
9. Kagawa A. Studies on the inheritance of flower inductive habits in *Brassica* crops. Res Bull Fac Agric Gifu Univ. 1971; 31:41–62.
10. Opeña RT, Kuo CG, Yoon JY. Breeding and seed production of Chinese cabbage in the tropics and subtropics. Shanhua: Asian Vegetable Research and Development Center, 1988.
11. Lang A. Physiology of flower initiation. In: Ruhland W, ed. Encyclopedia of Plant Physiology, Vol. 15, Part I. Berlin: Springer-Verlag, 1965: pp. 1380–1536.
12. Kahangi EM, Waithaka K. Flowering of cabbage and kale in Kenya as influenced by altitude and GA application. J Hort Sci. 1981; 56:185–188.
13. Suge H. Re-examination on the role of vernalization and photoperiod in the flowering of *Brassica* crops under controlled environment. Jpn J Breed. 1984; 34:171–180.
14. Amagasa T, Takahashi H, Suge H. Effects of vernalization and photoperiod on the flowering of *Brassica oleracea* var. *alboglabra*, *B. campestris* var. *chinensis* and their amphidiploid. Rep Inst Agric Res Tohoku Univ. 1987; 36:9–19.

15. Ali A, Machado VA. Use of gibberellic acid to hasten flowering in rutabaga. Can J Plant Sci. 1982; 62:823–826.
16. van Marrewijk NPA. Artificial cold treatment, gibberellin application and flowering response of kohlrabi (*Brassica oleracea* L. var. *gongylodes* L.). Sci Hort (Amst). 1976; 4:367–375.
17. Opeña RT, Kuo CG, Yoon JY. Breeding for stress tolerance under tropical conditions in tomato and heading Chinese cabbage. In: Chang WN, ed. Improved vegetable production in Asia. Taipei: Food and Fertilizer Technology Center Book Series, No. 36. 1987: pp. 88–109.
18. Sakamoto S, Takagi H, Kuo CG. *Ipomoea batatas* (L.) Lam. In: Westphal E, Jansen PCM, eds. Plant resources of south-east Asia. Wageningen: Pudoc, 1989: pp. 166–171.
19. Lardizabal RD, Thompson PG. Hydroponic culture, grafting, and growth regulators to increase flowering in sweet potato. HortScience. 1988; 23:993–995.
20. Suge H. Promotion of flowering in sweet potato by gibberellin A_3 and A_7. Jpn J Breed. 1977; 27:251–256.
21. Asian Vegetable Research and Development Center. Progress report for 1978. Shanhua, 1979.
22. Kuo CG, Chen HM, Chen HC. Plant hormones in tomato fruit-set and development at high temperatures. Taipei: Food and Fertilizer Technology Center Book Series, No. 34. 1986: pp. 53–70.
23. Kuo CG, Tsai CT. Alternation by high temperature of auxin and gibberellin concentrations in the floral buds, flowers, and young fruit of tomato. Hort Science. 1984; 19:870–872.
24. Kuo CG, Chen HM, Chen HC. Relationship between hormonal levels in pistils and tomato fruit set in hot and cool seasons. In: Green SK, ed. Tomato and pepper production in the tropics. Shanhua: Asian Vegetable Research and Development Center, 1989: pp. 138–149.
25. Sawhney VK. Gibberellins and fruit formation in tomato: A review. Sci Hort (Amst). 1984; 22:1–8.
26. Sawhney VK. The role of temperature and its relationship with gibberellic acid in the development of floral organs of tomato (*Lycopersicon esculentum*). Can J Bot. 1983; 61:1258–1265.
27. El-Abd SO, El-Beltagy AS, Hall MA. Physiological studies on flowering and fruit set in tomatoes. Acta Hort (The Hague). 1986; 190:389–396.
28. Satti SME, Oebker NF. Effects of benzyladenine and gibberellin ($GA_{4/7}$) on flowering and fruit set of tomato under high temperature. Acta Hort (The Hague). 1986; 190:347–354.
29. Bunger-Kibler S, Bangerth F. Relationship between cell number, cell size and fruit size of seeded fruits of tomato (*Lycopersicon esculentum* Mill.), and those induced parthenocarpically by the application of plant growth regulators. Plant Growth Regul. 1983; 1:143–154.
30. Sjut V, Bangerth F. Induced parthenocarpy—a way of changing the levels of endogenous hormones in tomato fruits (*Lycopersicon esculentum* Mill.) 1. Extractable hormones. Plant Growth Regul. 1983; 1:243–251.
31. Wittwer SH, Bukovac MJ. Quantitative and qualitative differences in plant response to the gibberellins. Amer J Bot. 1962; 49:524–529.
32. Khudairi AK. The ripening of tomatoes. A molecular ecological approach to the physiology of ripening. Amer Scientist. 1972; 60:696–707.

33. Bangerth F, Sjut V. Induced parthenocarpy—a tool for investigating hormone-regulated physiological processes in fruits. Acta Hort (The Hague). 1978; 80:169–174.
34. Asian Vegetable Research and Development Center. Progress report 1985. Shanhua, 1987.
35. Asian Vegetable Research and Development Center. Progress report 1986. Shanhua, 1988.
36. Thomas TH. Hormonal control of assimilate movement and compartmentation. In: Bopp M, ed. Plant growth substances. Berlin: Springer-Verlag, 1985: pp. 350–359.
37. Sasaki T, Kainuma K. Control of starch and exocellular polysaccharides biosynthesis by gibberellic acid with cells of sweet potato cultured in vitro. Plant Cell Rep. 1984; 3:23–26.
38. Anonymous. Vegetable cultivation in China. Beijing: Agriculture Publications, 1987.
39. George RAT. Vegetable seed production. London: Longman, 1985.
40. Nishi S. Applications for vegetables. In: The miracle of plant hormones. Tokyo: Kyowa Hakko, 1980: pp. 271–330.
41. Thomas TH. Vegetable crops. In: Thomas TH, ed. Plant growth regulator potential and practice. Section II. Croydon: BCPC Publications, 1982: pp. 109–122.
42. Thomas TH. Plant growth regulators in the production and storage of outdoor and glasshouse vegetables. In: Menhenett R, Jackson MB, eds. Growth regulators in horticulture. Monograph 13, British Plant Growth Regulator Group. 1985: pp. 29–42.

CHAPTER 36

Gibberellin-Induced Flowering and Morphological Changes in Taro Plants

N. Katsura, K. Takayanagi, T. Sato, T. Nishijima, and H. Yamaji

1 Introduction

Gibberellins (GAs) are the only known substances that induce flower formation in numerous plants under strictly noninductive conditions.[1]

In general, Araceae plants show at best very small photoperiodic responses for flower induction.[2] Some of them show juvenility or size dependency for flowering. For example, *Amorphophallus* plants, grown from new cormels, do not flower. About 4 years in the vegetative phase are required before flowering can occur.[3] In these Araceae plants, GA application stimulates flowering markedly.[3-6] Itoh[3] found that these responses occur after one application of GA, often at levels exceeding 100 ppm.

Tropical countries and Japan produce taro plants *Colocasia esculenta* Schott (including dasheens and eddoes). They are cultivated widely in these areas, and many cultivars with various morphological characteristics have been maintained by vegetative propagation. There is no cultivar bred by artificial cross-breeding. One of the reasons is the difficulty of flowering of these cultivars. Therefore, improving the induction of flowering of taro plants would increase the possibility of cross-breeding. As mentioned above, GA application can induce flowering in taro plants. However, the mechanism of GA action on the flowering in these plants is not clear yet. Thus, we have examined the effect of GAs on flowering and morphogenesis of taro plants.

2 Materials and Methods

2.1 Plant Materials

Taro plants were used as materials. Secondary cormels were planted in our experimental field between May and June every year except in specially noted cases. After each treatment in the field, they were grown until October. During summer, flower formation (strictly, inflorescence formation)

and other morphological changes were observed. In October, the corms and cormels were harvested.

In our first experiment, we used 13 cultivars and studied the different responses among them after GA treatment. For analyzing GA action on morphogenesis, cv. Eguimo was used. For the comparison of activity between various kinds of GAs, cv. Ishikawawase was used because the plants have round-shaped cormels relative to those of cv. Eguimo.

2.2 Treatment

In all our experiments, a solution of GA_3 was injected close to the meristem of each plant with a syringe. Water containing 0.1% acetone was injected into the control group. The other GA solutions were applied in the same way. The concentrations of GA used are given in Tables 1 and 3–6. A 500-ml solution of uniconazole, an inhibitor of GA biosynthesis,[7] was applied to each plant with soil drench 10 days after planting.

3 Results

Many cultivars of taro plants sporadically flower in Japan under natural conditions. Seven lines out of a total of fifty cultivars (including lines) flowered.[5] The flowered lines belonged to two of the eight cultivar groups used. These results remained unchanged during our 3-year study. A single application of GA_3 stimulated flower formation in these cultivars. Almost all cultivars showed some notable response to GA_3 (Table 1). About 40 days after the treatment, tillering first started at the basal parts of each petiole. Thereafter, a flower stalk with spadix, covered with spathe, emerged at about 60–70 days after the treatment. In some cultivars such as Tounoimo which flowers reluctantly,[8] flower stalks without spadix (Fig. 1) preceded true inflorescences (Fig. 2). In cv. Takenokoimo, which flowers eventually,[9] abnormal flowers were produced, characterized by branched or doubled spathe. These results suggest, generally, that the more readily a cultivar flowers, the easier it is for GAs to induce flowering.

Next, the action of GA_3 on the morphogenesis of taro plants was investigated using cv. Eguimo, a naturally flowering cultivar.[5] In fact, this cultivar flowered without any treatment in this experiment. Table 2 shows the flowering response of the plants from corms and cormels of cv. Eguimo. Only plants originating from the corms flowered. This result suggests that the previously reported flowering of this cultivar[5] was caused by contamination by corms in seed cormels.

As mentioned above, GA application induced flowering (Table 3). The treatment produced simultaneously several morphological changes. Gibberellin A_3 induced tillering in nontillering cultivars (Table 3). A corm of taro plants is made at the basal part of plants with one meristem and sever-

Table 1. Flowering induction by GA_3 in Japanese taro cultivars

		70 days after GA			90 days after GA		
Group of cultivars	Cultivar	Veg. (%)	Inter. (%)	Flower. (%)	Veg. (%)	Inter. (%)	Flower. (%)
Eguimo	Eguimo-3	87	0	13	43	0	57
Ishikawawase	Ishikawawase	74	13	13	25	0	75
Dotare	Aichiwase	100	0	0	57	0	43
	Sansyuu	57	0	43	29	0	71
	Dotare	86	0	14	86	0	14
Akame	Daikichi	50	33	17	50	0	50
Tounoimo	Ebiimo-39	14	86	0	14	43	43
	Ebiimo-41	24	63	13	24	13	63
	Ebiimo-42	40	60	0	40	20	40
	Tounoimo	12	75	13	12	25	63
Yatsugashira	Yatsugashira	100	0	0	100	0	0
Takenokoimo	Takenokoimo	33	50	17	17	0	83
Migashiki	Migashiki	75	25	0	33	0	67

Thirteen cultivars were planted in pots and grown in a greenhouse. Nineteen days after planting, GA_3 solution (0.1 ml of 1000 ppm GA_3 in 0.1% acetone) was applied close to a meristem of each plant with a syringe. Six to eight plants were used except for Migashiki and Yatsugashira (five plants were used). Seventy and 90 days after GA_3 treatment, flowering was observed from outside. Values in columns "Veg." and "Flower." are percentages of vegetative plants and flowered plants, respectively. Values in column "Inter" are percentages of plants that produced only flower stalk without spadix. Plants without GA_3 applications did not make any flower buds, judging from microscopic observation

Fig. 1. A spadixless flower induced by GA_3 application in taro plants (cv. Tounoimo). A solution of GA_3 (1000 ppm) was injected as described in the footnote to Table 1

Fig. 2. An inflorescence of taro plants

Table 2. Effect of seed corms on flowering of taro plants (cv. Eguimo)

Material	Flowering (%)	Number of buds
Corm	37.5	(flowered) 1
		(vegetative) 1
1st cormels	0	1
2nd cormels	0	1

Table 3. Effect of uniconazole and GA_3 on flowering and morphogenesis of taro plants (cv. Eguimo)

Treatment		Flowering (%)	Max. petiole length (cm)	Number of buds on a corm	Shape index of a cormel[c]
Uni[a]	GA[b] (ppm)				
−	0	0	82.0 ± 3.4	1.0	3.2 ± 0.6
−	1000	80	66.6 ± 8.9	8.6 ± 3.2	7.2 ± 1.6
+	0	0	40.4 ± 3.0	1.0	1.3 ± 0.2
+	30	0	42.7 ± 3.5	1.0	1.3 ± 0.2
+	100	0	49.5 ± 6.3	4.5 ± 2.6	1.7 ± 0.5
+	300	0	56.5 ± 4.4	3.4 ± 2.6	1.9 ± 0.5
+	1000	20	43.6 ± 4.2	8.8 ± 3.8	2.9 ± 1.2

[a] 500 ml of 50-ppm uniconazole solution for each plant was applied to the soil at the day of planting
[b] 0.1 ml of GA_3 solution was injected as close as possible to an apical meristem
[c] Shape index is the length/diameter ratio of the first cormels. The largest five cormels taken from each corm were measured. Six to ten plants were used for each treatment

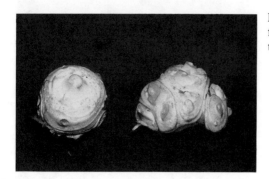

Fig. 3. Gibberellin A_3-induced fused corm. *Left*, Control; *right*, treated with 1000 ppm of GA_3

al lateral buds. These lateral buds were transformed into cormels. The tillering induced by GA_3 generated fused corms that had many apical buds (Fig. 3). The shapes of these fused corms approximated those of corms of cv. Yatsuga-shira, a naturally tillering cultivar. It should be noted that flowered plants derived from corms without GA_3 also showed tillering before the elongation of the flower stalk. However, they never produced such Yatsugashira-like corms.

On the other hand, GA_3 failed to increase the length of the petiole. However, application of uniconazole inhibited the growth of petioles and the inhibition was partially reversed by GA_3. Increased elongation was observed in cormels. Not all cormels were elongated by GA_3. The bigger ones showed more significant elongation. The shape of secondary cormels that were formed on cormels attached on corms did not respond to these chemicals. Therefore, GA_3 applied at the early time of the growth of lateral buds of corms is thought to be effective in inducing excess elongation of cormels. Uniconazole suppressed the elongation and produced round-shaped cormels. Gibberellin A_3 reversed the trend partially.

These results show that GA_3 stimulates not only flowering, but also accelerates vegetative growth, including the elongation of cormels. However, GA_4 and GA_7 induced only flowering. However, flowering was accompanied by tillering and the subsequent fusing of corms (Table 4). The action of GA_1 was the same as that of GA_3 (Table 5). It is conceivable, therefore, that flowering and elongation in taro plants depend on different modes of action of GAs.

4 Discussion

Traditionally, studies concerning the GA action on flowering has focused on plants requiring long day and/or low temperature. These conditions enable a "switching on" of the flowering under defined conditions. Gibberellin A_3 application on the Araceae flower group has been known to

Table 4. Effect of various GAs on flowering and morphogenesis of taro plants (cv. Ishikawawase)

Treatment	Flowering response				No. of tillers	Shape index of cormels
	Veg.	Tiller.	Inter.	Flower		
Control	5	0	0	0	1.0	1.3 ± 0.2
GA_3						
10 ppm	3	2	0	0	1.5	1.2 ± 0.2
100 ppm	0	1	0	4	3.4	1.8 ± 0.5
300 ppm	1	3	1	0	3.3	2.0 ± 0.6
GA_4						
10 ppm	3	2	0	0	1.8	1.3 ± 0.2
100 ppm	0	5	0	0	2.4	1.3 ± 0.2
300 ppm	2	2	0	1	2.2	1.2 ± 0.2
GA_7						
10 ppm	2	3	0	0	1.8	1.2 ± 0.2
100 ppm	0	4	0	1	2.6	1.3 ± 0.2
300 ppm	0	2	0	3	3.5	1.3 ± 0.2

Each concentration of GAs was injected close to a meristem of taro plants (cv. Ishikawawase). Five plants were used for each treatment. Veg. means plants under complete vegetative stage with an apical meristem. Tiller. means plants showed emergence of tillers. Inter. means plants having spadixless inflorescence. Flower means flowered plants. Shape index of cormels is the length/diameter ratio of cormels. Largest 5 cormels taken from each corm were measured, and the average for 5 plants is shown

Table 5. Effect of GAs on flowering and morphogenesis of taro plants (cv. Ishikawawase)

Treatment	Flowering (%)	Days to appearance of inflorescence	Number of buds in a corm	Shape index of a cormel
Control	0	—	1.0	1.6 ± 0.3
GA_1	80	69 ± 6	4.8 ± 1.0	2.1 ± 0.9
GA_3	60	70 ± 5	4.7 ± 0.6	2.0 ± 0.1
GA_7	60	66 ± 1	2.8 ± 0.5	1.6 ± 0.3

Gibberellins (500 ppm) were injected with a syringe as close as possible to an apical meristem; 0.1 ml of GA solution was used for each plant. Ten plants were used for each treatment. Days to appearance of inflorescence from the time of GA treatment were shown

stimulate or induce flowering.[3,10-13] However, the mechanism of action of GAs involved remains unclear. Many Japanese cultivars of taro plants used in the present study did not flower like those of African varieties[13] and Malaysian varieties.[14] Treatment with GA induced flowering in these plants; under the same conditions, there was no flowering response in the control group. In the cv Eguimo, plants originating from corms showed a spontaneous flowering response without exogenous GA. Gibberellin A_3 produced no difference in flowering rate when cormels of various weights were used (data not shown). These results suggest that flowering in the cv. Eguimo is affected by the physiological stage of the corm's meristem. The GA treatment first induced tillering and then heavy fused corms. Plants without GA treatment showed tillering only when corms were used as seeds, and fused corms failed to form in them. Therefore, GA treatment accelerated growth of tillers and caused the subsequent fusion of corms.

Although GA_3 failed to promote elongation of petioles and leaf blades significantly, it succeeded in reversing the retardation of petioles caused by the application of uniconazole. Moreover, it stimulated the longitudinal growth of cormels. Interestingly, GA_3 and GA_1, which contain a 13-hydroxyl group, stimulated both flowering and elongation of cormels. However, GA_4 and GA_7, which do not contain a 13-hydroxyl group, stimulated only flowering.

These results suggest that GAs play important roles in the morphogenesis of taro plants, especially flowering. This flowering mechanism remains unexplained. However, our findings increase the possibility of cross-breeding through a new system. Such a system could generate a larger amount of information on the GA action on flower induction, because we can switch on at will a flowering process by GA treatment under noninductive environments or at noninductive stages of the plant's growth. In turn, a greater volume of information can provide more insights into the flowering process and, thus, the cross-breeding of the Araceae plants.

5 Conclusions

In taro plants, flowering was induced when 0.1 ml of GA_3 solution (500–1000 ppm) was applied close to the meristem with a syringe. Almost all cultivars responded to the treatment. The treatment first induced tillering and provided corms that had many apical buds, as seen in the cv. Yatsugashira without GA_3 treatment. Corms produced under each tiller fused, and budding continued. Untreated plants never showed such phenomena. In the cv. Eguimo, an easily flowering cultivar, plants with flowers did not produce such corms under natural conditions. The budding on the corms, therefore, was thought to be a result of GA_3 treatment itself, not a result of flowering. The GA_3 treatment did not promote the elongation of petioles. However, the elongation growth of corms was stimulated by GA_3. Treatment with uniconazole, an inhibitor of GA biosynthesis, inhibited the elongation of both organs, and the elongation was recovered upon GA treatment. These results suggest that the endogenous GAs play some role in the growth of these organs. Gibberellins A_1 and A_3 induced the flowering of taro plants and the elongation of cormels simultaneously, whereas GA_4 and GA_7 stimulated only flowering. From these results, it is concluded that GAs play important roles in the morphogenesis of taro plants, especially in flowering through some still unidentified modes of action.

References

1. Zeevaart JAD. Phytohormones and flower formation. In: Letham, D.S., Goodwin, P.B., Higgins, T.J.V. eds. Phytohormones and related compounds —a comprehensive treatise Vol. 2. Elsevier/North-Holland Biomedical Press Amsterdam, 1978: pp. 29-325.
2. McDavid CR, Alamu S. Effect of daylength and gibberellic acid on the growth and promotion of flowering in tannia (*Xanthosoma sagittifolium*). Trop Agric (Trinidad). 1979; 56:17-23.
3. Itoh H. Studies on the utilization of growth regulators to horticulture (VIII). On the effect of floral induction by gibberellin treatment upon cormlets of *Amorphophallus konjac* K. Kock. Bull Fac Educ Kobe Univ. 1960; 24:65-72 (in Japanese).
4. Alamu S, McDavid CR. Promotion of flowering in edible aroids by gibberellic acid. Trop Agric (Trinidad). 1978; 55:81-86.
5. Katsura N, Takayanagi K, Sato T. Gibberellic acid induced flowering in cultivars of Japanese taro. J Jpn Soc Hort Sci. 1986; 55:69-74.
6. Miyazaki S, Tashiro Y, Kanazawa K, et al. Promotion of flowering by the treatment of seed corms and young plants with gibberellic acid in taro plants (*Colocasia esculenta* Schott). Jpn Soc Hort Sci. 1986; 54:450-466 (in Japanese).
7. Izumi K, Yamaguchi I, Wada A, et al. Effect of a new plant growth retardant (E)-1-(4-chlorophenyl)-4,4-dimethyl-2-(1,2,4-triazol-1-yl)-1-penten-3-ol (S-3307) on the growth and gibberellin content of rice plants. Plant Cell Physiol. 1984; 25:611-617.
8. Matsumoto T, Nakao Y. Promotion of flowering by gibberellin treatment in taro plants. Agriculture and Horticulture. 1984; 59:351-352 (in Japanese).
9. Odawara C. Nishimura K. Hitaka Y. Studies on the breeding of "taro". Kyusyu Agric Res. 1965; 27:231 (in Japanese).
10. McDavid CR, Alamu S. Promotion of flowering in tannia (*Xan-thosoma sagittifolium*) by gibberellic acid. Trop Agric (Trinidad). 1976; 53:373-374.
11. Harbaugh BK, Wilfret GT. Gibberellic acid (GA_3) stimulates flowering in *Caladium hortulanum* Birdsey. HortScience. 1979; 14:72-73.
12. Henny RJ. Gibberellic acid (GA_3) induces flowering in *Dief-fenbachia maculata* 'Perfection'. HortScience. 1980; 15:613.
13. Wilson JE. Effects of formulation and the method of applying gibberellic acid on flower promotion in cocoyam. Exp Agric. 1981; 17:317-322.
14. Gahni FD. Preliminary studies on flowering in *Colocasia esculenta* cultivars in Malaysia. Previsional report No. 11, Regional Meeting of Edible Aroid. International foundation for Science, Stockholm, 1981: pp. 336-339.

CHAPTER 37

Antheridiogens of Schizaeaceous Ferns: Structures, Biological Activities, and Biosynthesis

H. Yamane

1 Introduction

In the life cycle of ferns, germinated spores develop to prothallia, on which archegonia and anteridia are formed. Fertilization of eggs in the archegonia with sperms from the antheridia results in the production of young sporophytes. In 1950, Döpp[1] demonstrated that mature *Pteridium aquilinum* prothallia produced hormonal substance(s) inducing antheridial formation in many species of the ferns Polypodiaceae. Since then, more than ten fern species have been reported[2-4] to produce such antheridium-inducing substances, which were designated as antheridiogens.[5] Though it has been shown,[6] based on cross-testing of biological activities, that there are several different types of fern antheridiogens, only four have been characterized.[7-11] All the antheridiogens characterized were derived from schizaeaceous ferns, and they are all gibberellin (GA)-related compounds. The purpose of this article is to review chemical structures and biological activities of the antheridiogens and to discuss their biosynthetic relationships.

2 Chemcial Structures of Antheridiogens

2.1 Antheridic Acid from *Anemia phyllitidis* and the Other Three *Anemia* Species

A major antheridiogen in *A. phyllitidis*, tentatively named A_{An}, was isolated in a yield of 18 mg from 57 liters of culture medium of 57-day-old prothallia and characterized as **1a**, based on spectroscopic analyses.[7] In 1985, Corey and Myers[12] synthesized racemic compounds **1a** and **1b** (Fig. 1) by total synthesis and found that compound **1a** exhibited a proton nuclear magnetic resonance (PMR) spectrum different from that of naturally derived A_{An}. On the other hand, the PMR spectrum of synthetic compound **1b** was identical with that of A_{An}. This constitutes strong evidence for the

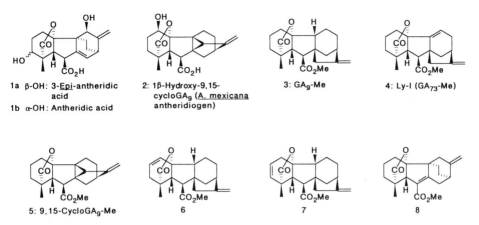

Fig. 1. Chemical structures of antheridiogens and related compounds

assignment of structure **1b** to A_{An}, although the final proof could not be obtained because of the lack of natural A_{An}. Recently, however, natural A_{An} was reisolated, and direct comparison of physicochemical properties and biological activities of natural A_{An} with those of synthetic compound **1b** was performed, with the result that structure **1b** can be unambiguously assigned to A_{An}. It was proposed that A_{An} be called antheridic acid, and the parent hydrocarbon antheridane.[8] Furber and Mander[13] succeeded in converting GA_7 into antheridic acid. Since the optical rotatory dispersion curve of antheridic acid synthesized from GA_7 corresponded to that of the natural material, the absolute configuration of the carbon skeleton of natural antheridic acid was shown to be same as that of *ent*-gibberellane. Antheridic acid was also shown to be the major antheridiogen in *A. hirsuta*,[14] *A. rotundifolia*, and *A. flexuosa*.[4]

2.2 *Anemia mexicana* Antheridiogen

A major antheridiogen in *A. mexicana* was isolated by preparative thin-layer chromatography (TLC) of an ethyl acetate extract of the acidified medium from 3-month-old cultures of the prothallia. The antheridiogen was subjected to combined gas chromatography–mass spectrometry (GC–MS), high-resolution MS, and combined GC–Fourier transform infrared spectrometry (GC–FTIR). The spectral data suggested that the antheridiogen was a monohydroxy-C_{19}-GA-like compound, such as GA_5, GA_7, GA_{31}, and GA_{62}, although it did not correspond to any known GA.[15] However, accumulation of more *A. mexicana* antheridiogen enabled its PMR spectrum to be measured. In the PMR spectrum, slight but unusual upfield shifts of exocyclic methylene resonances were revealed, whereas no signal due to an olefinic proton was observed, suggesting the presence of one more additional ring system in the molecule. Based on these spectral data, we proposed a putative carbon skeleton, *ent*-9,15-cyclogibberellane, for the *A. mexicana* antheridiogen.

Further characterization of the *A. mexicana* antheridiogen was largely due to synthetic work by Furber et al.[11] The chemical shifts of exocyclic methylene in the PMR spectrum of compound **5**, synthesized as a model compound, were very close to those of the *A. mexicana* antheridiogen. However, in the PMR spectrum of the *A. mexicana* antheridiogen, downfield shifts of C-5 and C-15 methine protons were observed. These downfield shifts and a signal due to the carbinyl proton at δ 4.12 could be explained when a hydroxyl is located at the C-1 β position.

Based on the above speculation, 1β-hydroxy-9,15-cyclo-GA_9 (**2**) was synthesized from GA_7 as a plausible candidate for the *A. mexicana* antheridiogen. The synthetic compound **2** and the natural *A. mexicana* antheridiogen gave identical PMR spectra, and their methyl ester trimethylsilyl derivatives afforded identical mass spectra and Kovats retention indices. Thus, the *A. mexicana* antheridiogen was unambiguously formulated as **2**.[11]

2.3 Methyl Esters of GA_9 and GA_{73} from *Lygodium japonicum*

Though *Lygodium japonicum* belongs to the same family (Schizaeaceae) as the genus *Anemia*, the chemical properties of the *L. japonicum* antheridiogen were considered to be partly different from those of the *A. phyllitidis* antheridiogen on the basis of cross-testing of biological activities; that is, the *Lygodium* antheridiogen never induced antheridial formation in *A. phyllitidis*,[16] although antheridic acid and GAs induced antheridial formation in *L. japonicum*.[17,18] Besides, the *L. japonicum* antheridiogen was thought to be neutral. However, we found that the methyl esters of some GAs (GAs-Me) showed high activity in inducing antheridial formation in *L. japonicum*. Their activities were one to two orders of magnitude greater than those of the corresponding free acids, although GAs-Me were generally inactive in most of the bioassay systems using intact higher plants.[19] It was also noted that in silica-gel (Si-gel) TLC the *L. japonicum* antheridiogen migrated very similarly to GA_9-Me. In a preparative GC, the retention time of the antheridiogen was almost identical to that of GA_9-Me. Gibberellin A_9-Me was tentatively identified by combined GC–selected ion monitoring (GC–SIM) analysis of the bioactive fraction purified by Si-gel TLC of a neutral ethyl acetate fraction from the culture medium.[9] Later, the definitive identification of GA_9-Me was carried out by full-scan GC–MS,[20] the concentration of GA_9-Me in the culture medium being 7.5–22.5 ng/liter, thus demonstrating the occurrence of GA_9-Me as an antheridiogen of *L. japonicum*. However, the total activity in the culture medium could not be attributed to GA_9-Me alone, and the presence of some other, more active, substance, tentatively named Ly-I, was strongly suggested.[9]

The neutral ethyl acetate fraction from the culture medium of 6-week-old prothallia of L. japonicum was prepurified using a Sep-Pak [octadecylsilane (ODS)] cartridge and a Si-gel short column and was then subjected to Si-gel-high-performance liquid chromatography (HPLC) followed by ODS-HPLC. Gibberellin A_9-Me was separated from Ly-I by the Si-gel-HPLC. The bioactive fraction from the ODS-HPLC was analyzed by capillary GC–MS, and the mass spectrum obtained for Ly-I suggested that Ly-I was a didehydro-GA_9-Me-like compound. An intensive peak at M^+-44 due to the loss of γ-lactone as CO_2 indicated a double bond (or a double-bond-equivalent functional group) in an α, β or a β, γ relationship to the C-10 carbon. Considering the occurrence of antheridic acid[4,8,14] and the A. mexicana antheridiogen[11] in the related Anemia species, the co-occurrence of GA_9-Me in L. japonicum, and the mass spectrum of Ly-I, the series of compounds 4–8 (Fig. 1) were synthesized as the Ly-I candidates. Ly-I was compared directly with the synthesized compounds 4–8 by capillary GC–MS, resulting in the unambiguous identification of Ly-I as 9,11-didehydro-GA_9-Me (4).[10] Since compound 4 is a new GA, a new GA number, GA_{73}, was assigned to the corresponding free acid. The concentration of Ly-I (GA_{73}-Me) in the culture medium was estimated to be approximately 3–4 ng/liter. Since GA_{73}-Me was highly active in inducing antheridial formation in L. japonicum, as described in Section 3 below, most antheridium-inducing activity of the culture medium can be considered to be derived from GA_{73}-Me.

3 Biological Activities

In A. phyllitidis, natural antheridic acid showed significant activity in the induction of antheridial formation at 8.6×10^{-9} M and spore germination in darkness at 8.6×10^{-10} M, as shown in Tables 1 and 2. Since the level of activity of ($\pm$)-antheridic acid was about half that of natural antheridic

Table 1. Effects of natural antheridic acid, ($\pm$)-antheridic acid, and ($\pm$)-3-epi-antheridic acid on induction of antheridial formation in light-grown prothallia in A. phyllitidis.[21] Reprinted with permission from Phytochemistry 26, K. Takeno et al., "Biological activity of antheridic acid, an antheridiogen of Anemia phylliditis," Copyright 1987, Pergamon Press PLC.

Sample	Antheridial formation (%) ± SE at concentration (M):				
	2.9×10^{-9}	8.6×10^{-9}	2.9×10^{-8}	8.6×10^{-8}	2.9×10^{-7}
Natural antheridic acid	0 ± 0	21 ± 5	90 ± 2	96 ± 2	93 ± 3
($\pm$)-Antheridic acid	0 ± 0	0 ± 0	72 ± 6	92 ± 1	98 ± 1
($\pm$)-3-epi-Antheridic acid	0 ± 0	0 ± 0	0 ± 0	38 ± 2	95 ± 0
Control		0 ± 0			

Table 2. Effects of natural antheridic acid, (±)-antheridic acid, and (±)-3-*epi*-antheridic acid on induction of spore germination in darkness in *A. phyllitidis*.[21] Reprinted with permission from Phytochemistry 26, K. Takeno et al., "Biological activity of antheridic acid, an antheridiogen of *Anemia phyllitidis*," Copyright 1987, Pergamon Press PLC.

Sample	Germination (%) ± SE at concentration (M):					
	8.6×10^{-10}	2.9×10^{-9}	8.6×10^{-9}	2.9×10^{-8}	8.6×10^{-8}	2.9×10^{-7}
Natural antheridic acid	3 ± 1	17 ± 2	58 ± 6	90 ± 1	92 ± 1	86 ± 2
(±)-Antheridic acid	0 ± 0	1 ± 0	15 ± 1	66 ± 1	88 ± 2	94 ± 2
(±)-3-*epi*-Antheridic acid	0 ± 0	0 ± 0	1 ± 0	38 ± 3	85 ± 2	91 ± 3
Control			0 ± 0			

Table 3. Effects of (±)-antheridic acid, (±)-3-*epi*-antheridic acid, GA_{73}, GA_9, 9,15-cyclo-GA_9, and GA_3 on elongation of the second leaf sheath in *Oryza sativa* L. cv. Tan-ginbozu[21,23]

Sample	Mean of the second-leaf-sheath lengths (% of control) ± SE at dosage (ng/plant):					
	0.1	1.0	10	100	1000	Control[a]
(±)-Antheridic acid	99 ± 3	100 ± 3	92 ± 4	108 ± 4	134 ± 6	a
(±)-3-*epi*-Antheridic acid	101 ± 9	103 ± 7	111 ± 5	139 ± 2	259 ± 10	a
GA_{73}	—	120 ± 1	151 ± 11	272 ± 5	328 ± 11	b
GA_9	—	125 ± 3	153 ± 9	246 ± 13	311 ± 12	b
9, 15-Cyclo-GA_9	—	122 ± 4	160 ± 4	274 ± 11	305 ± 3	b
GA_3	131 ± 7	191 ± 4	280 ± 8	357 ± 8	350 ± 10	a

The assay was done according to the microdrop method
[a] a: 15.7 ± 0.6 mm, b: 14.1 ± 0.6 mm

acid, the mirror image of natural antheridic acid is likely to be inactive. (±)-3-*epi*-Antheridic acid was approximately one order of magnitude less active than (±)-antheridic acid in both bioassays.

In most bioassays using intact plants, 3α-hydroxy-GAs were almost inactive, while 3β-hydroxy-GAs such as GA_1, GA_3, GA_4, and GA_7 were highly active.[22] Therefore, it is quite interesting that (±)-antheridic acid is more active than (±)-3-*epi*-antheridic acid in assays of induction of both antheridial formation and spore germination in darkness in *A. phyllitidis*, although in the dwarf rice assay, (±)-3-*epi*-antheridic acid was one order of magnitude more active than (±)-antheridic acid, as shown in Table 3. It should also be noted that 3-*epi*-GA_4 (3α-hydroxy-GA_9) was less active than GA_4 (3β-hydroxy-GA_9) in both antheridium-inducing and dark spore germination assays in *A. phyllitidis* (K. Takeno, personal communication).

In *L. japonicum*, GA_{73}-Me induced antheridial formation in dark-grown protonemata at 10^{-14} M, induced spore germination in darkness at 10^{-11} M, and inhibited archegonial formation at 10^{-11} M, as shown in Tables 4, 5, and 6, respectively. Gibberellin A_9-Me was two to four orders of magnitude less active than GA_{73}-Me in the above three assays. It is noteworthy that antheridic acid and its methyl ester were less active than their 3-epimers in the antheridium-inducing assay in *L. japonicum* (Table 4). This result is in marked contrast to the relationship of antheridic acid and its 3-epimer in the antheridium-inducing assay in *A. phyllitidis*. The *A. mexicana* antheridiogen and its methyl ester exhibited almost the same activities in the antheridium-inducing assay in *L. japonicum*. The activity of compound **5** was one order of magnitude or more greater than those of *A. mexicana* antheridiogen and its methyl ester (Table 4).

Details of biological activities of the *A. mexicana* antheridiogen in the

Table 4. Effects of GA_{73}-Me, GA_9-Me, related compounds on antheridial formation in dark-grown protonemata in *L. japonicum*[9,23]

Sample	Antheridial formation (%) ± SE								
	Concentration (M)								
	10^{-15}	10^{-14}	10^{-13}	10^{-12}	10^{-11}	10^{-10}	10^{-9}	10^{-8}	10^{-7}	10^{-6}
GA_{73}-Me	0 ± 0	32 ± 5	98 ± 1	100 ± 0	—	—	—	—	—	—
GA_9-Me	—	—	—	—	—	5 ± 3	60 ± 12	99 ± 0	99 ± 0	100 ± 0
9,15-Cyclo-GA_9	—	—	—	—	—	—	32 ± 8	96 ± 2	—	—
1β-Hydroxy-9,15-cyclo-GA_9	—	0 ± 0	—	—	—	—	0 ± 0	21 ± 2	99 ± 0	—
1β-Hydroxy-9,15-cyclo-GA_9-Me	—	0 ± 0	—	—	—	—	0 ± 0	16 ± 4	82 ± 2	—

Sample	Concentration (M)				
	2.9×10^{-9}	2.9×10^{-8}	2.9×10^{-7}	2.9×10^{-6}
(±)-Antheridic acid	—	0 ± 0	4 ± 2	92 ± 1
(±)-Antheridic acid-Me	—	0 ± 0	4 ± 1	87 ± 2
(±)-3-*epi*-Antheridic acid	0 ± 0	73 ± 9	97 ± 2	—
(±)-3-*epi*-Antheridic acid-Me	0 ± 0	0 ± 0	62 ± 12	—
Control		0 ± 0		

Table 5. Effects of GA_{73}-Me and GA_9-Me on induction of spore germination in darkness in *L. japonicum*[23,24]

Sample	% Spore germination ± SE at concentration (M):							
	10^{-13}	10^{-12}	10^{-11}	10^{-10}	10^{-9}	10^{-8}	10^{-7}	10^{-6}
GA_{73}-Me	0 ± 0	0 ± 0	12 ± 2	70 ± 2	—	—	—	—
GA_9-Me	—	—	—	0 ± 0	2 ± 2	3 ± 1	53 ± 1	89 ± 2
Control				0 ± 0				

Table 6. Effects of GA_{73}-Me and GA_9-Me on inhibition of archegonial formation in light-grown prothallia in *L. japonicum*[9,23]

Smaple	% Inhibition at concentration (M):						Control[a]
	10^{-13}	10^{-12}	10^{-11}	10^{-10}	10^{-9}	10^{-8}	
GA_{73}-Me	5	16	61*	79*	—	—	a
GA_9-Me	—	—	—	2	36*	73*	b

*Significant inhibition.
[a] Archegonial formation ± SE in the control was 82 ± 3% (a) and 85 ± 5% (b).

induction of antheridial formation and spore germination in darkness in *A. mexicana* have not been reported.

In the dwarf rice assay, GA_9, GA_{73}, and 9,15-cyclo-GA_9 all showed almost the same level of activity (Table 3). The dwarf rice does not seem to discriminate between GA_9 and GA_{73} (9,11-didehydro-GA_9), although introduction of the 9,11-double bond into GA_9-Me caused a drastic increase in antheridium-inducing activity in *L. japonicum*, as described above. The activity of 9,15-cyclo-GA_9 in the dwarf rice assay could be explained by the following possibilities: (1) 9,15-cyclo-GA_9 is converted to an active GA, (2) 9,15-cyclo-GA_9 is converted to an active *ent*-9,15-cyclogibberellane compound, or (3) 9,15-cyclo-GA_9 is active per se.

4 Biosynthesis

Given the occurrence of three skeletal types, antheridane, *ent*-9,15-cyclogibberellane, and *ent*-gibberellane in the schizaeaceous ferns, their biosynthesis is considered to be closely related. An *ent*-9,15-cyclogibberellane derivative could be a precursor of an antheridane derivative and/or an *ent*-gibberellane derivative; alternatively, a C-9 cationic intermediate could be a common precursor of three classes of antheridiogens, as shown in Fig. 2.

The following information has been obtained concerning the biosyn-

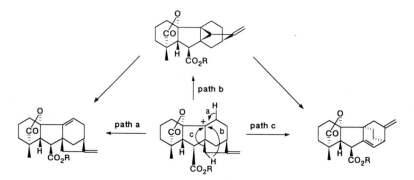

Fig. 2. Speculations on biosynthetic relationships of antheridiogens in the schizaeaceous ferns

thesis and metabolism of the antheridiogens. (1) It was confirmed by GC–MS that 17-d_2-9,15-cyclo-GA$_9$ was converted to 17-d_2-antheridic acid by *A. phyllitidis* prothallia (T. Yamauchi, H. Yamane, L.N. Mander, et al., unpublished results). (2) In *L. japonicum* prothallia, exogenously applied GA$_9$ was converted to GA$_9$-Me[9] and then hydroxylated at C-12α, C-12β, and C-13 to give GA$_{70}$-Me, GA$_{69}$-Me, and GA$_{20}$-Me,[25] respectively. 12-Hydroxylation of GAs-Me by *L. japonicum* prothallia has also been noted in GA$_4$-Me,[26] GA$_{12}$-Me, and GA$_{14}$-Me.[27]

5 Conclusions

Three skeletal types (antheridane, *ent*-9,15-cyclogibberellane, and *ent*-gibberellane) of compounds have been characterized as antheridiogens in the schizaeaceous ferns. They are assumed to constitute a new group of compounds with a closely related biogenetic origin.

The physiological roles of the antheridiogens are discussed in the following chapter by Takeno.

Acknowledgments. The author is grateful to Professor N. Takahashi (Department of Agricultural Chemistry, The University of Tokyo) and Professor L.N. Mander (Research School of Chemistry, Australian National University) for critical reading of the manuscript and to Dr. K. Takeno (Laboratory of Horticultural Science, Tohoku University) for supplying unpublished results.

References

1. Döpp W. Eine die Antheridienbildung bei Farnen fördernde Substanz in den Prothallien von *Pteridium aquilinum* L. (Kuhn). Ber Dtsch Bot Ges. 1950; 63:139–147.

2. Näf U, Nakanishi K, Endo M, On the physiology and chemistry of fern antheridiogens. Bot Rev. 1975; 41:315–359.
3. Emigh V, Farrar DR. Gemmae: A role in sexual reproduction in the fern genus *Vittaria*. Science. 1977; 198:297–298.
4. Yamane H, Nohara K, Takahashi N, et al. Identification of antheridic acid as an antheridiogen in *Anemia rotundifolia* and *Anemia flexuosa*. Plant Cell Physiol. 1987; 28:1203–1207.
5. Näf U. On dark-germination and antheridium formation in *Anemia phyllitidis*. Physiol Plant. 1966; 19:1079–1088.
6. Näf U. On separation and identity of fern antheridiogens. Plant Cell Physiol. 1968; 9:27–33.
7. Nakanishi K, Endo M, Näf U, et al. Structure of the antheridium-inducing factor of the fern *Anemia phyllitidis*. J Amer Chem Soc. 1971; 93:5579–5581.
8. Corey EJ, Myers AG, Takahashi N, et al. Constitution of antheridium-inducing factor of *Anemia phyllitidis*. Tetrahedron Lett. 1986; 27:5083–5084.
9. Yamane H, Takahashi N, Takeno K, et al. Identification of gibberellin A_9 methyl ester as a natural substance regulating formation of reproductive organs in *Lygodium japonicum*. Planta. 1979; 147:251–256.
10. Yamane H, Sato Y, Nohara K, et al. The methyl ester of a new gibberellin, GA_{73}: The principal antheridiogen in *Lygodium japonicum*. Tetrahedron Lett. 1988; 29:3959–3962.
11. Furber M, Mander LN, Nester JE, et al. Structure of a novel antheridiogen from the fern *Anemia mexicana*. Phytochemistry. 1989; 28:63–66.
12. Corey EJ, Myers AG. Total synthesis of ($\pm$)-antheridium-inducing factor (A_{An}, 2) of the fern *Anemia phyllitidis*. Clarification of stereochemistry. J Amer Chem Soc. 1985; 107:5574–5576.
13. Furber M, Mander LN. Conversion of gibberellin A_7 into antheridic acid, the antheridium inducing factor from the fern *Anemia phyllitidis*: A new protocol for controlled 1,2-bond shifts. J Amer Chem Soc. 1987; 109:6389–6396.
14. Zanno PR, Endo M, Nakanishi K, et al. On the structural diversity of fern antheridiogens. Naturwissenschaften. 1972; 59:512.
15. Nester JE, Veysey S, Coolbaugh RC. Partial characterization of an antheridiogen of *Anemia mexicana*: Comparison with the antheridiogen of *A. phyllitidis*. Planta. 1987; 170:26–33.
16. Näf U. On the control of antheridium formation in the fern species *Lygodium japonicum*. Proc Soc Exp Biol Med. 1960; 105:82–86.
17. Näf U. Control of antheridium formation in the fern species *Anemia phyllitidis*. Nature. 1959; 184:798–800.
18. Schraudolf H. Die Wirkung von Phytohormonen auf Keimung und Entwicklung von Farnprothallien. I. Auslösung der Antheridienwirkung und Dunkelkeimung bei Schizaeaceen durch Gibberellinsäure. Biol Zentralbl. 1962; 81:731–740.
19. Hiraga K, Yamane H, Takahashi N. Biological activity of some synthetic gibberellin glucosyl esters. Phytochemistry. 1974; 13:2371–2376.
20. Satoh Y. Studies on plant growth regulators controlling the life cycle of pteridophytes. Ph.D. thesis, The University of Tokyo, 1984.
21. Takeno K, Yamane H, Nohara K, et al. Biological activity of antheridic acid, an antheridiogen of *Anemia phyllitidis*. Phytochemistry. 1987; 26:1855–1857.
22. Brian PW, Grove JF, Mulholland TPC. Relationships between structure and

growth-promoting activity of gibberellins and some allied compounds, in four test systems. Phytochemistry. 1967; 6:1475–1499.
23. Takeno K, Yamane H, Yamauchi T, et al. Biological activities of the methyl ester of gibberellin A_{73}, a novel and principal antheridiogen in *Lygodium japonicum*. Plant Cell Physiol. 1989; 30:201–205.
24. Sugai M, Nakamura K, Yamane H, et al. Effects of gibberellins and their methyl esters on dark spore germination and antheridium formation in *Lygodium japonicum* and *Anemia phyllitidis*. Plant Cell Physiol. 1987; 28:199–202.
25. Sato Y, Yamane H, Kobayashi M, et al. Metabolism of GA_9 methyl ester in prothallia of *Lygodium japonicum*. Agric Biol Chem. 1985; 49:255–258.
26. Yamane H, Yamaguchi I, Kobayashi M, et al. Identification of ten gibberellins from sporophytes of the tree fern, *Cyathea australis*. Plant Physiol. 1985; 78:899–903.
27. Murofushi N, Nakayama M, Takahashi N, et al. 12-Hydroxylation of gibberellins A_{12} and A_{14} by prothallia of *Lygodium japonicum* and identification of a new gibberellin, GA_{74}. Agric Biol Chem. 1988; 52:1825–1828.

CHAPTER 38

Antheridiogen, Gibberellin, and the Control of Sex Differentiation in Gametophytes of the Fern *Lygodium japonicum*

K. Takeno

1 Introduction

Pteridophytes can be classified into two groups in terms of spore morphology, the heterosporous and the homosporous ferns. The sex of the gametophytes of heterosporous ferns is genetically determined.[1] The gametophytes that develop from macrospores produce archegonia, and those from the microspores produce antheridia. Intergametophytic matings always occur in these heterosporous ferns. On the other hand, homosporous ferns produce only one kind of spore, and all gametophytes from these spores are potentially bisexual; that is, they can produce both antheridia and archegonia. In natural and laboratory conditions, both unisexual female and unisexual male gametophytes are often observed in a single population of the gametophytes.[1]

Antheridial differentiation in many homosporous fern species is known to be induced by antheridiogen.[2] However, the factors controlling archegonial differentiation are not well known. The aim of this article is to document the control of sex differentiation by antheridiogens in a homosporous schizaeaceous fern, *Lygodium japonicum*.

2 Sex Differentiation of Gametophytes in *Lygodium japonicum*

The gametophytes of *Lygodium japonicum* (Thunb.) Sw. were aseptically cultured on 1/10-strength simplified Murashige and Skoog[3] mineral salts solution solidified with 0.3% agar in a petri dish (3 cm in diameter) at 25 °C under continuous white light. The gametophytes cultured under such conditions have the potential to produce reproductive organs of both sexes (Fig. 1). In a population, however, most of the large gametophytes are female and small ones male, bisexual gametophytes being few in number[4] (Fig. 2). It is apparent that the sex of the gametophyte is correlated with the width of the gametophyte, that is, growth stage of the gametophyte. The minimal width of female gametophytes is around 1.5 mm, regardless

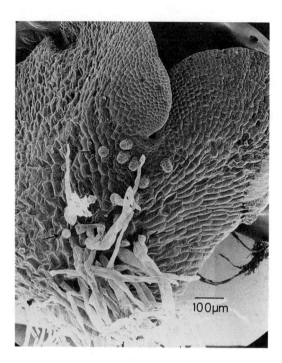

Fig. 1. Archegonia (circled with broken line) and antheridia (arrow) formed on the ventral side of gametophyte of *L. japonicum*

of culture conditions: culture period, photoperiod, temperature, and concentration of sucrose added to the medium (Table 1). If the gametophytes are cut into pieces, each regenerated fragment forms archegonia after reaching a width of about 1.5 mm.[4] These facts indicate that archegonial formation depends on the stage of growth of the gametophyte. Some physiological changes must occur to trigger the differentiation of archegonia in the gametophyte at this critical growth stage.

3 Induction of Antheridia

Sex differentiation of the gametophyte is not genetically fixed in *L. japonicum*. In fact, every gametophyte can form antheridia if it is treated with the *Lygodium* antheridiogens, methyl esters of gibberellins A_9 and A_{73} (GA_9-Me and GA_{73}-Me)[5,6] (see the preceding chapter by Yamane), or with other gibberellins (GAs),[7] especially as their methyl esters.[5]

4 Inhibition of Archegonia

It is difficult to study the factors controlling archegonial differentiation since a population of gametophytes consists of different growth stages. To avoid this problem, uniformly sized gametophytes were selected at an early

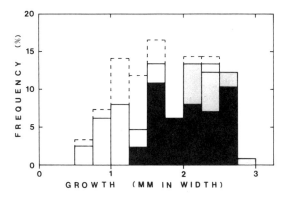

Fig. 2. The correlation between sex and growth in gametophytes of *L. japonicum* cultured under continuous light at 25 °C for 3 weeks. Frequency distribution of width of gametophyte and percentages of female (closed column), male (open column), and bisexual (stippled column) gametophytes and those with no reproductive organs (broken-lined column) belonging to each class of width are indicated (Modified from Takeno and Furuya[4])

Table 1. Minimal size of female gametophytes in populations of *L. japonicum* cultured under different conditions (From Takeno and Furuya[4])

Minimal size (width of gametophyte in mm)	Number of cases[a]
<1.00	0
1.00–1.25	3
1.25–1.50	10
1.50–1.75	14
1.75–2.00	9
2.00–2.25	2
>2.25	0

Gametophytes were grown under continuous light at 25 °C for 1, 2, 3, 4, 5, or 6 weeks, under continuous light at 30 °C for 6 weeks, under photoperiods of 8, 16, or 24 h at 25 °C for 4 or 6 weeks, or under continuous light at 25 °C on media supplemented with 0, 1, or 5% sucrose. Two or three replicated dishes in each condition were examined
[a] Width of smallest female gametophytes was measured in 38 populations

stage of growth and grown at low population density. In such a system, gametophytes grow with good synchrony, and the gametophytes form archegonia when they reach 1.5 mm in width 18 days after spore inoculation.

When this selected population was used as an experimental system, it was found that GAs inhibited archegonial differentiation.[8] The inhibition by GA_3 of archegonial differentiation was not the result of growth inhibition, although the growth was suppressed to some extent at high concentrations of GA_3. Inhibition of archegonial differentiation by GA_3 was observed when the gametophytes were treated on day 14 after spore inoculation or earlier, that is, 4 or more days before archegonial primordia became detectable. The inhibitory effect was nullified when the gametophytes were transplanted onto GA-free medium; they then formed archegonia 6 days after transplantation, regardless of the length of treatment with GA_3.

Filtrates of medium on which L. japonicum gametophytes were cultured as well as organic solvent extracts from the gametophytes inhibit archegonial differentiation in the same manner as does GA_3.[9] The natural inhibitors of archegonial differentiation were found to be identical with the Lygodium antheridiogens and identified as GA_9-Me[5] and GA_{73}-Me.[6] It was found that GA_9-Me and GA_{73}-Me had the dual function of inducing antheridial formation and inhibiting archegonial differentiation.[5,10]

5 Production of Antheridiogen and Sensitivity of the Gametophyte to Antheridiogen

The amount of antheridiogen secreted into the medium by the gametophytes of different ages was measured by bioassay in terms of both antheridial-inducing activity and archegonial-inhibiting activity.[4] The antheridial-inducing activity became detectable on day 14 after spore inoculation, and the archegonial-inhibiting activity, on day 16 (Fig. 3). The difference in the amount of antheridiogens detected as antheridial-inducing activity and as archegonial- inhibiting activity is probably due to the difference in the sensitivity of the two bioassays used. The Lygodium anther-

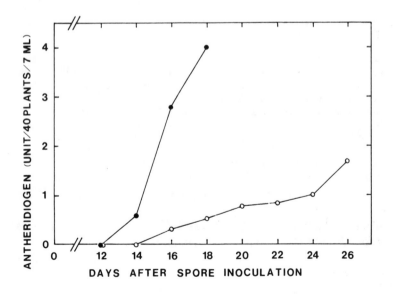

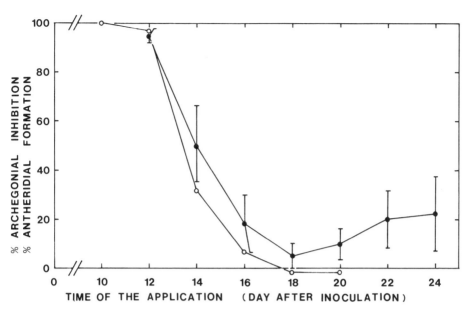

Fig. 4. Response to antheridiogens of different-aged gametophytes in *L. japonicum*. Ten to 24-day-old gametophytes were transplanted onto media with a methanol extract from 3-week-old gametophytes at 33 (for archegonial-inhibition assay; ○) or 3.3 (for antheridial-induction assay; ●) mg fr wt of tissue equiv/ml (Modified from Takeno and Furuya[4])

idiogen apparently induces antheridia at a lower concentration than that required to inhibit archegonia.[10] Even with a higher population density of donor gametophytes, no antheridiogen activity was detected in the medium before day 14.

Almost all the gametophytes produced antheridia and archegonial differentiation was almost completely inhibited when the gametophytes were treated with antheridiogens on day 12 after spore inoculation or earlier. However, if they were treated at day 14 or later, the antheridial-inducing effect and archegonial-inhibiting effect became weaker[4] (Fig. 4).

The results described above indicate that gametophyte sensitivity to the

←

Fig. 3. Time courses of the secretion of antheridiogens by *L. japonicum* gametophytes. The spores were inoculated in a dish containing 7 ml of agar medium, and forty uniformly grown gametophytes were selected and cultured for 12 to 26 days. The conditioned media prepared in this way were assayed at serial dilutions for archegonial-inhibiting (○) and antheridial-inducing activities (●). The amount of antheridiogen which induced antheridia or inhibited archegonia in 50% of assay gametophytes was defined as 1 unit (Modified from Takeno and Furuya[4])

antheridiogens is lost before the gametophyte secretes effective amounts of antheridiogen.

6 Model of Sex Differentiation

The above result means that only younger gametophytes, coexisting with older ones, can respond to the antheridiogens in terms of inhibition of archegonial initiation and formation of antheridia. Thus, the antheridiogens work as pheromones.

Based on this information, sex differentiation observed in a population can be explained as follows.[4] When gametophytes in a population are at different stages of growth, those that first reach the critical size form archegonia. Thereafter, they begin to produce antheridiogens. However, these larger gametophytes, which are differentiating archegonia, are not sensitive to the antheridiogens. As a result, the larger gametophytes are unisexual female only. On the other hand, retarded gametophytes in the population are sensitive to the antheridiogens secreted by larger gametophytes. As a result, archegonial formation is inhibited and antheridia are produced to give unisexual male gametophytes. Thus, a gametophyte population consists of large female gametophytes and small male gametophytes.

7 Mode of Fertilization and Sporophyte Formation

The unisexuality induced in individual gametophytes in a population may decrease the chance of intragametophytic selfing, which would increase genetic risk in terms of production of the next generation.[11,12] Another disadvantage of intragametophytic selfing is a lowered evolutionary potential.[13] Experiments were designed to compare ability of sporophyte formation by intragametophytic selfing and intergametophytic mating in *L. japonicum*.[14]

To induce bisexuality, gametophytes were selected for uniformity in width and transplanted onto fresh medium supplemented with 10^{-4} M GA_3 on day 12 after spore inoculation. They were cultured on the GA_3 medium for 8 days to initiate antheridial formation and then transferred onto fresh medium lacking GA_3 to allow archegonial formation. The time courses for the development of antheridia and archegonia in these gametophytes are shown in Fig. 5. Archegonial initiation was detected 6 days after transplantation to the GA_3-free medium; archegonia were mature 4 days later. Antheridia induced by the GA_3 treatment were also mature on the same day. Matured antheridia ruptured to release spermatozoids, and matured archegonia ruptured to attract and accept the spermatozoids in the presence of water.

To induce gametophytes to produce archegonia only, gametophytes

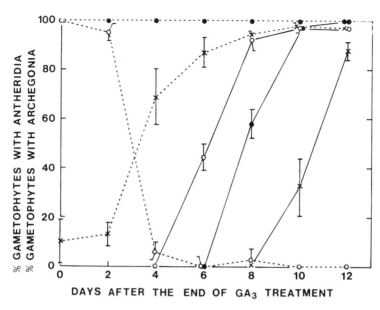

Fig. 5. Time courses of development of archegonium and antheridium in *L. japonicum*. Twelve-day-old gametophytes were transplanted onto medium with 10^{-4} M GA_3 and cultured for 8 days and then transplanted onto GA_3-free medium. Percentages of gametophytes that formed primordia (○), morphologically completed (●), and matured (×) archegonia (———) and antheridia (— — —) were determined every 2 days after the end of GA_3 treatment (Modified from Takeno and Furuya[14])

were selected for uniformity of width and transferred singly to fresh media on day 12 after spore inoculation. These gametophytes produced archegonia only. The archegonia in these unisexual gametophytes were discernible on day 18 after spore inoculation, and the time sequence of their development was the same as that for archegonia in the bisexual gametophytes. To induce gametophytes with antheridia only, gametophytes were transplanted onto medium supplemented with 10^{-4} M GA_3 on day 10 after spore inoculation and kept on the GA_3 medium for 12 days. These gametophytes produced antheridia only. The antheridia from these unisexual gametophytes reached maturity on day 22 after spore inoculation.

To allow self-fertilization only, bisexual gametophytes were isolated on day 30 after spore inoculation. Female and male gametophytes were paired, and pairs were isolated on day 22 to allow intergametophytic fertilization between female and male. Archegonia and antheridia of the bisexual gametophytes and those of the female and male gametophytes were mature and ready for fertilization on the day they were isolated and paired. A small amount of sterilized water was added to the gametophytes follow-

Table 2. Sporophyte formation in bisexual gametophyte and in unisexual female gametophyte in *L. japonicum* (Modified from Takeno and Furuya[14])

Sexuality of gametophyte	Mode of fertilization	% Sporophyte formation (No. of gametophytes observed)	
		Exp. I	Exp. II
Bisexual	Intragametophytic selfing	0 (16)	0 (95)
Female paired with male	Intergametophytic mating	37.5 (8)	26.3 (95)
Bisexual paired with male	Intergametophytic mating	39.6 (48)	19.2 (52)
Female paired with bisexual	Intergametophytic mating	20.0 (50)	17.0 (47)

ing the isolation and pairing. Gametophytes were examined on day 27 after the isolation and pairing for the presence or absence of sporophytes.

The data for sporophyte formation by self-fertilization in bisexual gametophytes and by intergametophytic fertilization in unisexual females paired with males are shown in Table 2. About 30% of the female gametophytes paired with males produced sporophytes, and no bisexual gametophytes produced sporophytes. The result indicates that intergametophytic mating was the only source of sporophyte formation.

Apogamous sporophyte formation was not observed in any gametophytes. It is concluded that the sporophytes observed on female gametophytes (Table 2) were the result of intergametophytic fertilization only. If a bisexual gametophyte was paired with a male or with a female, both the bisexual gametophytes paired with males and the female gametophytes paired with bisexuals produced sporophytes (Table 2).[14] These results indicate that the eggs and spermatozoids of the bisexual gametophytes possessed normal functions for producing zygotes which could develop sporophytes. Therefore, the lack of sporophyte formation by isolated bisexual gametophytes is due to the inability to form zygotes by selfing or failure of the resulting homozygotes to develop normally.

8 Conclusion

The *Lygodium* antheridiogens have two different functions: they induce antheridial initiation and inhibit the initiation of archegonia. Antheridiogens are produced only by gametophytes that have lost sensitivity to antheridiogens. The antheridiogens produced by older gametophytes act as pheromones on younger gametophytes. These properties of *Lygodium* antheridiogens can explain the sex differentiation of gametophytes through the action of antheridiogens at different growth stages. The sex differentiation regulated by the antheridiogens in *L. japonicum* is a system to favor intergametophytic mating.

The model of sex differentiation discussed here cannot be applied to other fern species, because the dual function of antheridiogens has yet to be observed in other species of ferns. Sex differentiation is known as a general trend in homosporous ferns, and so the same regulation mechanism as that found in *L. japonicum*, or a similar one, may exist in other ferns.

References

1. Miller JH. Fern gametophytes as experimental material. Bot Rev. 1968; 34:361–440.
2. Takeno K, Furuya M. Antheridiogens in ferns. Chemical Regulation of Plants. 1979; 12:16–28 (in Japanese).
3. Murashige T, Skoog F. A revised medium for rapid growth and bioassays with tobacco tissue cultures. Physiol Plant. 1962; 11:728–746.
4. Takeno K, Furuya M. Sexual differentiation in population of prothallia in *Lygodium japonicum*. Bot Mag Tokyo. 1980; 93:67–76.
5. Yamane H, Takahashi N, Takeno K, et al. Identification of gibberellin A_9 methyl ester as a natural substance regulating formation of reproductive organs in *Lygodium japonicum*. Planta. 1979; 147:251–256.
6. Yamane H, Satoh Y, Nohara K, et al. The methyl ester of a new gibberellin, GA_{73}: the principal antheridiogen in *Lygodium japonicum*. Tetrahedron Lett. 1988; 32:3959–3962.
7. Takeno K, Furuya M. Bioassay of antheridiogen in *Lygodium japonicum*. Dev Growth Differ. 1975; 17:9–18.
8. Takeno K, Furuya M. Inhibitory effect of gibberellins on archegonial differentiation in *Lygodium japonicum*. Physiol Plant. 1977; 39:135–138.
9. Takeno K, Furuya M, Yamane H, et al. Evidence for naturally occurring inhibitors of archegonial differentiation in *Lygodium japonicum*. Physiol Plant. 1979; 45:305–310.
10. Takeno K, Yamane H, Yamauchi T, et al. Biological activities of the methyl ester of gibberellin A_{73}, a novel and principal antheridiogen in *Lygodium japonicum*. Plant Cell Physiol. 1989; 30:201–205.
11. Voeller B. Developmental physiology of fern gametophytes: Relevance for biology. BioScience. 1971; 21:266–270.
12. Klekowski EJ Jr. Genetic load in *Osmunda regalis* populations. Amer J Bot. 1973; 60:146–154.
13. Klekowski EJ Jr. Reproductive biology of the pteridophyta. II. Theoretical considerations. Bot J Linn Soc. 1969; 62:347–359.
14. Takeno K, Furuya M. Sporophyte formation in experimentally-induced unisexual female and bisexual gametophytes of *Lygodium japonicum*. Bot Mag Tokyo. 1987; 100:37–41.

CHAPTER 39

Synthetic Pathways to Fern Antheridiogens from Gibberellins

M. Furber, L.N. Mander, and G.L. Patrick

1 Introduction

Following the discovery of an antheridium-inducing substance in gametophytes of the bracken fern, *Pteridium aquilinum*,[1,2] several discrete compounds (for which the term antheridiogen has been coined) have been isolated from a number of other fern species.[3] These compounds appear to play an important role in promoting cross-fertilization and thereby maintaining genetic diversity.[4] The very small quantities that have been obtained (nanograms to micrograms in most cases) have rendered structural analysis extremely difficult, however. An important clue to the constitution of these intriguing substances in the case of the family Schizaeaceae was the observation that they possessed gibberellin (GA)-like reactivity[5] and, conversely, that GAs had antheridium-inducing properties.[6,7] Nakanishi et al. were eventually able to arrive at formula **I** for antheridic acid,[8] the major antheridiogen from *Anemia phyllitidis*, a member of the family Schizaeaceae. The structure of antheridic acid was later refined to **II**,[9] following total syntheses of the racemates of **I** and **II**.[10] Antheridic acid (**II**) has also been shown to be a natural antheridiogen in further members of the *Anemia* genus, i.e., *A. hirsuta*,[11] *A. rotundifolia*, and *A. flexuosa*,[12] but could not be detected in *A. mexicana*.[13] However, a new GA-like substance was obtained from this last species, for which structure **III** was initially suggested, but which was ultimately shown to be **IV**.[13,14]

Structures of two antheridiogens from the related genus *Lygodium japonicum* have also been elucidated. The more abundant of these was found to be GA_9 methyl ester (**V**),[15] while the minor, but more potent, substance was shown to be 9,11-didehydro-GA_9 methyl ester (GA_{73}-Me) (**VI**).[16]

The range of GA structural types occurring naturally among the higher plants is surprisingly limited.[17] The basic skeleton is constant apart from the presence or absence of C-20, and even then the C_{20} GAs are present primarily as antecedents to the biologically operative C_{19} derivatives. It is as though the standard C_{19} system with the associated γ-lactone function has evolved as the optimal arrangement, with variations in hydroxylation

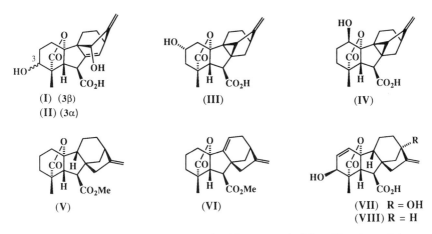

Fig. 1. Structures of known fern antheridiogens and gibberellins A_3 and A_7

patterns providing a degree of fine tuning for specific functions in different species. The occurrence of this new collection of compounds, **II**, **IV**, and **VI**, represents a significant departure from the structural homogeneity associated with the higher-plant gibberellins and leads to a number of fascinating conjectures concerning the evolutionary aspects of the biosynthesis and biological roles of the GAs in the plant kingdom. It also leads to the enticing prospect that this structural variability will be extended among other fern species. Given the tiny quantities of material that are likely to be available, it is clear that any search for new compounds will be arduous and that the role of synthesis in this quest will be crucial, as has already been demonstrated for compounds **IV**[14] and **VI**.[16] We have therefore mounted a major effort to establish efficient methods of access to these substances by synthetic conversions from the fungal GA_3 (**VII**) and GA_7 (**VIII**).[18] Aspects of this program are described in this paper.

2 First-Generation Syntheses of *Anemia* Antheridiogens from Gibberellins

Following an involvement in GA synthesis which had spanned two decades,[19] we became intrigued by the possibility of converting an appropriate GA derivative into antheridic acid (**II**). We first attempted this by means of a 1,2-bond shift initiated by the Lewis acid catalyzed opening of a 9α, 10α-epoxide as illustrated in Fig. 2.

This kind of transformation had been proposed as a possible basis for the biogenesis of **II**,[8] but we were not surprised to find that we were unable to duplicate this hypothetical biosynthetic conversion on such a complex system. In anticipation of a negative result, we had planned a more secure

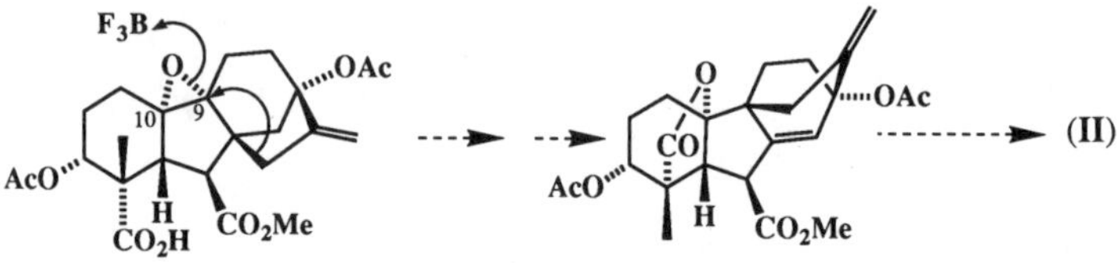

Fig. 2. Attempted rearrangement of a GA epoxide to an antheridane derivative

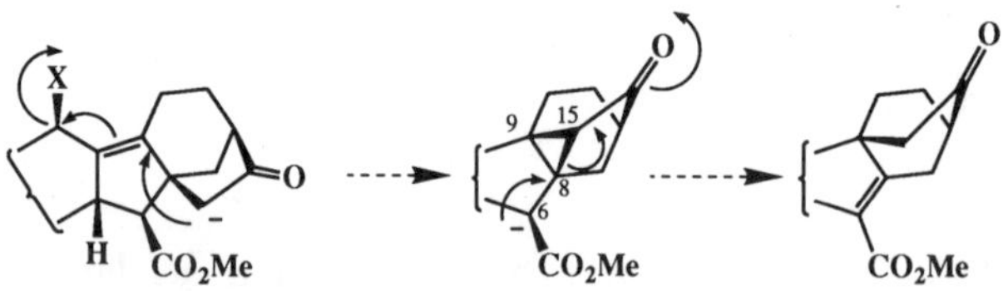

Fig. 3. Proposed plan for a controlled 1, 2-bond shift of C-15 from C-8 to C-9 in a GA derivative

sequence based on the formation of a 9,15-cyclopropyl ketone, followed by fragmentation of the 8,15-bond, which we expected could be initiated by enolization of the methoxycarbonyl function attached to C-6 (Fig. 3). These considerations were eventually translated into the preparation of **II** from GA_7 (**VIII**) as outlined in Fig. 4.[18] As well as establishing good access to this important compound, the conversion shows that antheridic acid has the same absolute configuration as the GAs.

Just as we were bringing this synthesis to a successful conclusion, our attention was drawn to the tentative assignment of structure **III** to a new antheridiogen from *Anemia mexicana* (N. Takahashi and H. Yamane, personal communication). With intermediates in hand possessing the same basic skeleton, we were able quickly to probe the validity of the hypothesis and deduce the correct structure to be **IV**.[14] This conclusion was then confirmed by the synthesis outlined in Fig. 5.[20] One especially useful feature of this new sequence was the direct isomerization of the allylic A-ring lactone function (step 7), which could be effected in one simple operation by diphenyl boron bromide, whereas this process had required four steps in our previous synthesis of **II** (Fig. 4).

3 Second-Generation Syntheses of *Anemia* Antheridiogens from Gibberellins

Although our respective syntheses of **II** and **IV** facilitated access to the *Anemia* antheridiogens, there were a number of features which we found to be unsatisfactory and believed would inhibit future investigations. In

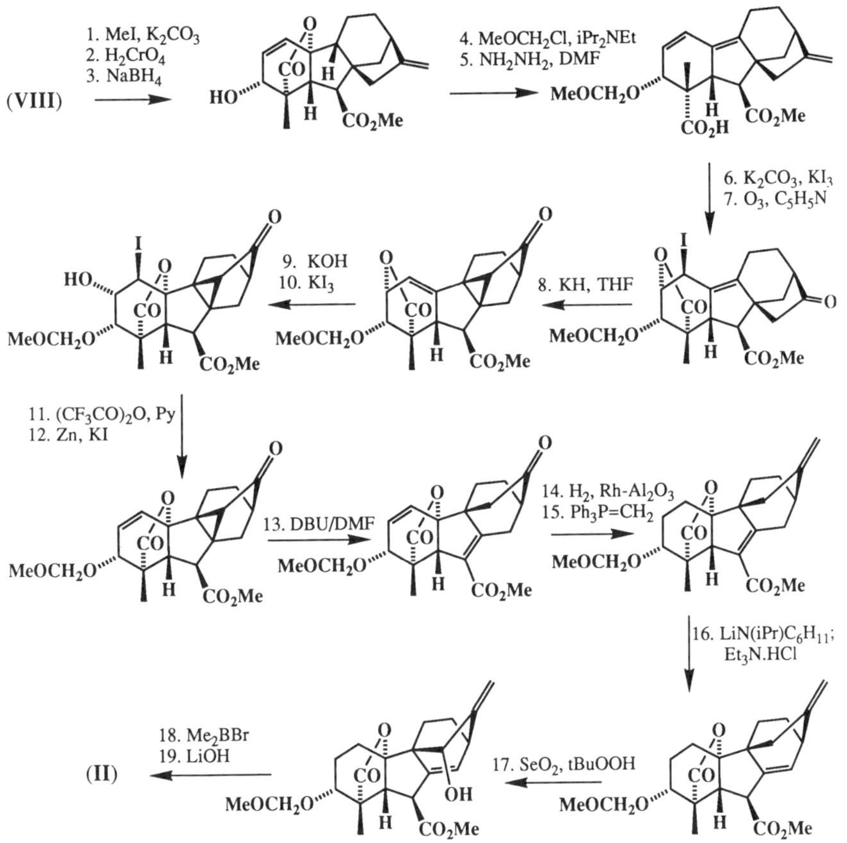

39. Synthetic Pathways to Antheridiogens 401

Fig. 4. Synthetic conversion of GA_7 into antheridic acid

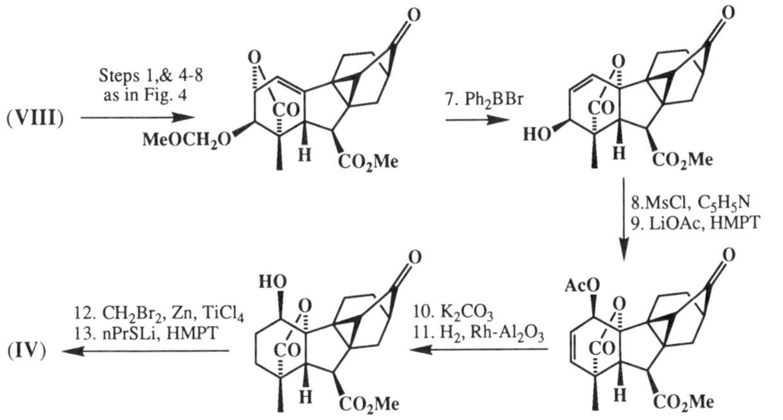

Fig. 5. Synthetic conversion of GA_7 into the antheridiogen from *A. mexicana*

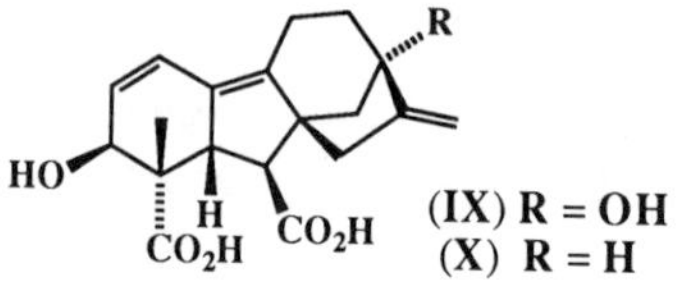

Fig. 6. Structures of gibberellenic acid and its 13-desoxy analogue

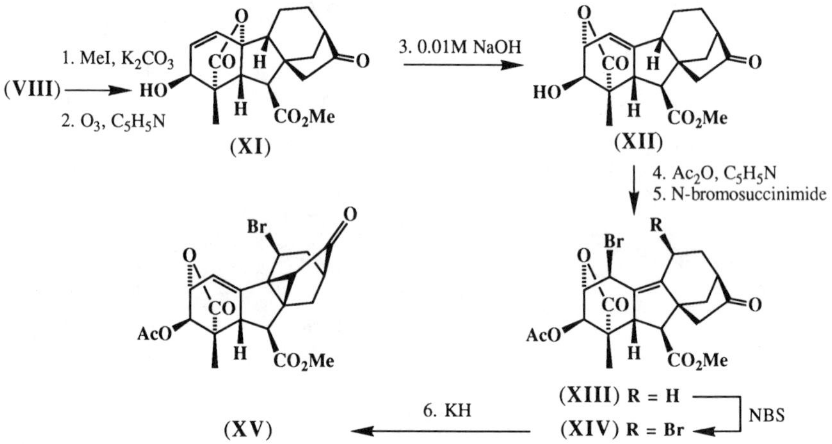

Fig. 7. Conversion of GA_7 into a valuable new precursor for antheridiogen synthesis

particular, the yield of the triene half ester with hydrazine (step 5, Fig. 4) was only 20%, while the pivotal intramolecular alkylation initiated by potassium hydride (step 8, Fig. 4) was rather capricious. The successful use of this reagent depends very much on the source, and reports of similar difficulties are commonplace.[21] Fortunately, it has been our experience to date that the reaction either proceeds in high yield or returns starting material.

We had initially chosen the hydrazine reaction[22] since it appeared to be the only secure method for introducing a double bond into the C-9–C-10 location. Although GA_3 (**VII**) and GA_7 (**VIII**) both afford acceptable 40% yields of gibberellenic acid (**IX**) and the 13-desoxy analogue (**X**), respectively (Fig. 6), we had difficulty in distinguishing between the carboxyl groups in these products, and it had been necessary to attempt the equivalent transformations on the methyl esters, for which the conversions were only half as efficient.

It was possible to bypass this obstacle as indicated in Fig. 7, however. Isomerization of the allylic A-ring lactone moiety with dilute sodium hydroxide in GAs like GA_7 (**VIII**) was known to afford the $\Delta^{1(10)}$-ene-2,19-

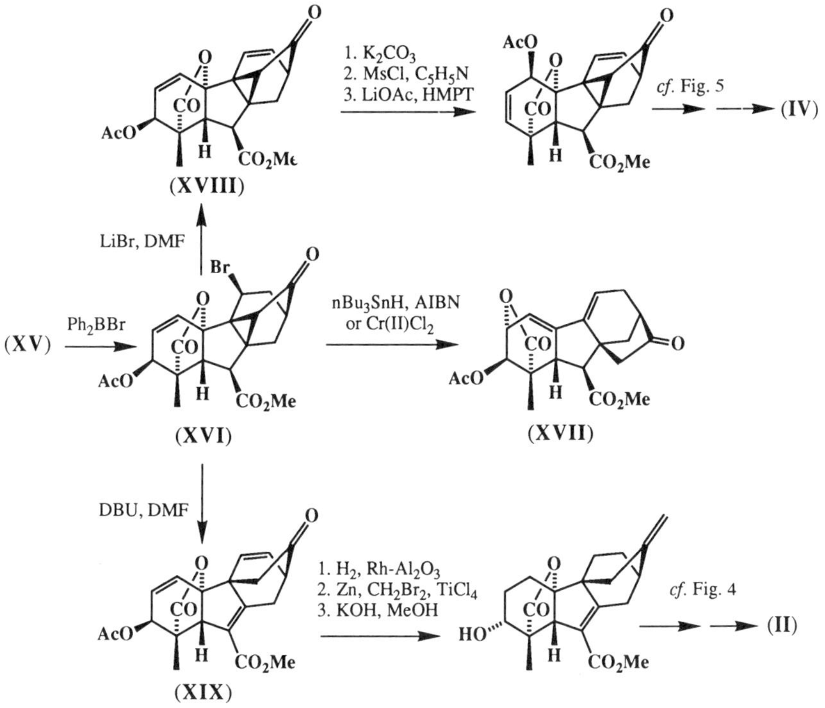

Fig. 8. Second-generation syntheses of the Anemia antheridiogens

lactones,[23,24] and we were able to carry out the analogous conversion of GA_7-17-nor-16-one (XI) into the isomer XII without difficulty. We envisaged that allylic bromination of the 3-acetate derived from this intermediate would take place with migration of the alkene bond and give rise to XIII, which could then serve as an alternative substrate to the allylic iodides utilized previously in the intramolecular alkylation steps. Unfortunately, it was difficult to obtain XIII in high yield because of further bromination to the 1,11-dibromide (XIV). We therefore allowed the reaction to progress to the latter stage and then proceeded to examine the feasibility of continuing the sequence with this compound in the expectation that the 11-bromo substituent could be removed subsequently. In the event, treatment of XIV with potassium hydride afforded the cyclopropyl ketone XV in high yield. [It should be noted, however, that this reaction only proceeded with selected batches of reagent obtained from Fluka Chemie, as had been found in the original sequence (step 8, Fig. 4).[18]]

After isomerization of the A-ring allylic lactone with diphenyl boron bromide to give XVI, attempts were made to remove the bromo substituent by reductive methods (e.g., n-Bu_3SnH, chromous salts), but this led to cleavage of the newly formed C-9–C-15 bond as well as rearrangement

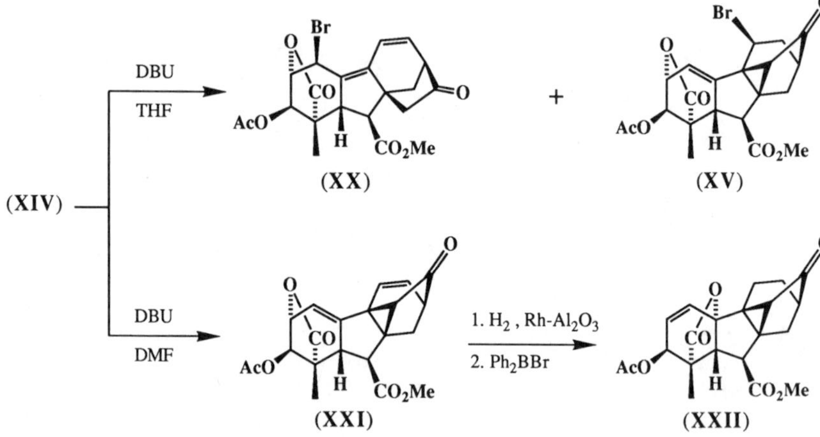

Fig. 9. Improved preparation of intermediates for antheridiogen synthesis

of the A-ring lactone function to give **XVII**. However, it was possible to effect elimination of the 11-bromo substituent with bromide ion in dimethylformamide (DMF) as the solvent to afford **XVIII**. This product could then be utilized as a precursor to the *A. mexicana* antheridiogen (**IV**) by means of a simple variation to the route used earlier (Fig. 5), as indicated in Fig. 8. Alternatively, elimination of bromide ion from **XVI** with a stronger base, 1,8-diazabicyclo[5.4.0]undec-7-ene (DBU), resulted in concomitant fission of the C-8–C-15 bond and afforded triene **XIX**, which was then used in a second synthesis of the *A. phyllitidis* antheridiogen (**II**). These reactions are also outlined in Fig. 8.

We also investigated the feasibility of removing the 11-bromo substituent from **XIV** prior to the intramolecular alkylation step with potassium hydride. Treatment with DBU in tetrahydrofuran afforded a 41% yield of the dienyl bromide **XX**, accompanied by a significant quantity of the cyclopropyl ketone **XV** (19% yield). The formation of the latter compound under these conditions provided a valuable lead that enabled us to bypass the troublesome potassium hydride-mediated process. Thus, we found (Fig. 9) that the reaction of **XIV** with DBU in DMF effected both cyclopropyl ketone formation and elimination of HBr in the C ring to furnish **XXI** in good yields. Hydrogenation followed by isomerization with diphenyl boron bromide then gave the very useful intermediate **XXII**.

The *Anemia* antheridiogens can thus be prepared more reliably and in higher overall yields by virtue of these various modifications.[25] Moreover, there are several opportunities to introduce isotopic labels at a wide range of sites.

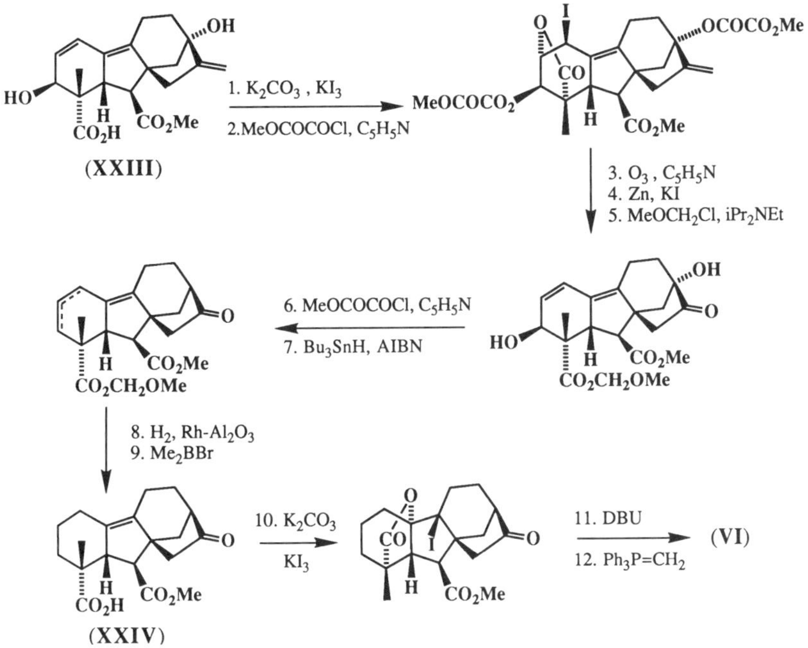

Fig. 10. Synthesis of the more potent antheridiogen from *L. japonicum*

4 Synthesis of the *Lygodium* Antheridiogen, GA_{73} Methyl Ester (VI)

The synthesis of the $\Delta^{9(11)}$ GA structure (**VI**) that had been deduced for the less abundant but more potent antheridiogen from *Lygodium japonicum*[16] appeared to be a much simpler task than that involved in the synthesis of the two *Anemia*-derived substances, but this view proved to be deceptive. The allylic lactone function in **VI** turned out to be quite labile, and, as a consequence, a number of early approaches were abortive. The first successful preparation was based on the iodolactonization of the 9-ene carboxylic acid **XXIV**, which was prepared from the diene half ester **XXIII** as outlined in Fig. 10. Anticipating poor yields in the preparation of this type of intermediate, we began with the more easily obtained GA_3 (**VII**), and although this decision imposed the additional task of removing the 13-hydroxyl group, we had assumed that this step could be combined with deletion of the 3-hydroxyl by means of a stannane reduction of a 3,13-bis(methyl oxalate) (step 7).[26] The oxalate functions were introduced early in the sequence so that they might also serve as protecting groups during this phase of the synthesis, but although they were adequate for

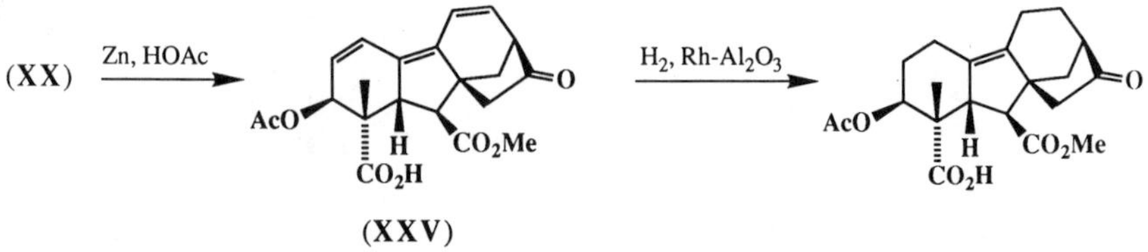

(XX) $\xrightarrow{\text{Zn, HOAc}}$ [structure] $\xrightarrow{\text{H}_2,\text{ Rh-Al}_2\text{O}_3}$ [structure]

(XXV)

Fig. 11. Revised synthesis of a key intermediate in the conversion of GAs to the more potent *L. japonicum* antheridiogen

(XVI) $\xrightarrow{\text{H}_2,\text{ Rh-Al}_2\text{O}_3}$ (XXVI) + (XXVII) $\xrightarrow{\begin{array}{l}1.\ \text{K}_2\text{CO}_3,\ \text{MeOH}\\2.\ \text{MsCl, Et}_3\text{N}\\3.\ \text{DBU}\\4.\ \text{H}_2,\ \text{Rh-Al}_2\text{O}_3\\5.\ \text{Ph}_3\text{P=CH}_2\end{array}}$ (VI)

Bu$_3$SnH, AIBN

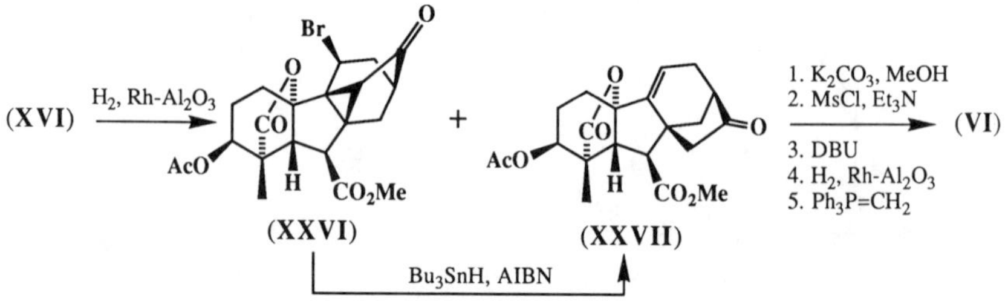

Fig. 12. Improved synthesis of the more potent *L. japonicum* antheridiogen

this purpose, they were very easily hydrolyzed and it was necessary to reinstate them at step 6.

The preparation of the antheridiogen in this way served the important purpose of confirming the tentative structure assignment, but in view of a pressing need for further supplies of **VI**, as well as isotopically labeled derivatives, we turned again to the isolactone approach (cf. Fig. 7) in a search for more effective access to a suitable 9-ene intermediate. It appeared that it should be feasible to convert the dienyl bromide **XX** into triene **XXV** by reductive elimination and then selectively to hydrogenate the double bonds in the A and C rings (Fig. 11).

While this sequence was successfully completed, the net yields were not sufficiently encouraging to persevere with this alternative approach. Instead, we examined the hydrogenation of lactone **XVI** (Fig. 8) with the intention of subsequently effecting the stannane-induced reductive cleavage of the cyclopropyl bromide moiety with formation of the $\Delta^{9(11)}$-olefinic bond in analogy with the formation of **XVII**. We were agreeably surprised to observe the direct formation of lactone **XXVII**, however, as well as a moderate amount of the expected intermediate **XXVI** (Fig. 12). This mixture was treated further with tributylstannane to complete the desired conversion and afford **XXVII** as a homogeneous product. Removal of the oxygen function from C-3 was not straightforward, and reasonably direct methods[27] based on stannane reduction of a thiocarbonate or thiocarbamate derivative led to concomitant hydrogenolysis of the allylic lactone function. We therefore resorted to elimination of a 3-methanesulfonate function to afford a 2,9(11)-diene, followed by selective hydrogenation of the Δ^2-olefinic bond as indicated (Fig. 12).

Although the preparation of the *Lygodium* antheridiogen by this last route appears to be unnecessarily elaborate, it is much more efficient and reliable than the earlier approaches. Moreover, all three antheridiogens, **II**, **IV**, and **VI**, may now be obtained from the common, readily prepared intermediate **XVI**.

5 Conclusion

The availability of these synthetic compounds has facilitated several biological and biochemical investigations that would not otherwise have been possible. We have also begun to establish an extensive spectroscopic data base, which should simplify the task of determining the structures of further growth substances of this general type. The biogenetic origins of these compounds are of special interest, and it is tempting to assume that they are formed from a common GA precursor (or set of precursors) that possesses a cationic center at C-9, at least in a formal sense. The speculation that the 9,15-cyclogibberellin structure contained in the *A. mexicana* antheridiogen (**IV**) might be implicated in the biosynthesis of antheridic acid (**II**)[14,18] has been given considerable credence by very recent results obtained by Takahashi and co-workers. Aspects of these studies are described by Yamane in Chapter 37 of this volume.

Acknowledgments. We are grateful to the assistance and cooperation provided by many of the principal protagonists in the gibberellin and antheridiogen field, especially Professors MacMillan, Pharis, Schraudolf, and Takahashi, and Drs. Nester, Takeno, and Yamane. The skilled assistance of Bruce Twitchin and the generous provision of fungal gibberellins by Abbott Laboratories have also been crucial to the success of this work.

References

1. Döpp W. Eine die Antheridienbildung bei Farnen fördernde Substanz in den Prothallien von *Pteridium aquilinum* L. Kuhn. Ber Dtsch Bot Ges. 1950; 63:139–147.
2. Döpp W. Über eine hemmende und eine fördernde Substanz bei der Antheridienbildung in den Prothallien von *Pteridium aquilinum*. Ber Dtsch Bot Ges. 1959; 72:11–24.
3. Näf U, Nakanishi K, Endo M. On the physiology and chemistry of fern antheridiogens. Bot Rev. 1975; 41:315–357.
4. Schraudolf H. Phytohormones and Filicinae: Chemical signals triggering morphogenesis in Schizaeaceae. In: Bopp M, ed. Plant growth substances. Berlin: Springer-Verlag, 1985:p. 270–274.

5. Sharp PB, Keitt GW Jr, Clum HH, et al. Activity of antheridiogen from the fern *Anemia phyllitidis* in three flowering plant bioassays. Physiol Plant. 1975; 34:101–105.
6. Schraudolf H. Relative activity of the gibberellins in the antheridium induction in *Anemia phyllitidis*. Nature. 1964; 201:98–99.
7. Voeller B. Gibberellins: Their effect on antheridium formation in fern gametophytes. Science. 1964; 143:373–375.
8. Nakanishi K, Endo M, Näf U, et al. Structure of the antheridium-inducing factor of the fern *Anemia phyllitidis*. J Amer Chem Soc. 1971; 93:5579–5581.
9. Corey EJ, Myers AG, Takahashi N, et al. Constitution of antheridium-inducing factor of *Anemia phyllitidis*. Tetrahedron Lett. 1986; 27:5083–5084.
10. Corey EJ, Myers AG. Total synthesis of (±)-antheridium-inducing factor (A_{An},2) of the fern *Anemia phyllitidis*. Clarification of stereochemistry. J Amer Chem Soc. 1985; 107:5574–5576.
11. Zanno PR, Endo M, Nakanishi K, et al. On the structural diversity of fern antheridiogens. Naturwissenschaften. 1972; 512.
12. Yamane H, Nohara K, Takahashi N, et al. Identification of antheridic acid as an antheridiogen in *Anemia rotundifolia* and *Anemia flexuosa*. Plant Cell Physiol. 1987; 28:1203–1207.
13. Nester JE, Veysey S, Coolbaugh RC. Partial characterization of an antheridiogen of *Anemia mexicana*: Comparison with the antheridiogen of *A. phyllitidis*. Planta. 1987; 170:26–33.
14. Furber M, Mander LN, Nester JE, et al. Structure of an antheridiogen from the fern *Anemia mexicana*. Phytochemistry. 1989; 28:63–66.
15. Yamane H, Takahashi N, Takeno K, et al. Identification of gibberellin A_9 methyl ester as a natural substance regulating formation of reproductive organs in *Lygodium japonicum*. Planta. 1979; 147:251–256.
16. Yamane H, Satoh Y, Nohara K, et al. The methyl ester of a new gibberellin, GA_{73}: The principal antheridiogen in *Lygodium japonicum*. Tetrahedron Lett. 1988; 29:3959–3962.
17. Takahashi N, Yamaguchi I, Yamane H. Discovery of gibberellins and their occurrence in nature. In: Takahashi, N, ed. Chemistry of plant hormones. Boca Raton, Florida: CRC Press, 1986:p. 57–151.
18. Furber M, Mander LN. Conversion of gibberellin A_7 into antheridic acid, the antheridium inducing factor from the fern *Anemia phyllitidis*: A new protocol for controlled 1,2-Bond shifts. J Amer Chem Soc. 1987; 109:6389–6396.
19. Mander LN. Synthesis of gibberellins and antheridiogens. Nat Prod Reports. 1988; 5:541–579.
20. Furber M, Mander LN. Synthesis and confirmation of structure of the antheridium-inducing factor from the fern *Anemia mexicana*. J Amer Chem Soc. 1988; 110:4084–4085.
21. MacDonald TL, Natalie KJ Jr, Prasad G, et al. Chemically modified potassium hydride. Significant improvement in yields in some oxy-Cope rearrangements. J Org Chem. 1986; 51:1124–1126.
22. Grove JF, Mulholland TPC. Gibberellic acid. Part XII. The stereochemistry of allogibberic acid. J Chem Soc. 1960; 3007–3022.
23. Cross BE, Grove JF, Morrison A. Gibberellic acid. Part XVIII. Some rearrangements of ring A. J Chem Soc. 1961; 2498–2515.
24. Kirkwood PS, MacMillan J, Sinnott ML. Rearrangement of the lactone ring of

gibberellin A_3 in aqueous alkali; participation of the ionised 3-hydroxy-group in an *anti* S_N2' reaction. J Chem Soc Perkins Trans 1. 1980; 2117–2121.
25. Furber M, Mander LN. An improved synthesis of antheridic acid, the antheridium inducing factor from the fern *Anemia phyllitidis*. Tetrahedron Lett. 1988; 29:3339–3342.
26. Dolan SC, MacMillan J. A new method for the deoxygenation of tertiary and secondary alcohols. J Chem Soc Chem Commun. 1985; 1588–1589.
27. Barton DHR, McCombie SW. A new method for the deoxygenation of secondary alcohols. J Chem Soc Perkins Trans 1. 1975; 1574–1585.

APPENDIX
Structures of Free Gibberellins

Appendix

GA₁(F,P) GA₂(F) GA₃(F,P)

GA₇(F,P) GA₈(P) GA₉(F,P)

GA₁₃(F,P) GA₁₄(F) GA₁₅(F,P)

GA₁₉(P) GA₂₀(P) GA₂₁(P)

GA₂₅(F,P) GA₂₆(P) GA₂₇(P)

GA₃₁(P) GA₃₂(P) GA₃₃(P)

GA₃₇(F,P) GA₃₈(P) GA₃₉(P)

Appendix 413

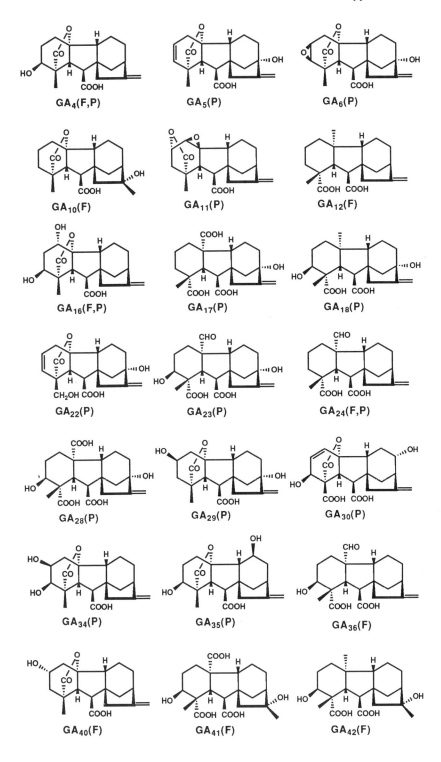

GA₄₃(P) GA₄₄(P) GA₄₅(P)
GA₄₉(P) GA₅₀(P) GA₅₁(P)
GA₅₅(F,P) GA₅₆(F) GA₅₇(F)
GA₆₁(P) GA₆₂(P) GA₆₃(P)
GA₆₇(P) GA₆₈(P) GA₆₉(P)
GA₇₃(p) GA₇₄(p) GA₇₅(P)
GA₇₉(P)

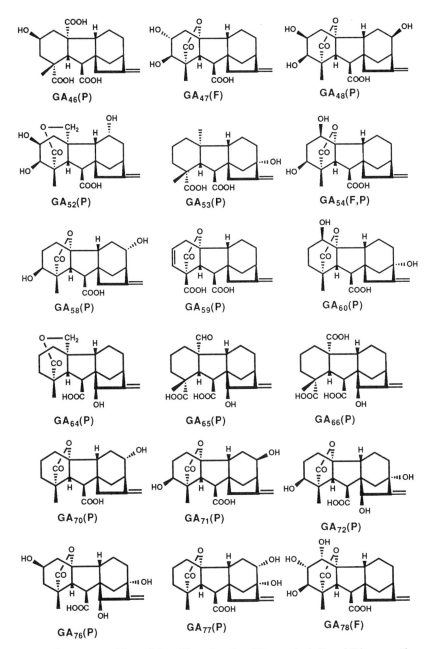

Fig. 1. Structures of free gibberellins, A_1–A_{79}. The symbols P and F in parentheses denote plant and fungus origin, respectively

Index

Abscisic acid (ABA)
 α-amylase gene transcription and, 130-133
 lettuce seed germination and, 289, 290, 291
 uniconazole treatment and, 334
 content in germinating wheat, 128-129
Abscission, of pedicels, 365
Acetyl coenzyme A carboxylase, 100
Adenosine triphosphate (ATP), calcium transport and, 110
Affinity chromatography, 143-145
Affinity probes, for GA receptors, 137
Agricultural applications, 301-305, 361-367
Agrobacterium rhizogenes, 241
Agropine, 241
Agrostemma githago, 273
Aleurone
 α-amylase synthesis and, 106-107, 111-112, 115-116
 in *Avena fatua*, GA receptors in, 136-145
 in barley, calcium transport and, 106-112
 hydrolytic enzyme production and, 126
 mRNA levels in, 130
 Sepharose-immobilized GAs and, 138-145
 in wheat, GA production and, 125-133
Alkaloid synthesis, 241-247
1-Aminocyclopropane-1-carboxylic acid (ACC), 336
α-Amylase gene transcription of, 120-122, 130-133, 136
 multigenetic expression of, 118-120
α-Amylase synthesis
 calcium transport and, 106-107, 111-112
 in rice, 114-122
 Sepharose-immobilized GAs and, 138-145
AMO-1618, 188, 196, 274, 298
Amorphophallus, flowering of, 370
Ancymidol, 58, 229-239, 299-305, 313-314, 325
Anemia species, 378-386, 398-407
Antheridiogens

 in *Anemia*, 378-386
 in *Lygodium japonicum*, 380, 389-397
 synthesis of, 398-407
Antheridane, 397
Anthers, in rice, GA biosynthesis and, 11, 17-18
Anthesis, 10
Anthocyanins, 300
Antibodies, anti-GA, specificities of, 159-163
Apical meristem, 199
Apricot seed, 352
Arabidopis thaliana, fruit and seed development of, 181-183
Archegonia, inhibition of, 390-392
Ascorbate, 65, 68, 78, 103
ATP, calcium transport and, 110
Atropa belladonna, 242-243, 247
Auxin, 211, 241
Avena fatua, 136-145

Bacterial infection, hairy root growth and, 241
Bacteriophage T4, 146
Bakanae, 2-4
Barley
 aleurone, GA-regulated biochemistry and, 106-112
 GA concentrations in, 95-96
 growth retardants and, 101
BAS 106, 366
BAS 111..W, 58, 299-305
Bentazon, 305
Biosynthesis, of gibberellins
 in *Cucurbita* endosperm, 51-60
 during seed development, 180-181
 enzymes and regulation of GA concentration, 94-103
 genetic loci for, 40-42
 inhibitors of, 296-305
 in maize, 22-31

417

wheat germination and, 125-133
Brassica, flowering of, 362-363
Brassinosteroid levels, uniconazole and, 339-347
BTS 44 584, 298
BX-112 and derivatives, 311-318

C-13 hydroxyl group, *see* 13-Hydroxylation pathway gibberellins; *specific gibberellins*
C_{19} gibberellins, immunogenic conjugates of, 149-155; *see also specific gibberellins*
C_{20} gibberellins, immunogenic conjugates of, 155-158; *see also specific gibberellins*
C-20 hydroxylase, 51, 59-60
C-20 oxidation, 53-54, 59, 88, 316-318
Cadmium ion, 76
Calcium transport, 106-112, 117
Calibration curve, for radioimmunoassay, 147
Canopy structure, plant growth regulators and, 302
Carrier protein, conjugate GAs and, 149-158
Castasterone, 341-347
CCC, *see* Chlormequat chloride
cDNA clones, 130
Cell elongation, 200-201, 211-217, 220
Cell-free extract, for GA metabolic studies
from *Cucurbita maxima*, 51-59
from *Phaseolus coccineus*, 83-91
from *Phaseolus vulgaris*, 62-69, 72-81
from *Marah macrocarpus*, 67-68
from rice, 18-20
from spinach, 276-278
Cell wall
sugar composition of, root growth and, 235
yield threshold of, 211, 216
Cellulose microfibrils, 211, 220
Chinese cabbage, flowering of, 362-363
Chlormequat chloride (CCC), 296, 298-304, 366
4'-Chloro-2'-(α-hydroxybenzyl)isonicotin- anilide, *see* Inabenfide
Chlorphonium chloride, 298
Chlorophyll content, growth retardants and, 330, 331-332
Chlortoluron, 305
Citrus, endogenous GAs in, 204
Climatic stress, improving resistance to, 304
Cofactors, 65, 67-69, 76; *see also specific cofactors*
Cobalt ion, 76
Colchicine, 222, 225-226
Colocasia esculenta, GA-induced flowering in, 370-376
Conifers
endogenous GAs in, 203-206

exogenous GAs and, 200-203
GA-promoted flowering of, 166-171
Conjugated GAs
biological activity of, 258
of C_{19} GAs, 149-155
of C_{20} GAs, 155-158
endogenous GA concentrations and, 96
enzymatic formation of, 254-255
enzymatic hydrolysis of, 255-258
in fruits of *Phaseolus coccineus*, 252-254
naturally occurring, 249-250
physiological functions of, 260
synthesis and analysis of, 250-252
wheat germination and, 125, 126-128
in woody plants, 203
in xylem sap, 205
Conjugation
biochemical and physiological aspects of, 249-260
immunoassays and, 146, 149-158
Copper ion, 76
Cortical microtubules
cell elongation and, 211-217
onion bulb development and, 220-227
Cotton, growth retardant and, 302, 313
Cotyledons
2ß-hydroxylase and, 97
3ß-hydroxylation in, 62-69
in vitro GA_4 metabolism in, 89-91
Cowpeas, 46, 280-287
Cremart, 220, 225-226
Cropping efficiency, GA effects on, 350, 354-359
Cross-reactivities, of anti-GA antibodies, 159-163
Cucumber, growth retardant effects on, 313, 332
Cucurbita maxima
endosperm, GA biosynthetic pathways in, 51-60
ent-kauradienoic acid conversion in, 84
inabenfide and, 320-328
Cupressaceae, flowering in, 167-171
Cupressus arizonica, 203
ent-9,15-Cyclogibberellane, 385
Cyclohexanedione and derivatives, 100-104
Cyclohexanetriones, 58-59, 300, 301, 302
Cytochrome P-450, plant growth regulators and, 300, 305
Cytochrome c reductase (CCR), 110
Cytokinin, 241, 333-336
Cytosol, calcium homeostasis in, 107-109

Daminozide, 296
Datura innoxia, 242-247

Deoxygibberellin C, 3ß-hydroxylase activity and, 76-78
Deuterium labeled GAs
  enzyme kinetics, 66
  metabolism of, in rice, 264-271
  as internal standards for GC-MS, 33, 281-287, 292
Dicyclohexylcarbodiimide (DDC), 155
Dioxygenases, 68-69, 300
  inhibitors of, 100-104, 318
DNA mobility shift, 121-122
DOCHC, 311-318
Drought stress, improving resistance to, 304
Dwarf mutants
  Arabidopis, seed development of, 181-183
  cell elongation studies using, 212-217
  end-of-day FR response and, 47
  exogenous GA and, 202-203
  GA levels in, 14-18, 35-36, 94-96
  in maize, GA biosynthesis studies, 22-31, 32-38
  of rice, 10-20
  root growth and, 229-239
dx gene, in rice, GA biosynthesis and, 17
Dye loading method, 107
dy gene, in rice, GA biosynthesis and, 17-18

Early-13-hydroxylation pathway, 17, 20 22, 28, 32, 42, 48, 62, 84-85, 126, 275, 280
  see also 13-Hydroxylated gibberellins
Early-non-hydroxylation pathway, 17, 20, 32, 84, 127
Ears, of maize, 37
EL 500, 300
Embryo preparations, GA_{12} conversions in, 56-58
End-of-day far-red response, 46-48, 280-287
Endoplasmic reticulum, in barley aleurone cells, calcium transport and, 106-112
Endosperm
  GA_7 biosynthesis in, 63
  GA_{19} content of, 128
  of pumpkin, GA biosynthetic pathways in, 51-60
  of wheat, GA metabolism in, 125-133
Enzymatic hydroxylation, 62-69, 72-81; *see also specific hydroxylation entries*
Enzyme(s), GA concentration regulation and, 94-103; *see also specific enzymes*
Enzyme cofactors, 65, 67-69, 76
Enzyme-linked immunosorbent assays (ELISAs), 146, 149, 158-159
Enzyme tracers, preparation of, 158
Epoxidation, 78, 399-400
Ethephon, 296

Ethylene, cell expansion and, 212
Ethylene evolution, uniconazole and, 335, 336, 339

Far-red (FR) response
  GA metabolism and, 280-287
  gibberellin mutants and, 46-48
  lettuce seed germination and, 289-294
Fast protein liquid chromatography (FPLC), 87-89
Female inflorescence, of maize, 32
Ferns, antheridiogens of, 378-386, 389-397, 398-407
Ferrous ion, 65, 67-69, 76, 78, 103
Fertilization mode, for ferns, 394-396
Flower formation, 303
  of Chinese cabbage, 362-363
  of conifers, GA-promoted, 166-171
  of sweet potato, 363-364
  GA_3 treatment in, 350, 352-354
  in taro plants, 370-376
  uniconazole and, 331
Fluorescent dyes, 107
Flurprimidol, 300
Frost damage, improving resistance to, 304
Fruit development, GA effects on, 179-184, 350-359
Fruit set, of tomato, at high temperature, 364-365
Fruit trees, growth regulation in, 303
Fungal diseases, 2-4
  chemical resistance to, 305, 339
Fusaric acid, 5
Fusarium, 2-3

Gametophytes, sex differentiation of, 389-390, 394-397
Garden pea, 40
GC-MS (gas chromatography-mass spectrometry), 10, 18-19, 25, 33-34, 37, 42, 52, 73, 85, 95, 128, 281, 292, 340-346, 381
GC-SIM (gas chromatography-selected ion monitoring), 10, 25, 27, 33-34, 42, 95-96, 170, 266, 281, 292, 321, 334, 340-346
Genetic influences
  for α-amylase synthesis, 118-122
  exogenous GAs and, 43-44, 202-203
  Ri plasmid and, 241
  in sweet pea, 40-48;
  see also Mutants
Germination
  phytochrome-mediated, of lettuce seeds, 289-294

of wheat, GA metabolism and, 125-133
Gibberella fujikuroi, 3, 84
ent-Gibberellane, 385
Gibberellin
  agricultural applications of, 361-367
  antibody cross-reactivities of, 160-163
  as antibulbing hormone, 223
  conjugates, *see* Conjugated gibberellins
  discovery and early studies of, 5-7
  mutants, *see* Mutants
  naming conventions of, 7
  open lactones, 84-86
  structures of, 412-415
Gibberellin A_1
  α-amylase gene transcription and, 130-133
  concentration, determinations of, 94
  conjugates of, 250
  conversion of, from GA_{12}-aldehyde, 84-87
  conversion of, from GA_{20}, 22-23, 28-30, 40, 42, 62-68
  discovery of, 7
  in dwarf mutants, concentrations of, 95-96
  end-of-day FR response and, 46-48
  far-red treatments and, 284-287
  florigenic activity of, 166, 174, 374-376
  growth retardants and, 101, 314-318
  2β-hydroxylation of, 97-100
  3β-hydroxylase and, 22-23
  inabenfide and content of, 328
  interaction of photoperiod and, 201
  in lettuce seed, 292-294
  in maize seed, 34-37
  in maize shoots, 22, 28-29
  in *Phaseolus*, 73, 84-89
  production in endosperm, 55-56
  in rice, 10-20
  in spinach, 275, 278
  in sweet pea, 40-48
  uniconazole action and, 343-347
  wheat germination and, 125-133
  in woody plants, 203, 205, 206
3-*epi*-Gibberellin A_1
  in lettuce seed, 292
  in spinach, 275
Gibberellin A_2
  discovery of, 7
Gibberellin A_3
  α-amylase biosynthesis and, 120-122
  antheridiogen synthesis from, 405-407
  archegonial differentiation and, 391-392
  in barley, calcium transport and, 106-112
  Chinese cabbage flowering and, 363
  conjugates of, 250, 252-255
  conversion to, from GA_5, 29-30, 62-70
  discovery of, 7
  dwarfish genotypes and, 203
  enzymatic conversion to, 67-70

florigenic activity of, 171-174, 353
growth retardants and, 100-104
hairy root growth and, 242-247
hardwood stem elongation and, 200
maize sex differentiation and, 37-38
maize shoots and, 22
microtubule orientation and, 220-227
root growth and, 229-239
S-3307 (uniconazole) and, 224
shoot elongation and, 94
short-day induced growth cessation and, 201
sour cherry fruit set and, 350-359
substitution of, for red light effects, 290
taro plant flowering and, 371-376
tomato fruit set and, 364-365
virus expression and, 354
in woody plants, 203
Gibberellin A_4
  conifers and, 168-172, 200-201
  conjugates of, 250
  conversion to from GA_{12}-aldehyde, 84-88
  discovery of, 7
  florigenic activity of, 168-172, 374-376
  2β-hydroxylase and, 97-100
  in maize seed, 34-47
  in *Phaseolus*, 83-91
  interaction of photoperiod and, 202
  in rice, isotope studies of, 264-271
  in rice, 11-20
  Sepharose immobilization technique of, 137-145
  in spinach, 275
  sweet potato flowering and, 364
  in woody plants, 203
  in wheat aleurone tissue, 125-126
  wheat germination and, 127-128
Gibberellin A_5
  conjugates of, 250
  conversion to from GA_{12}-aldehyde, 84
  conversion to from GA_{20}, 29-30
  enzymatic conversion of, 62-68
  epoxidation of, 78
  3β-hydroxylation and, 62-69, 76-78
  in maize shoots, 29
  in *Phaseolus*, 73, 84
  in spinach, 275
  sweet potato flowering and, 364
  tomato fruit set and, 364
Gibberellin A_6
  conversion to, from GA_{12}-aldehyde, 84
  epoxidation of GA_5 to, 78
  in *Phaseolus*, 73, 84
Gibberellin A_7
  antheridiogen synthesis from, 380, 400-405
  conifers and, 168-172, 200-201
  florigenic activity of, 168-172, 374-376

Index   421

interaction of photoperiod and, 202
in spinach, 275
sweet potato flowering and, 364
tomato fruit set and, 364
Gibberellin A_8
conjugates of, 250, 252, 255-258
in dwarf mutants, concentrations of, 95
in maize seed, 34
in maize shoots, 28-29
in *Phaseolus coccineus* cell-free system, 86-87
in spinach, 275
in sweet pea, 42
uniconazole action and, 343
wheat germination and, 126-128
Gibberellin A_9
conifer flowering and, 168-171
conjugates of, 250
DOCHC and, 316
endosperm conversion of, 63
enzymatic activity and, 66
2ß-hydroxylase and, 97-100
3ß-hydroxylase activity and, 76-78
in maize seed, 34-37
methyl esther as antheridiogen, 380, 384-385, 390, 398
interaction of photoperiod and, 202
in rice, 11, 19
in woody plants, 203, 205
1 ß-hydroxy-9,15-cyclo-Gibberellin A_9, 380-386
1ß, 2ß, 3ß-trihydroxy-Gibberellin A_9, 126-127
Gibberellin A_{12} and derivatives
conversion of, to C_{19} GAs, 84-89
DOCHC and concentration of, 316
inabenfide and, 324-325
metabolism of, in embryo preparations, 56-58
metabolism of, in endosperm preparations, 51-54
from mevalonic acid, 84
in rice, 11, 19
in spinach, 275
Gibberellin A_{14}
conversion to, from GA_{12}-aldehyde, 85
Gibberellin A_{15}
conversion to, from GA_{12}-aldehyde, 84-88
DOCHC and concentration of, 316
inhibition of 3ß-hydroxylase activity by, 76-78
Gibberellin A_{17}
conversion to, from GA_{12}-aldehyde, 84
growth retardants and concentration of, 101
in maize seed, 34
in rice, 11
in spinach, 275

Gibberellin A_{19}
conversion to, from GA_{12}-aldehyde, 84-86
in dwarf mutants, concentrations of, 95-96
far-red treatments and concentration of, 284-287
growth retardants and concentration of, 101, 314-318
inabenfide and concentration of, 328
in lettuce seed, 292-294
in maize seed, 34
in maize shoots, 28-29
interaction of photoperiod and, 201
in rice, 10-20
in spinach, 275-278
uniconazole action and content of, 343-345
wheat germination and content of, 126-128
in woody plants, 203
Gibberellin A_{20}
conversion to GA_1, 22-23, 28-30, 40, 42
conversion to GA_5, 29-30
conversion to, from GA_{12}-aldehyde, 84-87
DOCHC and concentration of, 316
in dwarf mutants, concentrations of, 95-96
far-red treatments and concentration of, 284-287
growth retardants and concentration of, 101, 314-318
2ß-hydroxylase and, 97-100
3ß-hydroxylase and, 22-23, 62-69, 72-81
inabenfide and concentration of, 328
in lettuce seed, 292-294
in maize seed, 34-37
maize shoot growth and, 22-31
interaction of photoperiod and, 201
in rice, 11
in spinach, 275, 277, 278
uniconazole action and content of, 343-347
wheat germination and content of, 126-128
in woody plants, 203, 205
Gibberellin A_{20}-conjugate, 125, 127
Gibberellin A_{24}
conversion to, from GA_{12}-aldehyde, 84-88
DOCHC and metabolism to, 316-317
growth retardants and metabolism to, 59
in rice, 11
Gibberellin A_{25}
conversion to, in embryo preparations, 56
conversion to, in pumpkin endosperm, 53
Gibberellin A_{26}
conjugates of, 250
Gibberellin A_{27}
conjugates of, 250
Gibberellin A_{28}
conversion to, in endosperm, 53, 55-56
Gibberellin A_{29}
conjugates of, 250
enzymatic conversion to, 62-68

growth retardants and concentration of, 101
in maize seed, 34
in maize shoots, 28–29
in rice, 11
in spinach, 275
in sweet pea, 42
uniconazole action and content of, 343–345
2-*epi* Gibberellin A_{29}, 42
Gibberellin A_{32}
florigenic activity of, 175
in sour cherry seed, 352
Gibberellin A_{34}
in maize seed, 34
in *Phaseolus* germinating seeds, 89
in rice, 11, 19
in spinach, 275
wheat germination and concentration of, 127–128
Gibberellin A_{35}
conjugates of, 250
Gibberellin A_{36}
GA_{37} conversion to, 88–89
growth retardants and conversion of, 59
Gibberellin A_{37}
conjugates of, 250
DOCHC and concentration of, 316
in *Phaseolus coccineus*, conversion to, from GA_{12}-aldehyde, 84–89
Gibberellin A_{38}
conjugates of, 250
in *Phaseolus coccineus*, conversion to, from GA_{12}-aldehyde, 84
Gibberellin A_{43}
in pumpkin endosperm, 52
Gibberellin A_{44}
conjugates of, 250
conversion to, from GA_{12}-aldehyde, 84–88
growth retardants and concentration of, 101
inhibition of 3ß-hydroxylase activity by, 76–78
in maize seed, 34
in maize shoots, 28–29
metabolism to, in pumpkin endosperm, 53
in rice, 11, 19
in spinach, 275
uniconazole and content of, 343
Gibberellin A_{49}
in pumpkin endosperm, 52
Gibberellin A_{51}
in rice, 11
Gibberellin A_{53}
conversion to, from GA_{12}-aldehyde, 84
in endosperm, conversions of, 53–56
in maize seed, 34
in maize shoots, 28–29
in rice, 11, 19
in spinach, 275–278

Gibberellin A_{73}
methyl ester as antheridiogen, 380–385, 390, 398, 405–407
ß-Glucosidase, 255
Glucosyl conjugates, 249–260
Glucosyl transferase, 89
ß-Glycanase, 120
Glycoprotein, α-amylase as, 117
Golgi apparatus, 110, 117
Grasses, growth regulation of, 303, 313
Growth retardants
applications of 301–305
brassinosteroid levels and, 339–347
effects of, GA metabolism in *Cucurbita* endosperm, 58–59
effects of, hormone levels in rice 330–337
GA levels and, 100–104
growth of woody plants and, 201
hairy root growth and, 242–244
herbicidal activity of, 305
suppression of long-day responses by, 274
microtubule arrangements and, 223–225
mode of action of, 311–318
root growth and, 229–239
sites of action of, 320–328
sterol levels and, 339–347
survey of, 296–300
see also specific growth regulators

Heat tolerance, in tomato, 364–365
Height regulation, in ornamentals, 304
Herbicidal activity, plant growth regulators and, 305
Heterosis, 94
HOE 074784, 299, 300
Hydrazine reaction, 402
Hydrolysis, of GA conjugates, 255–258
Hydrolytic enzymes, in aleurone and scutellum, 126
6ß-Hydroxyhyoscyamine, 244–246
Hydroxylases, 56–58; *see also* 2ß-Hydroxylase; 3ß-Hydroxylase
2ß-Hydroxylase
DOCHC and, 316
growth retardants and, 101
in *Phaseolus*, 89–91, 97–100
purification of, 98
3ß-Hydroxylase
in *Cucurbita*, partial purification of, 59
epoxidation of GA_5 to GA_6 by, 78
maize internodes as source of, 23, 30–31
in *Phaseolus*, 72–81, 84
2ß-Hydroxylation, 63, 68
3ß-Hydroxylation
far-red light and, 46–47
genetic inhibition of, 40

of GA_5 and GA_{20}, 62-69
GA53 metabolism in endosperm and, 53-55
growth retardants and, 58-59, 101
in rice, organ-specificity of, 19-20
12α-Hydroxylation, in endosperm
 preparations, 52-56
13-Hydroxylation of GA_4, 89
13-Hydroxylated GAs
 BX-112 and activity of, 314-318
 in cereal dwarf mutants, concentrations of,
  95-96
 conversion to, from GA_{12}-aldehyde, 85-86
 cyclohexanetriones and metabolism of, 300
 growth retardants and activity of, 300,
  314-318
 in *Phaseolus*, 85-86, 89
 phytochrome and metabolism of, 280
 in rice, 11
 in spinach, 275
 in sweet pea, 42
 wheat germination and concentrations of,
  126
 in woody plants, 203
N-Hydroxysuccinimide, 149
l-Hyoscyamine, 244-246
Hyoscyamus niger, 242-243, 246, 247

IAA, *see* Indole-3-acetic acid
Immunoaffinity purification, 164
Immunoassays, 146-164
 for GAs in maize, 33, 35
 validation of, 163-164
Immunogens
 antibody specificities and, 159-163
 preparation of, 149-158
Inabenfide, 299, 300, 320-328
Indole-3-acetic acid (IAA), 193, 333, 335,
  351
Indo-1 dye, 107
Inhibitors, *see* Growth retardants
Inosine diphosphatase, 110
Internal standards, 10, 27, 29, 96, 101, 128,
  170, 284, 292, 340, 346
Internode elongation, in rice, GA_1 and, 11
Internode length mutants, in sweet pea,
  40-48, *see also* Dwarf mutants
Iron ions, 65, 67-69, 76, 78, 103
Isofucosterol, 347
Isotope effects, enzyme kinetics and, 66
Isotope ratios, for quantification of GAs,
  26-27
Isotope labeled GA(s), 18-19, 24-29, 51,
  54-55, 64-68, 72-79, 83-91, 95-99,
  128, 170, 192-193, 264-271,
  275-277, 284, 292, 312-317,
  321-326, 340-341 *see also* **Deuterium**
 labeled GA(s) and Tritium labeled
  GA(s)
Isozymes, of α-amylase, 114

Juniperus scopulorum, 203

ent-Kaurene and derivatives
 growth retardants and, 58-59
 inabenfide and oxidation of, 321-326
 from mevalonic acid, 84
 Pisum growth and development and,
  188-197
 in rice, 17
 seed development and, 179-184
Kinetic isotope effect, 66
Kinetics of 3ß-hydroxylase, 75-76
Kinetin, 212
Kurosawa, Eiichi, 3-5

LAB 117 682, 300
LAB 129 409, 300
LAB 130 827, 300
LAB 140 810, 298
LAB 150 978, 336
Lactuca sativa, 229-239, 289-294
Lathyrus odoratus L., gibberellin mutants in,
  40-48
Lettuce
 phytochrome mediated germination of,
  289-294
 root growth and, 229-239
Life cycle, of rice, 9-10, 16-17
Lodging control, 301-302, 311
Lolium temulentum, flowering of, 174-175
Long-day conditions
 onion bulb development and, 220
 photoperiodic after-effect and, 273
 spinach growth and GA metabolism under,
  273-279
Long-day plants
 flowering of, 174-175
 growth responses to exogenous GAs, 200
lv mutants, end-of-day FR response and,
  46-48
Lycopene, 364
Lygodium japonicum, 380-381, 389-397, 398

Magnesium ion, 110
Maize (*Zea mays*)
 cell elongation studies of, 212
 growth retardant effects on, 313
 GAs in reproductive tissue of, 32-38
 tissue-specific GA metabolism in, 22-31

Male inflorescence, of maize, 32
Manganese ion, 56-58, 67, 76
Marah macrocarpus, 63-64, 67-70
Membranes
 calcium transport and, 109-111
Membrane proteins, affinity chromatography of, 143-145
Mepiquat chloride, 296, 298
Mercury ion, 76
Messenger RNA (mRNA), 130, 136, 241
 seed development and, 179
Mevalonic acid (mevalonate)
 ent-kaurene synthesis from, 191
 GA_{12} biosynthesis from, 84
 preparation of GA_{12} from, 312, 321-324
Microsomal membranes, 109-111
Microtubules
 cell elongation and, 211-217
 onion bulb development and, 220-227
Monensin, 110, 117
Monoclonal antibody, 66, 146, 159-163
Multi-genetic expression, of α-amylase, 118-120
Mutants
 end-of-day FR response and, 46-48
 GA concentration in, 95-96
 GA-deficient, seed development and, 179-185
 of maize, GA metabolism in, 22-31
 in sweet pea, 40-48
 see also Dwarf mutants

Nickel ion, 76
p-Nitrophenol, 149
Norbornanodiazetines, 300

Oak, growth retardant effects on, 313
Onium compounds, 298-304
Open lactones, 84-86
Orchard trees, 303, 350-359
Organ-specific gibberellins, in rice, 9-20
Ornamentals, height regulation of, 304
Oryza sativa, see Rice
Osmotic potential, 211
Oxidases, of GA_{53} and GA_{19}, 277-278
2-Oxoglutarate
 cyclohexanetrione activity and, 300
 enzyme activity and, 65, 68, 78
 growth retardants and, 101-103
 2ß-hydroxylase and, 100
Oxygen, enzymatic activity and, 65, 78

Paclobutrazol, 201, 283, 299-305, 325, 327, 335

Panicle initiation, 9
Parthenocarpy, 351-352
Pea (*see Pisum sativum*)
 Pedicels, abscission of, 365
Peterson reaction, 155
Pharbitis nil
 flowering of, 172-174
 seed development of, 179, 180-181
Phaseolus coccineus
 cell-free preparations of, GA metabolism in, 83-91
 fruits of, GA conjugates in, 252-254
Phaseolus vulgaris
 2ß-hydroxylases from, 97-100
 3ß-hydroxylation and, 62-69, 72-81
 inabenfide and cell-free extracts from, 320-328
Phloem, 175, 366
Photoperiod
 GA metabolism and, 201-202, 273-279
 GA mutants and, 46-48
 spinach growth and, 273-279
Phytochrome
 in cowpea, GA metabolism and, 280-287
 gibberellin mutants and, 46
Picea, endogenous GAs in, 204, 205
Pinaceae, 167-168, 201, 204, 205
Pisum sativum
 effects of uniconazole in, 339-347
 ent-kaurene biosynthesis vs. growth of, 188-197
 GA mutants in, 40-48
 root growth, 229-239
Plant growth regulators, 296
Plasmid, Ri, 241-247
Pollen, of maize, 32, 36-37
Polyclonal antibodies, 146, 159
Prothallia, 378
Prunus cerasus L., fruit set enhancement in, 350-359
Pseudotsuga, 200-206
Pteridium aquilinium, 378
Pumpkin
 GA biosynthetic pathways in, 51-60
 inabenfide and, 320-328
Purification
 of brassinosteroids, 341
 of C-20 hydroxylase, 59-60
 of GAs, 24, 281, 340-341
 of 2ß-hydroxylase, 98
 of 3ß-hydroxylase, 72-74
 by immunoaffinity chromatography, 164
 of sterols, 341

Radioimmunoassay (RIA), 146-149, 158-159, 163

Receptors, for GAs, 120-122, 136-145
Red light effects, 289-294; see also Far-red light
Reproductive tissues
 in maize, 32-38
 in rice, GA biosynthesis in, 17-18
Rhododendron, effect of uniconazole on, 331
Rice (*Oryza sativa*)
 α-amylase action in, 114-122
 control of lodging in, 302
 cytokinin content, 333-336
 effects of growth retardants on endogenous phytohormone contents in, 312, 320, 330-337
 effects of growth retardants on growth of, 311-318, 320-328
 endogenous GAs in, 312, 320
 GA_4 metabolism in, 264-271
 life cycle of, 16-17
 Nihonbare, 10, 14-15, 264-271
 organ-specific gibberellins in, 9-20
 Tan-ginbozu,10, 14-15, 315, 383
 Tong-il, 10, 14-15
 Waito-C, 10, 14-15
Ri plasmid, 241-247
Roots
 growth of, GA requirements for, 229-239
 hairy, tropane alkaloid synthesis and, 241-247
 removal of, onion bulb formation and, 222-223
 of sweet potato, accumulation of assimilates by, 365-366

S-3307 (uniconazole), 223-225, 242-244, 291, 299-304, 325, 327
Salix, GA/photoperiod interactions in, 201-206
Schizaeaceous ferns, 378-386, 389-397, 398-407
Scopolamine, 245
Scutellum
 ABA content in, 129
 α-amylase action in, 114-117
 hydrolytic enzyme secretion from, 126
 mRNA levels in, 130
Seed(s)
 endogenous GAs and development of, 179-184
 of lettuce, phytochrome mediated germination of, 289-294
 of maize, 33-36
 of *Phaseolus coccineus*, GA metabolism in, 83-91
Seedlings
 chemical improvement of quality of, 304

effect of root removal in, 222-223
 endogenous phytohormone contents in, 333-335
 ent-kaurene biosynthesis in, 192-197
 root growth of, 229-239
Seed production, Chinese cabbage, GAs and, 362-363
Senescence, chemical delay of, 330, 332
Sepharose, GA_4 immobilization with, 137-145
Sex differentiation
 in conifers, 171
 in fern gametophytes, 389-390, 394-397
 GA control of, 37
 uniconazole and, 332
Shoot elongation
 cell elongation and, 211-217
 control of, *see* Growth retardants
 ent-kaurene biosynthesis and, 188-197
 in maize, 22, 23, 32
 of orchard, 357-359
 root growth and, 229-239
 stress resistance and, 304
 in woody plants, GA regulation of, 199-207
Short-day plants
 flowering of, 172-174
 growth responses to exogenous GAs, 200
Silene armeria, 273
Siliqua development, 182-183
Silk, of maize, 32
Single-gene dwarf mutants, *see* Dwarf mutants
Sitosterol, 347
Slender plants, 45-46, 95-96
Solanaceous plants, 242-247
Solubilization, of membrane proteins, 143
Sour cherry, fruit set enhancement of, 350-359
Soybean, growth retardant effects on, 313
Spermine levels, uniconazole and, 339
Spinach, GA photoperiod and GA metabolism in, 273-279
Sporophyte formation, in ferns, 394-396
Spur formation, 354
Stem elongation
 gibberellin mutants and, 40-48
 light-induced changes of, 196-197, 280-287
Sterol levels, uniconazole and, 339-347
Steviol, 314
Storage roots, 365-366
Stress resistance, 304
Succinic acid-2,2-dimethyl hydrazide (SADH), 336
Sugar beets, growth retardant effects on, 313
Sumiki, Yusuke, 5-6
Sweet pea, 40-48

Sweet potato
  flowering of, 363–364
  storage roots of, 365–366

Tall genotypes, 35–36, 94, 95–96
Taro plants, GA-induced flowering in, 370–376
Tassels, 32, 37
Taxodiaceae, flowering in, 167–168
T-DNA, 241
Tetcyclacis, 58, 299–305, 325, 336, 347
Tetrahydrophthalimide (THPI), 289, 290, 291
T4 bacteriophage, 146
THO (tritiated water), 62, 64, 66
Tillering, 9
  GA_3-induced, 371–373
Tissue specificity
  of α-amylase multigenetic expression, 118–120
  of *ent*-kaurene biosynthesis, 192–194
  of GA metabolism in maize, 22–31
  of GA metabolism in rice, 10–20
  of GA metabolism in sweet peas, 41
Tomato
  fruit set at high temperature in, 364–365
  Seed development in GA-deficient mutants of, 179, 181–185
Transcription, of α-amylase genes, 120–122, 130–133, 136
Triadimefon, 300
Triadimenol, 300
Triapenthanol, 58, 299–305
Triazoles, 223–225, 299–304, 330–337, 339–347; *see also* Growth retardants
Trimethylammonium iodides, 298
Tritium labeled GA(s)
  as substrate(s) for 2ß-hydroxylase and, 97
  as substrate(s) for 3ß-hydroxylase and, 72
Tropane alkaloid synthesis, 241–247
Tropical vegetable production, 361–367
Tunicamycin, 117
Turf growth, chemical regulation of, 303

UDP-glucose, 255
Uniconazole, 223–225, 299–304, 339–347, 325, 327, 373

Vacuolation, 109
Vegetable production, GA control of, 361–367
Vernalization, 362–363
Vigna sinensis L., 280–287
Viral disease, 354–356

Water uptake, 211
Wheat
  GA concentrations in, 95–96
  germination of, 125–133
  growth retardants and, 101, 313
  plant growth regulators and lodging control in, 301
Woody plants
  endogenous GAs in, 203–206
  flowering of, 171–172
  GA metabolism in, 205–206
  growth retardants and, 201
  reduction of vegetative growth in, 303
  see also Conifers

Xylem, 205, 303

Yabuta, Teijiro, 4–6
Yield
  plant growth regulators and, 302
  sweet potato storage root and, 366
Yield threshold, of cell wall, 211, 216

Zea mays, *see* Maize
Zeatin, 333–334, 339
Zinc ion, 76